Society in Prehistory

Also by the same author

From the Caves to Capital
The Making of Modern Japan: A Reader

Society in Prehistory

The Origins of Human Culture

TIM MEGARRY

First published 1995 by
MACMILLAN PRESS LTD
Houndmills, Basingstoke, Hampshire RG21 2XS
and London
Companies and representatives
throughout the world

ISBN 0–333–31117–5 hardcover
ISBN 0–333–31118–3 paperback

A catalogue record for this book is available
from the British Library.

10   9   8   7   6   5   4   3   2   1
04 03 02 01 00 99 98 97 96 95

Copy-edited and typeset by Povey–Edmondson
Okehampton and Rochdale, England

Printed and bound in Great Britain by
Mackays of Chatham PLC, Chatham, Kent

For
Katy and **Matty**
and for my father
Bill Megarry
who first introduced me to human evolution

Contents

	List of Illustrations	viii
	Acknowledgements	ix
	Introduction: Prehistory and Society	1
1	Evolution and Human Society	18
2	Culture and Evolution	45
3	Biology and Culture	64
4	Primate Societies	91
5	The First Hominids	113
6	Tools and Culture	154
7	Tools, Brains and Behaviour	182
8	A Foraging Economy	207
9	Man the Hunter?	226
10	Modern Humans and Human Behaviour	266
11	Sex and the Division of Labour	291
12	Sexuality and Social Life	312
	Glossary	347
	Further Reading	354
	Bibliography	356
	Index	387

List of Illustrations

FIGURES

1.1	A model of the origin of species	36
1.2	Evolutionary adjustment by natural selection	38
4.1	Taxonomic and phylogenetic relationships of living hominoids	94
5.1	Diagrammatic representation of the reciprocal feedback process involved in hominid evolution	124
6.1	Palaeolithic stone tools (see with Table 6.1)	163
6.2	A representation of relationships in the human fossil record	170
7.1	The procedural templates involved in making Oldowan and Acheulian tools	186
7.2	A possible configuration of the social behaviour of tool-making hominids two million years ago	203

TABLES

1.1	The geological timescale of human evolution	24
1.2	Timescale for human evolution and cultural development	25
6.1	Palaeolithic tool manufacture	162
7.1	Brain volumes of hominid species	189

Acknowledgements

Special thanks are owed to Paul Hirst, Stephan Feuchtwang and Henrietta Megarry for their help and encouragement over many years and to Bernard Campbell for encouragement and advice. I must also thank Janet and Richard Briffett, Brigitte Flock, Imogen Forster, George Hallam, John Hood-Williams, Mike Kelly, Joanne Robertson and Hazel Shave.

Acknowledgements and thanks are also due to authors and publishers for allowing the use of copyright material as follows: Barbara Isaac, Peter Andrews and Garland Publications, Bernard Campbell and Aldine de Gruyter, John Gowlett and Academic Press, and British Museum Publications.

Introduction: Prehistory and Society

Man is a vertebrate, mammalian, and primate animal. That is what man *is*, and the fact should never be lost from account.
(Service 1971: 22)

From the point of view of sociological anthropology it is inconceivable that mankind, perhaps to be more exact modern man, *homo sapiens*, could have come into existence and have continued to exist other than through society.   (Fortes 1983: 1)

In terms of the history of life the rise of humans and their ancestors during the last 5 to 7 million years is both recent and bizarre. This whole episode in evolution, which has involved at least 10 species of bipedal primate that are known, and others that can only be conjectured, has occurred in less than 0.2 per cent of the time that there has been life on Earth. If we are to restrict this relationship to our own species, which emerged in its fully sapient form together with modern human behaviour only 40 000 years ago, we represent a mere 0.001 per cent of the 3.5 billion years of natural history. Humanity exists in a brief moment that is little more than a bubble submerged in an ocean of deep time.

Though we lack the historical depth of other more stable species, it is also the case that our rate of development has been rapid: our brains have tripled in size within a record 3 million years. However it is the acquisition of a collection of social and cultural traits that are used as a unique means of adaptation – as the basis for a new way of life – that is the most extraordinary aspect of our evolution. Some of this development is echoed by other animals and we have no absolute monopoly of culture or even language abilities. But the power of our mode of cultural adaptation, which has allowed us to transform diverse environments while continually redefining our culture and social organisation, is unprecedented and represents a biological breakthrough which has opened a new chapter in the history of life. Our

1

emergence, like that of all other life forms, has been caused by
evolutionary processes, but our existence in its present fully
human form cannot be explained merely by a fortunate conjunction of biology and ecology. Natural selection has in our case
involved a reciprocal relationship between important new factors
like a growing reliance upon tools and social behaviour which
have interacted with our physical form. In this sense it is
reasonable to place humans and their societies in an evolutionary
context.

Social scientists have sometimes shown an awareness of this
relationship but, apart from notable exceptions (Festinger 1983),
they have neglected consideration of its implication for the rise of
human society. Humanity may have evolved physically, but what
this implies for society and for human social behaviour has rarely
been confronted. The social sciences have largely rejected theories
of social evolution proposed by nineteenth century sociologists.
But at times this legitimate dismissal seems almost to have echoed
popular mistrust and the hostility of religious sects towards all
forms of evolutionary explanation (Barker 1979). As a result
sociologists in particular have usually been suspicious of theories
which connect human behaviour with natural selection. Social
scientists have invested considerable effort in repudiating the
crass and deterministic ideas presented by one or other variety of
biologism that attempt to reduce our understanding of social
complexities to non-social agents like genes and instincts. Sociobiology, the current form taken by such theories, has quite
justifiably raised the hackles of a generation of social scientists by
its assertion that human nature and society can be explained by
evolutionary biology. Social behaviour here, in all its rich and
varied forms, is no more than an effect of natural design.

The sociological response to such affronts to its territorial
integrity has often been to vigorously reassert cultural determination, but the price paid here has been ignorance of the
evolutionary processes that were involved in eliciting the very
subject matter of sociology itself: human social life. Another
consequence has been a refusal to take evolution seriously that is
seen in the general reluctance of sociologists to investigate
continuities between ourselves and the rest of nature (Giddens
1989). This has frequently been allied to claims that no part of our
ordinary behaviour, feelings or motives can be ascribed to
anything but social causes. The use of cultural determinism as
an exclusive mode of social explanation has sometimes meant that

sociology, like some religious doctrines, feels confident in making an absolute distinction between nature and culture. Animals are thought to run by an instinctual clockwork while humans, being devoid of any evolutionary inheritance, are driven by cultural forces alone. Our behaviour is the product of social consciousness derived from the interaction of culture with historical development, while animals are mechanisms without tradition that plod along according to a predetermined programme. But both these propositions are in fact false. Learning is essential for many animals and many of our reflex actions and emotional responses contain non-cultural elements.

If natural selection and our primate heritage is to be taken seriously our continuity with nature should be acknowledged. For there to be nothing left from our evolutionary history but basic reflexes would be extraordinary, and to claim that we operate by culture alone is as absurd as asserting that our behaviour is the product of a genetic code. We are surrounded by examples of universal forms of human behaviour that are seen in children playing, mother–child interaction, sexual conduct, the expression of sorrow or aggression and the body language of grief and joy. While this behaviour is of course heavily enculturated and open to a spectrum of socially constructed meanings, it is nonetheless mutually comprehensible to people in widely different societies.

However the conviction that a radical separation remains between ourselves and nature has been a deeply entrenched part of the social science tradition for two centuries. Ironically both Marx and Engels, who held Darwin and the theory of natural selection in high esteem, upheld this dualism. Marx's distinction between animal and human labour, made in *Capital* in 1867, indicates a confidence in celebrating human properties that have no precedent among animals:

> A spider conducts operations that resemble those of a weaver, and a bee puts to shame many an architect in the construction of her cells. But what distinguishes the worst architect from the best of bees is this, that the architect raises his structure in imagination before he erects it in reality.   (Marx 1970: i: 178)

Engels was to take this distinction further in his *Dialectics of Nature* with the argument that labour, 'is the primary basic condition for all human existence, and this to such an extent that, in a sense, we have to say that labour created man himself' (Engels 1940: 279). In a chapter written in 1876 with the title, '*The Part Played by Labour in the Transition From Ape to Man*', he proposed that from the earliest phase of our ancestry, production was to be the cause of all subsequent cultural advancement and was responsible for special human features like language and a large brain. The labour process begins for Engels with the first tool-making and leads, through the complementary association of dexterity, language and the brain in social activity, to the achievement of increasingly complex operations and tasks.

It is now well established that brain evolution did not occur until well after a protracted period of tool-making, and Engels' achievement in presenting an alternative to contemporary idealistic theories of human origins, which proposed that the power of reason and intellectual capacities of our ancestors came before social and technical development, is not to be dismissed. Idealistic explanation was in fact to remain as an influential element of anthropological theory well into the twentieth century, since the realisation that the first bipedal tool-makers had brains no larger than apes' was not to prevail until over 80 years after the appearance of Engels' labour theory of culture. In this sense Engels can be said to have formulated a more comprehensively materialist theory of human evolution, together with a causal mechanism, than even Darwin himself who was then preoccupied with establishing man's place in nature and never directly confronted problems raised by the emergence and significance of culture (Trigger 1966). Engels' emphasis on an interaction between behaviour and physical and social change should also be accepted, but the assumption that labour can be isolated as a category which differentiates us from animals is doubtful. Those who follow Engels here rely upon the continuing validity of an animal–human dichotomy, now and in our evolutionary past, that assigns unique qualities like language and tool-making to our species alone (Woolfson 1982, Mandel 1968).

Can labour really be identified as a causal agent in human evolution? Do humans alone work while animals only behave? This last question, which is posed by Ingold (1983), can no longer be answered with the same certainty that was possible in the nineteenth century. If labour means a self-conscious interaction between an organism and its environment which embodies intentionality and purposiveness in a socially directed context, at least some animals may be said to work. Animals are not mindless automata that act randomly or remain unaware of what they are doing. Higher animals, such as our closest living primate relatives, have been shown to possess a subjective will. Some behaviour may be genetically based but may also be part of a conscious purpose in both humans and animals, and consciousness itself is present to varying degrees among many animals. Uniquely human abilities may be claimed to lie with our capacity to make symbolic representations of our conscious intentions – as in verbal instructions or drawings. But work can exist independently of symbolism and such abilities came very late in our evolution. Cultural behaviour is an established part of primate societies and at the time our ancestors began their first innovations in tool-use and tool-making no form of superior or distinctive mental powers, or subsistence activity that could be seen as labour, can be assumed.

Social scientists have also resisted the use of Darwinian ideas because of a lingering suspicion that any use of biological or evolutionary explanations of human behaviour will open the way to discriminatory models that might justify racial, sexual or class inequalities. This fear is of course well founded. All manner of theories that have championed ethnocentrist, nationalist, racist and xenophobic principles have used the biology of natural selection as their justification in the past century. Although it is just as well to remember, in parenthesis, that the social sciences have also at times been recruited to support such ideas, it is the reprehensible misuse of Darwin's theories that has been most responsible for supporting discriminatory doctrines.

Academically respectable racist theories have all but vanished from physical anthropology since World War II, although journals such as the *International Anthropological and Linguistic Review*, which published a paper by Raymond Dart on violent behaviour among australopithecines, continued this tradition well into the 1950s. But far more sinister and notable for their social effects were the ideas of some nineteenth century evolutionary biologists, such as Ernst Haeckel (1834–1919), who defended and popularised Darwin's works in Germany. Haeckel was a leading scientist whose contribution to biology was notable. In 1866 he invented the term 'ecology' as the science which dealt with "the household of nature" (Mayr 1982: 121). His depiction of man's place among the primates and the order of branching that led to humanity is close to contemporary models (Cartmill et al. 1986). But Haeckel's popular exposition of natural selection, which at times dismayed Darwin because of its acceptance of Social Darwinism (Bowlby

1991) was an abuse of evolutionary theory that led directly to the racial theories of the Nazis (Stein 1988). Haeckel was to fuel a current of European thought which held that society could be finally and scientifically transformed by a mixture of reform, education and racial hygiene (Peukert 1989). The fallacies of Social Darwinism, which were never endorsed by Darwin himself, are explored in the next chapter and it is appropriate here only to comment on some of the ideological results of this illegitimate use of evolutionary theory.

Haeckel was to revive and combine existing racial theories, which specified a hierarchy of racial types, the innate inferiority of some races and justification of Aryan superiority, with an apparently plausible scientific foundation provided by Social Darwinism (Banton 1977). Africans and native Australian peoples were considered to have evolved less and therefore to be closer to the apes. His belief that human races and societies could be ranked according to measurable achievements, on a scale of civilization, was connected to the idea that races and societies had engaged as whole units in a struggle for existence throughout world history. These propositions became married to a set of mystical ideals associated with romantic nationalism and Aryan superiority that were later to be used to justify conquest, permanent subordination and even extermination of 'inferior' races and the sick and disabled (Burleigh and Wippermann 1991). Such acts were fully consistent, it was thought, with natural biological law.

In this perversion of Darwinian evolution, Haeckel had erroneously substituted races and societies for single organisms as the unit of natural selection. Evolution had created human beings, who were seen purely as products of nature but were simultaneously denied any species unity. Like other biological determinists Haeckel also assumed society and culture to be no more than a reflection of natural laws without autonomy or independent effect in social life. Haeckel's misuse of science also consisted of drawing moral worth from what he supposed to be scientific facts. Science of course cannot be used to verify moral values. It is possible however to employ science to confirm or falsify the scientific basis of theories of the kind purported by Haeckel. No branch of anthropology today provides any evidence to support racist theories, and the facts of human unity that were first established in Darwinian evolutionary theory over a century ago have now been reconfirmed and given a better grounding by the sciences which contribute to prehistory.

Drawing connections between society and evolution in no sense implies a capitulation to biologistic forms of reductionism or a denial of the importance of culture. Nor is there any implication that our social affairs – political ideas, rules or conflicts – can be explained biologically. But much is to be gained from understanding that human social life rests upon a complex set of *integrated* social, psychic and biological mechanisms that have arisen during an evolutionary journey that is partly shared with other animals. This is to argue that the simple opposition between nature and nurture, that is so frequently encountered in the social sciences, is largely misconceived since no final distinction or boundary dividing nature from culture can be drawn. The contention that the greater part of human behaviour is learned, is not contradicted by the fact that learning occurs only by virtue of a set of biological and psychological mechanisms that have evolved with the development of human and pre-human society. But these abilities of body and mind, in both humans and in higher animals, do not exist independently from social interaction. Humans are animals whose behaviour is generally in accord with a cultural tradition, but we are also animals who are programmed to learn culturally and we show a compulsion to acquire new elements of a growing behavioural repertoire from a very early age. This implies that both the means of learning and the learned content are activated and developed only through a lengthy process of social association and action.

Human learning, in common with animal learning, is not however a haphazard affair. Specific learning dispositions, as they are termed in ethology – the comparative study of animal behaviour – direct the acquisition of behavioural capacities (Eibl-Eibesfeldt 1979). Behaviour patterns develop along lines laid down within a range of variation determined by the organism's inheritance. Children, for instance, feel impelled to accomplish the formidable task implied in learning speech, and later to fully develop their language skills. This is achieved with little effort and without formal teaching despite the difficulties involved in pronunciation, grammar and the acquisition of a 5000 word vocabulary each year for 16 years (Miller and Gildea 1987). Most language learning occurs unaided yet learning to read or write, which is not part of our inheritance, usually requires protracted teaching and arduous practice over many years.

Another claim made by ethology, that must be given wider consideration by the social sciences, is the idea that some aspects of human cultural behaviour start from preadaptations which are also present in closely related animals. Darwin began this line of inquiry with his work, *The Expression of the Emotions in Man and the Animals,* which was published in 1872. He writes:

> Infants scream from pain directly after birth, and all their features then assume the same form as during subsequent years. These facts alone suffice to show that many of our most important expressions have not been learnt; but it is remarkable that some, which are certainly innate, require practice in the individual, before they are performed in a full and perfect manner; for instance weeping and laughing . . . When, however, we turn to less common gestures in ourselves, which we are accustomed to look at as artificial or conventional, – such as shrugging the shoulders, as a sign of impotence, or the raising [of] the arms with open hands and extended fingers, as a sign of wonder, – we feel too much surprise at finding that they are innate. That these and some other gestures are inherited, we must infer from their being performed by very young children, by those born blind, and by the most widely distinct races of man.   (Darwin 1965: 351–2)

This aspect of Darwin's work has been ignored or played down by the social sciences despite an accumulation of supporting evidence from primate studies that indicate a continuity with human behaviour (Goodall 1971). Primate research has shown that apes and monkeys use soft embraces to convey security and affection, while chimpanzees and orangutans kiss and gorillas use stare threats with emotional motives that are similar to humans. Human–primate continuities have also been found in areas like mothering, bonding, sexual behaviour, learning and socialisation (Goodall 1976, Loy and Peters 1991). The sensory motor stages of development that are found in human infants are also present among gorillas and chimpanzees and occur at about the same age (Parker and Gibson 1979). Work by ethologists has confirmed that human facial expressions are mutually comprehensible across cultures. The emotions behind smiling or the expression of grief or anger, for instance, can be intuited by people in diverse societies and these emotions are perceived correctly because they emanate from fixed action patterns that coexist with cultures which designate them with a range of social meanings. Universal forms of human response are aroused by innate releasing mechanisms seen for example in the eliciting of care-giving behaviour when a small child is in the presence of many adults. Basic modes of social interaction in which a smile may be used to repair social bonds, or facial expressions like pouting serve as a means of temporarily suspending association to block aggression, follow similar and predicable patterns in all cultures. There are ritual elements of any friendly human social encounter that include a universal grammar of traits like eye contact, head-tossing, bowing, eyebrow-flashing, patting or kissing which comprise some of the basic mechanics of interpersonal relations. The fact that interaction among children shows the use of universal strategies, seen in establishing friendly contact, sharing, aggression or appeasement, argues strongly that our species has evolved with a set of mechanisms to facilitate social life (Eibl-Eibesfeldt 1979).

Further evidence of a subtle interplay between biological mechanisms and social life is provided by studies of closely observed baby–mother interaction and early development. It seems likely that a child progresses through an innate developmental programme in a series of stages that are controlled by brain growth and which are activated by social interaction from early infancy. Rather than simply being developed by society, Trevarthen (1979) proposes that children develop themselves as their motivations change. With growth, a child's focus of interest in people and things changes as does motivation which accordingly leads to new areas of cultural and emotional experience that the child is then ready to learn about. Initially a child strongly indicates the need for consistent care-giving in the first phase of bonding with a mother figure soon after birth. There is also a strong desire to react positively to affectionate care-givers by expression and gesture. Within weeks, long periods of apparently unproductive play, that increasingly involves the mother, is used as a means of engaging and practising social abilities. Trevarthen and Grant (1979) argue that different types of play indicate the commencement of new stages of development. Play itself is vital for psychological survival in society since it is through play in a risk-free environment that the child is able to absorb the basis of rules and conventions that underlie cultural life. A mother induces a readiness for cultural learning by helping the child to integrate cultural meanings with his growing imagination and self-consciousness. Mother–child interactions explore the meaning behind conventional facial expressions and, in subsequent types of play, social awareness is enhanced as symbolic acts like 'pretend' games or 'feeding' dolls occur.

Human social existence requires an ability to correctly perceive the motives and intentions of others in relation to those of the self, and frequently to turn these into a collective social objective. Early interaction helps to develop this perception and foster common goals. Learning and teaching are crucial parts of the culture acquisition process but it is the child who sets the learning agenda at each developmental stage by an inclination to learn different things in the course of play. Children clearly possess an overwhelming urge to play, to explore meaning and symbols and to become psychologically integrated with their social worlds. Evolution is replete with examples of developmental mechanisms that facilitate fitness for animals leading a particular way of life. Our species arose with a dependence on cooperation and later on symbolic communication as parts of a complex social life. Social integration in humans depends upon standardised yet complex forms of interaction and information exchange which require abilities that are likely to be provided by innate mechanisms as much as by learning.

These issues are familiar to social scientists but there is exciting new evidence provided by prehistory that casts light on the evolutionary roots of culture itself. What is offered here is the possibility of understanding that our nature as social beings can now be traced to rudimentary cultural origins that are some two million years old, and back even further to our distant primate ancestors. This interminably long phase of development, which is at last beginning to yield information about the origins of human social life, is also the period of our apprenticeship as culturally directed animals, and this era of emergence must be significant to the social sciences in general.

Our existence as a result of natural selection raises a fascinating series of questions about the forces that created our species. Some of these issues: why we have a propensity to live in mixed-sex groups containing permanent familial bonds, marriage and a division of labour, have long been a focus of anthropology. The question of our origins in terms of both our physical and behavioural natures has certainly been mused on in the social science tradition. But until recently the record of prehistory was so restricted that almost all theories of human origin were little more than uninformed speculation. Before 1940 anthropologists were forced to work with a meagre fossil record in times when techniques in biochemistry were primitive. Little was known of the behaviour of our living primate relatives the apes. Hunter-gatherer studies showed little interest in production and the social implications of foraging remained unknown. Archaeology of the early stone age, the Lower Palaeolithic, which has shown a 2 million year old cultural record in Africa, began in earnest only some 30 years ago. Since the 1960s however, the study of prehistory has advanced in a series of quantum leaps and there are now good prospects for gaining an understanding of at least some aspects of the human career. Genetics, primate ecology, the ethnography of past and present hunting societies, palaeontology and archaeology have all advanced to a point that allows us to make at least a tentative reconstruction of early social life.

The desire to know our origins as a species is often accompanied by a conviction that, should we gain access to knowledge of the earliest times in which humanity was formed, we would also understand our natural condition. This in turn, it is often thought, would allow us an unrivalled understanding of human nature itself (Ardrey 1961, 1976). But this turns out to be an over simple view for two reasons. An initial understanding of the complex interacting forces that lie behind and sustain human evolution only affirms the fact that during the long journey from primate status we have apparently gained the capacity for an inordinately flexible adaptive behavioural repertoire that makes simple ideas of a singular human nature unviable rather than illusive. Our cultural and social way of life is capable of overriding, obliterating and mediating vestigial elements of our primate or pre-human heritage.

A second consideration must be that the human evolutionary journey is itself anything but undeviating and contains along its uncertain route a series of innumerable twists, dead ends and false starts. Natural selection has of course constructed humans by a series of cumulative steps on the foundation of preadaptations that were present in primates and two-legged human ancestors. In this sense we were not created by pure chance alone since evolution can only select from what has evolved. But the evolution of humanity was in no sense inevitable. Random mutation, working from existing biological and behavioural structures, can lead to a whole series of possible outcomes along an evolutionary trajectory. The number of organisms, along with their unique attributes, that actually have ever existed are only a tiny proportion of those that could actually exist (Dawkins 1986). During our evolutionary 'progress', as we often mistakenly think of it, there was no necessary reason why we should have acquired some of the specialist traits that we most admire in our species. But if one conclusion can be drawn from the study of this journey it is that our cultural equipment – the human means of adaptation and the lens through which we perceive our environment – was formed in a piecemeal fashion as a number of separate accretions over at least the past 5 million years. Each part of this equipment, including our ability to extract objects from the environment for specific purposes, the use and making of tools and fire, the ability to engage in a social life that contained a division of labour, language and aesthetic feelings that were finally expressed as art and symbolic representation, self-consciousness, body-decoration and religious ideas, all arose at different times and in an order that is uncertain. As the evidence of prehistory is unrolled, it becomes clear that because our current behavioural abilities were pioneered by different species of human ancestor across tens of millennia, there is in reality no beginning, or pristine time which may be invoked as an origin. There never was a fixed state of nature from which we have departed and, however unsettling it may be, we have to accept the fact that the only original human condition is one of endless change and development.

Although culture is not exclusively a human invention, its use as the basis of a new way of life has led to a proliferation of human diversity. This implies a widening gulf of social difference between societies that have become increasingly distant. But is this entirely true? Is there nothing that we retain in common as a social species? There are, as we have seen, individual behavioural traits that are shared by humanity as a whole, and some even with our primate relatives, which in a vague way can be described as a human condition. This in itself raises another issue for the social sciences. On the one hand we are a species that is seen as having adapted through culture by creating countless ways of life in a myriad of unique societies that can be distinguished precisely by their diverse cultural contents. In this view, which is typical of sociology, we are seen as creatures conditioned by societies that are separated by voids of time and social distance. But, conversely, it should also be recognised that human societies, like individuals, contain sets of shared characteristics. If individual ways of feeling, being and experiencing the world are, at some levels, common to humanity as a whole, it is also the case that our social systems contain common features although the consequences which these have for social life differ widely. Barriers of language, belief or custom, while real and often divisive, are not an impediment to understanding that we share common human traits at both an individual and social level. Profoundly different and containing though our cultural structures are, there can also be real and meaningful points of contact between peoples of quite dissimilar cultures. If this were not the case cultural diffusion would be impossible and human societies would continue to develop along unique pathways until they reached divergent points across which no form of communication would be possible.

But this does not occur. Human beings are prevented from creating cultures that proliferate idiosyncratically, like receding galaxies, to become infinitely different and alien to each other, by being everywhere forced to address critical imperatives concerning production and social order, reproduction and death. Despite being enmeshed in our own cultural solutions to these basic issues, we also have the ability to empathise with the solutions that other cultures have applied to these dilemmas and reflect on them as part of a shared human condition. Mutual fascination with the art or philosophy of other cultures is an obvious example. If evolution is in part responsible for the pervasive qualities that are part of humans and their societies there are no deterministic or prescriptive lessons to be drawn. But the social sciences have nothing to lose by acknowledging that as a species we share a nature or even that some of our feelings and behaviour are not the products of culture alone.

Even if we have developed the use of culture to extraordinary lengths we remain animals and we have certainly not escaped from the condition imposed upon all life that we adapt and survive only through dependence upon other life forms. Human evolution has created a strange species of primate that is unique in being able to transform a range of different environments by its cultural activities. But whatever advantage this has given us over the rest of nature we remain the subjects of natural selection which will itself also cause our extinction as a distinct species. Inevitably we will either be slowly modified into post-humans or meet with catastrophe. While our destiny is uncertain there is no doubt that cultural life has itself had a devastating effect upon the natural world with which we coexist. We would do well to reflect that when the human career began, the Earth contained greater biological diversity than at any time in its history, while today

fewer species of plants and animals have survived than at any
time in the past 65 million years (Wilson 1989). One lesson of
prehistory is the extraordinary precocity of our species which
should induce humility. It is only since we acquired our fully
sapient qualities 40 000 years ago that we became able to engage in
social and economic transformations, and these had little real
environmental impact until large-scale agriculture began to
change world ecology about 5000 years ago. The acceleration of
our cultural progress since then raises serious questions about our
future. Cultural success, it seems, has come at considerable
environmental expense, and the pressures that we have placed
on nature will demand new cultural solutions.

This book is not intended to be a comprehensive guide to human
evolution or even less a review of the latest evidence. In the past 30
years new discoveries and techniques have led to fresh interpretations of prehistory that have engendered a series of debates
and controversies (Lewin 1987) that can only be given a brief
consideration here. Disputes concerning the rate of evolution, for
instance, have divided those who have claimed that natural
selection elicits relatively abrupt punctuated bursts of change, that
suddenly produce new species in a geologically short time-scale
(Gould and Eldredge 1977, Stanley 1981), from opponents who
envisage continual imperceptible or gradual modification (Dawkins 1986). Differences found among the stone tool-forms of
Neanderthal peoples have been seen by some authorities as a
reflection of cultural variation and by others as the result of
practical design for implements serving different functions
(Bordes 1968, Binford and Binford 1969). New understandings of
the processes by which fossils are created have arisen with the
development of taphonomy, a subordinate branch of palaeontology. This new science has led to fresh interpretations of fossil
evidence and has caused a revision of older theories (Hill 1988).
While there is no space to pursue these developments in any detail
it is important to follow new insights that have been provided on
questions as diverse as the speed of our evolutionary development, the 'humanity' of our ancestors, violence and cannibalism as
formative influences in our past, and hunting or scavenging as a
primary adaptation.
   Rather than review recent advances made by prehistory, the
intention of this book has been to provide a discussion of some of
the issues that are raised when we attempt to examine humanity

by integrating the usually separate realms of social science and prehistory. What aspects of our experience in prehistory constrain social organisation? How did the first human-like society begin? What were the forces and circumstances which maintained prehistoric societies and under what conditions were pre-human ancestors transformed into our own species?

Another intention is to justify the conviction that the social sciences have a need to incorporate an evolutionary understanding. This is to propose that society and our nature as humans, which are as much a product of evolution as our physical form, cannot be fully understood unless we examine the context and processes in which they emerged. This appeal is also addressed, in a reversed form, in another direction to some of the specialist disciplines of prehistory in which the growth of a reductionist approach has at times led to a rejection of culture as one of the driving forces of natural selection. Advances gained in understanding the ecological adaptations made by human ancestors have unfortunately also led to demoting the role of material culture and social relations as formative influences on our long-term development (Foley 1987). But ecology alone does not provide an adequate explanation. Biological phenomena have effects on social life, and behaviour in turn effects physical evolution. Beside the need to explain our dental configuration or manual dexterity, we also have the obligation to investigate some of the distinctive human abilities that are just as crucial a part of our adaptive system. We cannot explain attributes like tool-making, art, language or prolonged bonds between the sexes simply as adaptational by-products of changing ecological conditions. These traits were not just the results of a lucky combination of a feeding strategy with environmental change and they will remain mysterious unless we explore their proper cause which lies in an interaction between social and biological factors.

This book then is also intended as a defence of culture as a significant causal agent involved in the evolutionary process itself. This is in no sense an original approach, in fact it represents a reassertion of some of the ideas guiding the study of human evolution that can be traced to the last century which have more recently been refined by biologists and anthropologists such as Dobzhansky and Washburn.

The full cultural awakening of human ancestors was excessively prolonged and has a likely history of four to five million years. The earliest known stone tools appeared in north-east Africa between 2 and 3 million years ago, but a little after 2 million years ago there is good evidence of widespread manufacture and use throughout the East African region from Tanzania to Ethiopia. But the extraordinary monotony of the cultural record in these times shows that this behaviour was practised by distinctly pre-human ancestors. Tens of thousands of generations of tool-makers were to remain faithful to their cultural traditions by producing artifacts with almost no discernable change for hundreds of thousands of years. While a culture-based way of life is found here, it is clearly qualitatively distinct from anything that we associate with our own species. As intelligent problem-solvers the first tool-makers were little more advanced than chimpanzees or a five-year-old human child (Parker and Gibson 1979). These pre-humans were foragers but their social organisation and technology was quite different from that of modern hunter-gatherer peoples, and to assume that both are examples of a similar mode of production is as unhelpful as claiming that England in 1851 and Japan today are both industrial economies. At this stage in our pre-human career our forebears followed what Jelinek (1977) refers to as a 'paleocultural' way of life. Culture was a crucial and growing part of social life but, despite its significance for the future, it was lacking in most of the elements that mark out our species today. Apart from a rudimentary material culture, and the means to communicate this to the young, there is little or no evidence for fire, speech, art and kinship which all emerged later at separate times. This evidence indicates that the foundations of human society, our psychological and physical abilities, are not a singular united entity but consist of a collection of traits that have been assembled to form what can only be seen as a current but changing configuration of attributes.

Prehistory presents the social scientist with considerable difficulties. It is true of course that the study of human evolution stands at the junction between the social and natural sciences (Foley 1991). Understanding evolution and the sciences that comprise prehistory often involves the crossing of dangerous interdisciplinary boundaries, and it is easy to grasp why a conception of the human subject as a static entity has usually been the preferred option for most social scientists. Human culture is often taken for granted by sociology and there has been a reluctance to consider our present social and behavioural characteristics as anything but a permanent fixture. However sociology is also part of an academic tradition that has been prepared to confront large questions and in particular to examine major social transitions. Work produced in this vein has never been more than partially successful, since integration and generalisation from disparate specialist fields involves considerable risk. But the need for a synthesis which draws together issues and themes pursued in separate disciplinary areas is justified by both the recent advances made in prehistory and by the need of the social sciences for a deeper perspective.

1

Evolution and Human Society

Society envelops human existence, and the nature of this social organisation, rather than its origins, is primary for sociology. But what are human characteristics and how were they acquired? These two interrelated questions, which are often ignored by social scientists, form the basis of this book. This chapter outlines the varieties of evidence which may be used to explain the emergence of human society.

THE HUMAN SPECIES AND VARIABILITY

A zoological description of the human species includes a set of related anatomical characteristics as follows:

An erect two-legged posture, known as bipedalism, is fundamental. Since the limbs and trunk have become adapted for bipedal movement, legs are longer than arms and the vertebral column has an S-shaped curve. The prehensile or grasping hands have a large thumb that may be rotated through an angle of 45 degrees and opposed with any other finger to make possible both a precision grip, as in holding a pen or a pair of chop-sticks, and a power grip, seen in using a hammer. The major part of the body is hairless. The brain is particularly large in proportion to the body, with a mean capacity of 1350 cc, but a wide variation from this average is common. The face is short, almost vertical, and is dominated by a large forehead and domed skull. The jaws and chewing muscles are relatively small and hardly project from the face at all; the brow ridges are only slightly developed. The dental arcade is rounded and the canine teeth are no larger than the incisors or premolars.

These physical features take no account of individual or sexual variations. Nonetheless, on the basis of this bare morphology, a zoologist can distinguish and name modern humans *Homo sapiens sapiens*: a single animal species that is but one of some 180 other living species that together form the Order of Primates.

Dividing the human population into six or nine geographical sub-species or racial groups (Campbell 1975) has been a prevailing practice among anthropologists since the 1940s. Despite the apparent physical diversity presented by these groups it was recognised that our species possessed an overall unity, since ultimately humanity consists of a single breeding unit. The differences found in body-build, skin, hair or eye colour and facial characteristics were seen as the results of adaptations to ancient environments. These differences were known to be small in terms of genetics or comparative zoology, and the human population, it was stressed, was seen as a single species (Washburn 1963). The widely accepted modern definition of species as, 'groups of actually or potentially interbreeding natural populations, which are reproductively isolated from other such groups' (Mayr 1942: 120 and 1982: 273), was claimed to apply to humans as much as to any other life-form.

Biologists argued that human races could be said to be groups who genetically were only comparatively isolated and therefore possessed roughly common characteristics, although these were always variable between individuals. A race shared similar genetic or inherited patterns with other populations but had certain genes and gene combinations, known as 'gene pools', that were found more frequently within the race than among other races, and thus these features could be said to be typical of it (Dunn 1956, Barnett 1950). But because races are open genetic systems that grade into each other, there could be no certain boundary characteristics that might be used for classification: neither could a particular number of races be said to exist, nor could all individuals be accommodated in one or another group. As race was meaningful only at the level of whole populations, a single individual was, at best, only a comparative approximation to a collective physical type.

The comparative isolation of races has in fact never broken down into a state of separation or complete reproductive isolation. If this had occurred, more than one species of human would now exist and interbreeding among different racial groups would not be possible. Continual genetic exchange between dispersed human groups from the earliest times has thus made humanity a single but polytypical species represented by races whose physical characteristics and interrelations are constantly changing.

In this conception, racial distinctions were based upon small degrees of genetic difference. There is new evidence however, which, while strengthening this unitary view of the human species, questions the reality of this difference and therefore the validity of basing racial categories on genetic criteria. Anthropological attempts to classify races have frequently been absurd or have resulted in ambiguities since there is no final or reliable way of distinguishing one race from another (Rensberger 1987). But it is the idea that humans can be divided into a genetically distinct set of races that has been questioned by the 1991 Reith Lecturer, Steve Jones. Modern genetics has shown that there are no genetically distinct groups within humanity and that therefore the category of race, used as a biological means of identifying a discrete series of sub-species which divide humanity, has become unviable. If we were in fact divided into pure racial groups, argues Jones, races would be quite distinct from each other with regard to their *overall* genetic composition rather than in the genes for skin colour alone. There are in fact less than ten genes which determine skin colour and about eight additional genes responsible for 'racial' features among a likely total of a hundred thousand genes carried by each of us (Wilson and Cann 1992, Rennie 1993). Mapping genes across the world presents a very different picture to the one drawn by those who support the belief that we are divided into separate racial groups that may be typified by skin colour. Changes in skin colour are not found to be accompanied by changes in the bulk of our genes: in reality the two systems vary independently of each other. 'When gene geography is used to look at an overall pattern of variation it seems that people from different parts of the world do not differ much on the average. Colour does not say much about what lies under the skin' (Jones 1993: 192). Jones shows that genetic variation is likely to be far greater between individuals of the same 'racial group' than between countries or 'races'. He continues:

> around eighty-five per cent of total diversity . . . of genes, world-wide, comes from the differences between different individuals from the same country: two randomly chosen Englishmen, say, or two Nigerians. Another five to ten per cent is due to differences between nations; for example, the people of England and Spain, or those of Nigeria or Kenya. The overall genetic differences between 'races' (Africans and Europeans, for example) is not much greater than that between different countries within Europe or within Africa. Individuals – not

nations and not races – are the main repository of human variation for functional genes. A race, as defined by skin colour, is no more a biological entity than is a nation, whose identity depends only on a brief shared history.　　(Jones 1993: 192–3)

Jones' arguments have contributed to an ongoing debate in anthropology concerning the legitimacy of making racial classifications within the human species (Gould 1981). The rules of biology allow a species to be divided into sub-species on the grounds that overall genetic distinctions can be made with regard to form, physiology or behaviour and that sub-species occupy a part of the total geographic range of the species. Neither of these requirements can be satisfied in humans. Compared with other animals we are a genetically homogeneous species, probably because of our recent evolutionary emergence, and fixed, definite boundaries cannot be applied – now or in the past – to either human morphology or the area of the world we inhabit. It would be bogus to deny facts of physical variability among humanity or to pretend that even the superficial differences which distinguish human races are not of biological interest since they are the result of important evolutionary adaptations to climate which occurred before major episodes of mass migration. Nevertheless this recent genetic evidence strengthens the views of some anthropologists that race has much more significance as a social than a scientific category and is of very limited use as a concept which differentiates human populations (Fried 1987, Livingstone 1962).

No description of a species however can rest upon anatomical criteria alone. Behaviour is as much a defining property of a species as are the physical characteristics with which it interacts. There is a striking resemblance between humans and the modern apes which may in part be extended into the realms of behaviour. The relationship posed by such categories as 'man' or 'human' or 'animal', which this juxtaposition involves, raises a number of highly controversial and subjective elements for possible inclusion as human behavioural attributes. Problem-laden though these elements may be it is only by recourse to an examination of behaviour that a distinctive character for *Homo sapiens sapiens* can be established.

Some human behavioural characteristics are shared with other non-human primates even though these may be present to a much lesser extent. The capacity to make and use tools and to imitate, learn and integrate diverse patterns of behaviour into a way of life that is peculiar to a particular group is found in humans and among some apes and monkeys. However, the capacity for mental abstraction which has given rise to symbolic communication seen in language is uniquely practised by the human species, although some echo of this capacity has been found in chimpanzees. The ability and habitual tendency to make changes in the mode of human life, both in the technical sense and in terms of relationships between people, describes a cultural dimension that is not present in animal life. Human social interaction is regulated by a system of flexible but formal rules which are derived from abstract principles. Human societies have come to contain stark contrasts in their appearance precisely because they are organised upon cultural rules. Yet despite such contrasts it is possible to describe, in broad outline, human social practices that are universal, although culturally variable. A socially determined division of labour and a shared consumption of its results; a system of ethical ideals; regulated mating and kinship organisation, exist in a multiplicity of forms but are ever present. These features, with the addition of language, are often taken as the basis of human society and the experience of them from an early age is essential if an individual is to achieve a socially recognised human status.

Evidence gathered from children who had been isolated in infancy, and from children who had probably been mothered by wild animals, is complex and is not amenable to simple interpretation. Yet it is possible to observe from such cases that the effects of early isolation constitute social deprivation with consequences that are as much physical as they are psychological. The absence of a learned language and the access that this would have given these children to basic human social relations was matched by an absence of acceptable physiological attributes. Tear glands did not function; the object of sexual feelings remained unspecified, and the deportment of the body was bizarre by all conventions. These isolated children were barred from the symbolic and emotional life of their communities not merely because of an absence of learning but because the mental and physical processes which underlie social behaviour were not activated. The point to be taken here is that social relations rest upon a combination of both learned and physical properties which have been simultaneously shaped during the course of human evolution. With some caution it is possible to claim that art and self-adornment, disposal of the dead and transcendent ideas are also universal, though in different forms, in all cultures. It is, for instance, a general social practice to distinguish between sacred and mundane although what is implied by this distinction varies between societies.

All these cultural practices however have an origin and a history, and they should be seen, like associated technological developments which began very gradually with the application of tools and later fire to new means of gaining subsistence, as products of an immensely ancient accumulation. In this process the physical evolution of the human species has interacted with social and cultural practices. The theme of this book is the acquisition of both the physical and cultural properties of humans during prehistory. The description of a process of 'humanisation', in which cultural behaviour became increasingly dominant and a slow departure from the animal world began, is the major task of the prehistorian (J. G. D. Clark 1979). In this sense it can be argued that social science has a real interest in the problems of prehistory. A proper discussion of human evolution does not therefore merely describe the physical stages which have been represented by human ancestors, but allies such description with concomitant changes in social behaviour that led ultimately to a hunting and gathering society.

In this book evidence for the origin of human characteristics is taken from a variety of physical and social sciences. The evolutionary theories of Darwin must be examined together with the modern evidence for human evolution. There is also new evidence for pre-human cultural behaviour which comes from both the study of living non-human primates and from archaeology. Recent field studies of primates have contradicted assumptions long held by zoologists on matters as fundamental as diet, habitat and social systems of some apes and monkeys. Observation in both the field and laboratory have shown that both the chimpanzee and orangutan make and use crude tools and that, under special conditions, chimpanzees can be taught to use symbols in communication. Such findings must be seen as a challenge to the notion that culture is unique to the human species.

New archaeological evidence has become available from the earliest period in which a material culture was practised by the *Hominidae* or hominids – the zoological family which includes *Homo*, or man – and extinct species which were ancestral to humans. Inferences drawn from stone tools and bone refuse that accumulated as much as 2 million years ago in East Africa during the beginnings of the Lower Palaeolithic culture period, allow the

24

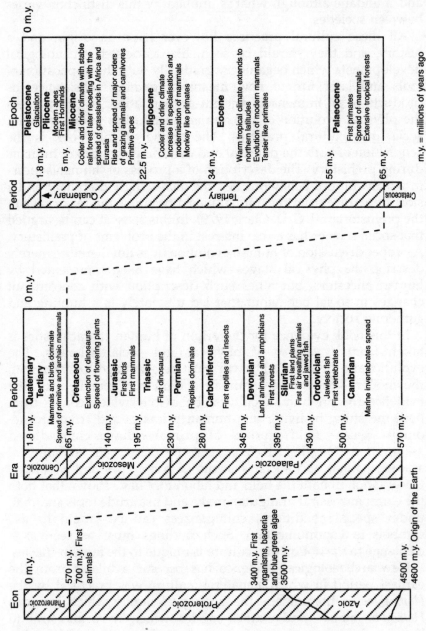

TABLE 1.1 The Geological Timescale of Human Evolution

SOURCE: Based on Jolly and Plog (1976), Simpson (1983) and van Eysinga (1978)

TABLE 1.2 *Timescale for Human Evolution and Cultural Development*

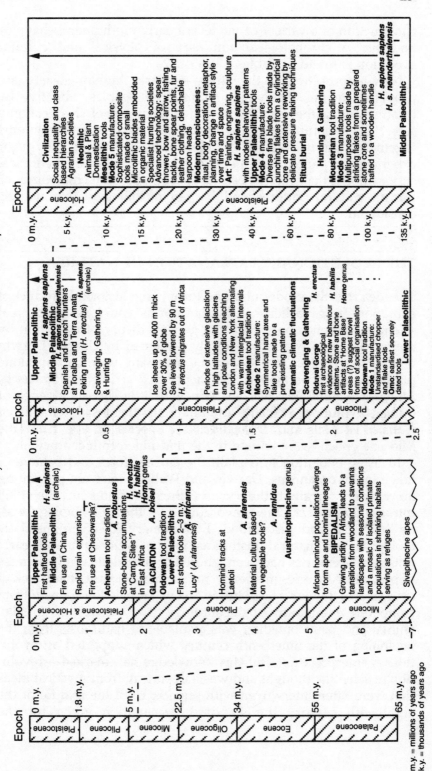

m.y. = millions of years ago
k.y. = thousands of years ago

SOURCE: Based on EHEP (1988), Foley (1987), Richards (1985) and Stanley (1989)

construction of a variety of models of early pre-human behaviour patterns. A chronological summary of physical and cultural evolution is found in Tables 1.1 and 1.2.

New approaches to the study of early human society are provided by the observation of modern peoples who live by hunting and gathering. The adaptive flexibility of this economy seen in its compatibility with utterly diverse environments together with its historical persistence, makes it the major experience of humanity. There are also a number of theoretical explanations that have been proposed for the emergence of human cultural patterns and human morphology that will be considered.

THE EVOLUTIONARY EXPLANATION OF NATURE

The idea that the modern human form is the historical product of evolution from earlier non-human forms could be given real scientific credence only with the general theory of evolution by natural selection which was jointly presented by Charles Darwin and Alfred Wallace in 1858. In the following year Darwin alone published a full account of the theory in *The Origin of Species by Means of Natural Selection*. Darwin had little direct evidence of human evolution available to him: only the fossil remains of *Neanderthal Man* and *Dryopithecus*, a species of ape from the Miocene epoch, were known in the mid-nineteenth century.

In using evolution to explain the current form and behaviour taken by all organisms, Darwin and Wallace were not presenting an entirely original theory. As Burrow notes in the 1968 introduction to *The Origin*, evolution as a theory in biology was, by 1859, old and discredited. Evolutionary ideas proposed by biologists and natural historians in the eighteenth and early nineteenth centuries had been ignored or ridiculed largely because of prevailing theories in science, rather than because of any lack of evidence in favour of evolution. New knowledge gained from comparative anatomy, palaeontology (the science of fossils), animal and plant breeding was available to naturalists from the beginning of the nineteenth century which supported an evolutionary interpretation. But this knowledge had marked a revolution in scientific thought and was a departure from mystical ideas that were often interwoven with science. Until the middle of the eighteenth century the standard explanation given for the existence of fossils was the 'vis plastica' theory of an Arab scholar Avicenna (980–1037). In this doctrine a 'wind of fertility' existed as a creative force in nature, and it was this which produced the images of living organisms from inanimate material and implanted them in stone. It was also widely believed that spontaneous creation could occur. Living creatures such as worms, flies, rats and other vermin could arise spontaneously from decaying matter. Even though these ideas had been overthrown, considerable intellectual barriers remained for evolutionary theories before Darwin's *Origin* appeared.

No realisation of geological time was present until the middle of the nineteenth century, and thus it appeared that an evolutionary process would have no historical depth in time in which to operate. The seventeenth century divine, James Usher, had proposed, on the authority of the Bible, that a mere 6000 years separated his own times from the creation of the world; and on the basis of further numerology drawn from the Old Testament, Dr Lightfoot, the Master of St Catherine's College, Cambridge, asserted that man had been created by the Trinity on October 23rd, 4004 BC, at nine o'clock in the morning. While such literalistic interpretations of the Bible were under attack by 1859, there were few who were inhibited from basing scientific theories on theological arguments. Scientists saw a duty to confirm, through science, the plan, design and even the predetermined end which The Creator had set for nature.

There were other reasons for the Victorian rejection of evolutionary theories. Until the appearance of *The Origin*, evolution had been closely associated with ideas of progress. A *scala naturae* or progressive succession of organisms through historical time formed the basis of many theories before Darwin. In biological terms this implied that a change from simple to more complex organisms had occurred. Since the fossil record as it was then known showed clearly that 'higher' or complex animals had emerged in some cases before 'lower' or simple animals, evolution seemed to be untenable. However in Darwin's arguments for evolution, and its mechanism of change natural selection, no ideas of progress were implied.

The evolutionary ideas proposed by the French naturalist Jean Lamarck in 1809 formed a consistent theory which drew upon philosophical and biological thinking that had been current for over a century. Lamarck postulated an innate drive to perfection within organisms and a series of stages in nature marked by increasing complexity which culminated in man. New organisms were supposedly the products of spontaneous generation that were brought forth by a capacity latent in nature that was conferred by God. Such new animals would either be perfected versions that exemplified progressive change or would represent adaptations to a changing environment that induced new needs which were themselves a motive for biological change. Finally, the inheritance of characteristics acquired during an organism's life was proposed. Lamarck had argued for the fact of evolution and had suggested, at least in part, that adaptation was a driving force. However none of the propositions which supported these claims could be confirmed. Above all, the study of extinct and modern animals demonstrated that no form of classification which implied progression by a *scala naturae* could be validated (Mayr 1982).

Before the appearance of *The Origin*, biologists remained unconvinced by evolutionary theories. Scientific thought in general was only able to conceive of a natural world as one containing fixed species. Like medieval astronomers, who believed in a cosmos of fixed stars, Victorian science could not accept change as a primary dimension of nature. The prevailing form of Natural Theology, which fed into theories in biology, was set against any possibility of a gradual transmutation of one species into another, despite the steady accumulation of evidence that later gave support to such a conclusion. Instead immutable species were seen to be appointed to a place on earth as part of a Plan. Such explanation had necessarily to involve itself in an appeal to supernatural intention which, in turn, involved a teleology, since organisms were thought to exist only to serve the purposes of such a Plan. All natural forms and their functions – the hump on camels or on zebu cattle, the process of digestion or the existence of mountains – were merely evidence of the intelligence and intention of a Divine Will.

The doctrine of fixed species was buttressed by two linked and mutually reinforcing varieties of theory in biology: essentialism and catastrophism. The essentialist thesis held that a real and permanent essence of all natural life-forms existed beneath any chance superficial variation that might exist between members of the same species. Since pure model-types were seen as fixed morphological entities, no flexibility or adaptation existed in nature under the essentialist argument. No form of transmutation of species could occur since these would be deviations from a true model.

The discovery of fossil strata containing extinct animal species, and the birth of palaeontology, provoked initial dismay among naturalists. Only by positing a catastrophe of all species followed by a subsequent new Creation would the geological record appear to make sense. As the content of fossil strata became better known and the complexity of the earth's history ever more clear, permutations of the catastrophist and creationist arguments attempted new forms of interpretation. The reconciliation of science with the new discoveries was to produce a series of grotesque arguments. Rather than one single catastrophe and a following creation, thirty, fifty and even a hundred extinctions all followed by new creations were proposed. As palaeontology showed the continuity of some species through strata and the disappearance of others, a further twist in explanation argued that the Creator, in making continual readjustments to His Plan, reshaped nature by the occasional creation of a new species while allowing the extinction of others.

Nevertheless the intellectual climate which Darwin joined had, by the 1830s, been invigorated by new scientific advances, and fundamentalist explanation was retreating with accumulating doubt. Geology, which was Darwin's main interest well into his four and three-quarter year circumnavigation on HMS *Beagle*, had already seriously questioned creationist ideas. Darwin's mentor Charles Lyell had published the first volume of his *Principles of Geology* in 1830 which accompanied Darwin on his voyage a year later. Lyell had shown that the present configuration of the earth's crust had been caused by volcanic and hydraulic activity: uniform processes that were open to rational understanding. The British landscape, it became realised, concealed evidence beneath it of epochs when familiar places had once been tropical swamp, glacier, desert or even sea. To men of science, mountain building and erosion were to become more plausible modes of constructing scenery than Divine catastrophe (Rudwick 1992). Lyell was later to argue that the human species was very much older than stipulated by Genesis, although he never fully embraced evolutionary ideas and in the 1860s he disappointed Darwin by his equivocal support. But Lyell's work established that the age of the earth was inordinately greater than biblical interpretations and that explanations of its nature and structure were more properly within the province of science than religion (Bowlby 1991).

The discovery of human antiquity was also promoted by the emerging science of archaeology, which was to grow more persuasive throughout the nineteenth century, and was to become another force which pushed against creationist doctrine. Human remains had been unearthed from deposits containing the fossil bones of extinct fauna that were foreign to Europe, such as elephants, hippos, large carnivores and apes from sites in France and England. Crude stone tools including hand-axes, that are now identified as belonging to cultures nearly half a million years old, had sometimes been discovered in the same context. Such findings, which showed that humans had been on earth well before the modern age, put Genesis in question and by the time *The Origin* appeared, the idea of reconciling biblical chronology with contemporary evidence had become absurd. By 1859 the antiquity of the human species had become widely accepted (Grayson 1983). It was the nature of our ancestry and the means of transmutation that were to become controversial.

Darwin's interpretation of nature overthrew theological causation, the fixity of species and the centrality of man in creation by advancing a general and abstract scientific thesis. All prior explanations in biology which relied upon unquestioned assertions drawn from Rational Christianity were made redundant under the process of evolutionary change governed by natural selection. It was Darwin's main intellectual achievement to realise that a void existed at the heart of biology. The theoretical vacuum that was implied by the absence of a conceptual framework meant that both similarities and diversity between species could not be explained. His greatest contribution to science was to provide an organising principle and an operating mechanism – natural selection – that was responsible for this continuity and difference.

Darwin refused to accept that the anatomical character of an animal could be interpreted on the assumption that it had been created only for the function it currently served. All organisms represented only a current stage in a series of modifications that had occurred under different conditions of existence; the most complex organs and instincts had been perfected '. . . by the accumulation of innumerable slight variations, each good for the individual possessor' (Darwin 1968: 435). Without this insight the structure and behaviour of numerous forms of life remained mysterious and anomalous: certain organs and instincts appeared to have no current function, and similarities in the anatomy of different animal species would seem to be the result of chance alone. However by assuming that species were capable of change and that whole groups of animals were related, the biological diversity evident in the contemporary and fossil records could be seen to possess a comprehensible form of organisation.

NATURAL VARIATION

The Origin opens with an initial observation that variation is a theme of nature. Individuals in any given species differ widely and vary in size, strength, health, fertility and longevity. Man has known of these simple facts for a long time and has made use of them in animal and plant breeding. Domestication of species is but one form of selection: 'The key is man's power of accumulative selection: nature gives successive variations; man adds them up in certain directions useful to him' (Darwin 1968: 90). Why such variations should always be present in a species population was unclear to Darwin and this omission was used against him by critics of evolution. Darwin was forced to assume that a 'blending' of characteristics occurred in sexual reproduction. This was an unsatisfactory explanation because such 'blending' would tend to cancel beneficial variations and produce a species that changed little.

The answer to this problem came in the isolation and under-standing of the genetic mechanism of inheritance which ensures that while offspring may closely resemble their parents they are never exact copies. Genes are units of highly complex chemical material, called DNA, which are present within each living cell of an organism. At the point of fertilisation half of the genes of each parent are combined. It is this unique inheritance, encoded in the first cell and thereafter copied as an exact chemical reproduc-tion in all subsequent cells, which together with environmental factors, will determine the characteristic form of the individual organism.

The complete genetic constitution inherited by the organism is called the *genotype*. This inheritance is fixed at the moment of fertilisation and remains unchanged during the organism's life span. During sexual reproduction the process of gene combination involves the creation of matched pairs of gene chains which form highly intricate and unique chemical structures. The complexity inherent in sexual reproduction is due to the shuffling of many tens of thousands of individual genes. It is this process of genetic combination which is in part responsible for the continual creation of new or variant genotypes.

It should be noted at this point that an individual organism cannot be seen only as the outcome of its genotype. All organisms are the products of an interaction between genes and environment. The *phenotype* is the totality of characteristics displayed by an organism from conception to death. The phenotype is not inherited and is only partly the result of the genotype. Organisms will encounter dissimilar environments in terms of climate, nutrition, disease, competition and social interaction which will give rise to radically different physical forms and behaviour. Important as these differences are, particularly in the study of human society, they should not be seen as variations caused by genetic combination.

A more fundamental source of genetic variation, however, lies in a mutation of the genotype caused by an imperfection in the reproduction of cells. Such 'mistakes' in cell replication are random and accidental modifications to the genetic structure which occur infrequently. Because mutations are chance events, which occur regardless of whether they are beneficial or harmful, they represent a 'disordering process' for the organism which can often be lethal. Dobzhansky comments:

> Occasionally, however, a newly arisen mutation may increase the adaptation of the organism . . . This in turn will depend upon the nature of the mutation and the nature of the environment . . . A mutation increasing the density of hair in a mammal may be adaptive in a population living in Alaska, but it is likely to be selected against in a population living in the tropics. Increased melanin pigmentation may be beneficial to men living in tropical Africa, where dark skin protects from the sun's ultraviolet radiation, but not in Scandinavia, where the intensity of sunlight is low, and light skin facilitates the synthesis of vitamin D.   (Dobzhansky 1977: 66)

The processes of sexual reproduction and mutation provide a continual source of variations which are offered to the environment for acceptance or rejection. These discoveries, which began with the genetic theories of Mendel (1822–1884) in the mid-nineteenth century, were consolidated by experimental work done in the 1930s. This and modern research has combined to form what has come to be known as the synthetic theory of evolution. In this theory Darwin's proposals are fully substantiated and a more complete understanding of the mechanisms by which nature introduces variation for selection by the process of evolution is provided.

NATURAL SELECTION AND THE ORIGIN OF SPECIES

Darwin's observation that nature is plastic leads to an explanation of why variations become incorporated within a species. The reproductive potential of a species is vastly greater than that required to maintain a constant level of population. Even the slow-breeding elephant has the potential of producing 15 million elephants in the short space of 500 years from a single pair. But this is not the case; there are not 15 million elephants in the world and nor do other more prolific species reproduce themselves at anything near their potential rate of increase. A very great number of reproductive cells and young individuals of all species never reach maturity. If this were not so, '. . . the earth would soon be covered by the progeny of a single pair' (Darwin 1968: 117).

Species are accustomed to a particular way of life – seen in their mode of acquiring food or in reproduction – that has a special relationship with the environment. A species therefore occupies a habitat in the environment, or an ecological 'niche'. This 'place in the polity of nature' (Darwin 1968: 152) is usually shared, at least in part, with other species. Each individual organism is therefore faced with competition from other species and from members of its own species. As the environment itself is subject to change, the niche occupied by a species will be prone to instability. Taken all together, the constantly shifting ecological balance of the environment between organic and inorganic forces imposes on species a 'Struggle for Existence'. In such a contest:

> There is no obvious reason why the principles which have acted so efficiently under domestication should not have acted under nature . . . [since] more individuals are born than can possibly survive. A grain in the balance will determine which individual shall live and which shall die, – which variety or species shall increase in number, and which shall decrease, or finally become extinct. [And] . . . Owing to this struggle for life, any variation, however slight and from whatever cause proceeding, if it be in any degree profitable to an individual or any species, in its infinitely complex relations to other organic beings and to

external nature, will tend to the preservation of that individual, and will generally be inherited by its offspring. The offspring, also, will thus have a better chance of surviving, for of the many individuals of any species which are periodically born, but a small number can survive. I have called this principle, by which each slight variation, if useful, is preserved, by the term of Natural Selection . . .   (Darwin 1968: 441/442/115)

Darwin provided a means of understanding why species were distributed as they were in an environment. The arrangement of plants on an entangled bank, or animals on a savanna, was not the result of chance but of a comprehensible process. The arrangement was in fact a temporary accommodation of various species to each other and their environment. More significantly Darwin argued that while natural selection worked extremely slowly, momentous changes had occurred in all organisms:

If we look to long enough intervals of time, geology plainly declares that all species have changed; and they have changed in the manner which my theory requires, for they have changed slowly and in a graduated manner. [This leads to the conclusion that:] All the individuals of the same species, and all the species of the same genus, or even higher group, must have descended from common parents; and therefore, in however distant and isolated parts of the world they are now found, they must in the course of successive generations have passed from some one part to the others. [And, finally:] . . . I should infer from analogy that probably all the organic beings which have ever lived on this earth have descended from some one primordial form, into which life was first breathed.   (Darwin 1968: 440/437/455)

Natural selection governs the process of 'speciation' which has led from the 'primordial form' to the present diversity of life. A species population, with a common gene pool, will contain variant individuals. Environmental change may alter the current niches of a species and may also open new niches. Those able to seize and exploit such novel ecological niches will be the possessors of the randomly distributed variations that favour adaptation to the changed environment. The relative 'success' of the variant group will be seen in its reproduction of a whole population which contains the variant characteristics as a matter of course. Despite the fragmentation of the original species, interbreeding between the two moieties may still occur, but should there be an intensification of environmental adaptation, two sub-species will appear. At this point both genetic and morphological differences between the two populations will become more marked and reproductive isolation will be usual. The two sub-species will now encounter a reproductive barrier presented by geography or behaviour: mating seasons, courtship patterns or physiological differences will tend to make the sub-species sexually incompatible, and this will prevent any further gene-flow between the two populations. The process is completed as each new species demarcates itself from its original stock by an alliance of its novel genotype with a major environmental opportunity. The whole of this process is summarised in Figure 1.1. Whole new orders of animal have arisen and proliferated by taking such opportunities. Amphibians advanced on to dry land, reptiles adopted flight and mammals entered the sea. Darwin saw natural selection operating gradually and unevenly but nevertheless with spectacular results:

> . . . I can see no limit to the amount of change, to the beauty and infinite complexity of the co-adaptations between all organic beings, one with another and with their physical conditions of life, which may be effected in the long course of time by nature's power of selection. (Darwin 1968: 153)

EVOLUTION AND PROGRESS

Dramatic and momentous as these episodes in evolutionary history are, they should not be taken as evidence of any progressive development in nature (Gould 1994). Natural selection merely explained the transmutation of species, it did not imply the direction that change might take and still less could it be the means by which a drive toward perfection was channelled. Evolutionary change was not perfectionist but conservative, since it allowed the preservation of an old life-style in the context of a changed environment. An example of this principle is provided by Romer (1958) in a discussion of the evolution of the first amphibian animals during the Devonian period, nearly 400 million years ago. The tempestuous seas in Devonian times

FIGURE 1.1　　*A Model of the Origin of Species*

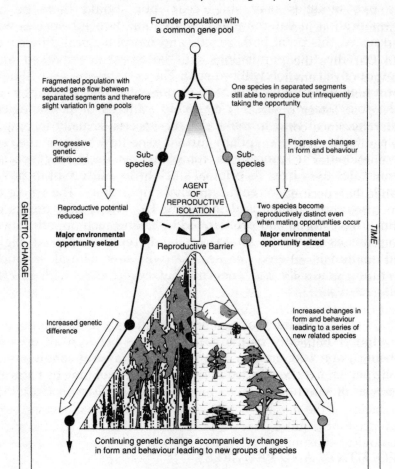

SOURCE:　Adapted from Napier (1971: 22)

frequently stranded fish in drying pools. Such experience selected for the development of lungs to supplement gills and later some 'lung fish' used their steering fins to return to the sea. The development of rudimentary limbs from fins did not imply a change in habitat or behaviour, because the first amphibians were still water dwellers who lived on the same food resources and ventured on land only when forced to: their '. . . legs were merely an accessory adaptation to aid them in staying in the water' (Romer 1958: 68). Locomotion by four feet was superimposed on a swimming mechanism to allow the first amphibians to 'swim on land'.

In Darwin's theory, all species were subject to the capricious changes of the environment and those best fitted to a new regime were mutations created by blind chance. There could be no 'higher' or 'lower' organisms placed upon a ladder of perfection or hierarchy since contemporary species, like their ancestors, were the results of an accommodation with natural forces: they were in no sense 'better'. Change under the terms of natural selection was not an 'advance' on the status quo, and neither could a direction of change be specified. The development of complex neurological structures could occur in the brains of mammals, while simultaneously bacteria and viruses developed more simple, yet more effective, means of living upon such hosts. Natural selection continually caused change in all organisms however simple or complex and no singular trend towards size or complexity was evident in nature. This process of constant evolutionary readjustment or homeostasis is illustrated by Figure 1.2.

Darwin's emphatic rejection of progress horrified his contemporaries: 'I believe . . . in no law of necessary development' (Darwin 1968: 348). The random and haphazard course taken by evolution denied the possibility of any goal for nature or of a plan which culminated in the emergence of the human species. All the intricacies of the natural world were the products of chance, not evidence for the fulfilment of a purpose. Darwin had avoided such teleological reasoning: leopards did not gain spots to become better hunters nor giraffes long necks to browse from tall trees. Rather these characteristics had been acquired because ancestors of these species with better camouflage and extended vertebral columns had tended to survive and concentrate these features in future populations.

Evolutionary theory is now at the heart of biological explanation and the validity of Darwin's theses has been reinforced by innumerable experiments and observations which illustrate the operation of natural selection. The fruit fly *Drosophila* and microscopic soil organisms have been bred in laboratories while being subjected to environmental change which has led to genetic adaptation in descendant populations. Wild rat populations have been known to develop immunities to man-made poisons after generations of exposure, as have viruses and bacteria to drugs.

FIGURE 1.2   *Evolutionary Adjustment by Natural Selection*

Broadly Stable Environment – Constant Selection Pressure

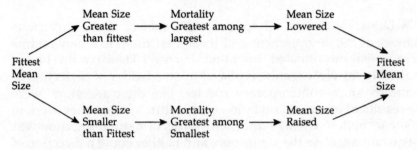

Long Term Environmental Change – Selection Pressure Alters

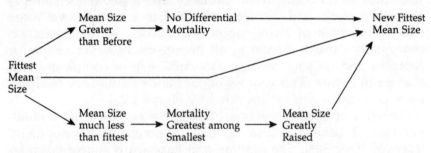

SOURCE: Adapted from Campbell (1975: 27)

Copper-tolerant plants have developed on spoil heaps left from mining copper ore that is normally toxic to all vegetation. During the process of industrial growth in the north of England the Peppered Moth, *Biston betularia,* has changed its colouring to match the bark of trees that are now stained by pollutants. Current debates in biology occur within the ambit of evolutionary theory. There are controversies which concern the rate of evolutionary change and the precise lineage or 'phylogeny' of organisms, but no scientific theory which would serve as an alternative to evolution exists. With regard to human evolution scientific opinion varies, as we shall see, upon such matters as the timing of the divergence between the primate stocks that were ancestral to *Homo sapiens* and to modern apes, on the significance of brain enlargement, on the reasons for bipedalism and on the behavioural characteristics of human ancestors. However, such differences do not question the main proposition that modern humans are products of evolution with an ancestry that may with confidence be traced to a prehuman state.

SOCIAL EVOLUTION

Since the eighteenth century, evolutionary ideas have been applied to the historical development of human society. We must briefly examine the validity of these ideas and show how they were used in sociology by contemporaries of Darwin.

Theories of social evolution were current *before* Darwin's *Origin* was published. They arose initially from a wish to derive sociological explanation from philosophical ideas of progress, but after 1859 new attempts were made to join these ideas with biological evolution. This is not the place to present a full discussion of the origins and development of social theories that deal with progress and evolution: however, a clear distinction between these ideas and Darwin's theory must be made.

Despite Darwin's clear rejection of progress and his insistence that moral or social values could not be drawn from nature, there were many nineteenth century popularisations of evolution which reproduced current obsessions and prejudices with an aura of scientific authority. As Burrow remarks, the fashion for 'Social Darwinism' was '. . . a mould into which, for a while, the most diverse and often incongruous ideas were pressed' (Darwin 1968: 48). Social Darwinism was little more than a rag-bag of contemporary preoccupations made into a pseudo science by a misalliance of social observation with biological analogy. As we have seen in the Introduction, Darwin's biological principle 'the struggle for existence' was made by others to serve racist and nationalist doctrines, which included colonial expansion, and even to sanctify free-market competition as a state natural to man. The key concepts of biological evolution, adaptation and natural selection, were either dropped or misrepresented in these theories, and notions of fitness or progress substituted. Darwin had provided a biological theory which became improperly joined to a pre-existing stream of ideas.

By the middle of the nineteenth century an assortment of social evolutionary theories had accumulated and become dominant. One product of the eighteenth century Scottish Enlightenment was a long-term view of history which conceived of a series of connected economic stages described as hunting, pastoral, agricultural and commercial. A natural propensity towards an increase in the division of labour was seen as the source of progress from one stage to the next. Such sequential development tended to give prominence to the history of Western society and to celebrate its existing nineteenth century configuration as a stage that was higher than all previous social forms. On this basis it was also claimed that such a model represented a goal of social progress and that the purpose and necessary direction of history was revealed in the struggles of 'lower' forms to achieve this end. The once benign attitudes of Europeans to Asian societies and cultures changed to contempt; these social systems were now seen as cases of arrested change.

Herbert Spencer's work *Progress: Its Law and Cause*, which was published in 1857, two years *before* Darwin's *Origin of Species*, best exemplifies Victorian social evolutionary theory. Spencer attempts to base his social theory on biological arguments: in fact a single causal mechanism is held to be responsible for both society and nature. Progress for Spencer was apparent '. . . in those changes of structure in the social organism . . .' (Spencer 1982: 526) which have led to an increase in social complexity. While Spencer wished to be relieved of any popular notion of progress, which he conceived of as '. . . the making of a great quantity and variety of the articles required for satisfying men's wants; in the increasing security of persons and property; in widening freedom of action' (Spencer 1982: 526) it is clear from later argument that such a state would represent a necessary outcome of greater complexity. But these achievements are not evidence of progress simply because such changes lead to an increase in human happiness – this would be a teleological argument.

In fact an underlying cosmic law of growth and change, that is applicable to biological and social phenomena, is conceived to be at work. Spencer takes growth from a simple homogenous primary 'germ' to a complex heterogenous structure by the process of 'differentiation', as an equivalent if not identical process, for both the growth of animals and plants and for the development of civilisations. What unites the two varieties of change is a law which Spencer states: 'Every active force produces more than one change – every cause produces more than one effect' (Spencer 1982: 531). Spencer was in no doubt as to the ubiquity of these forces '. . . this law of organic progress is the law of all progress. Whether it be in the development of the Earth, in the development of Life upon its surface, in the development of

Society, of Government, of Manufactures, of Commerce, of Language, Literature, Science, Art . . .' (Spencer 1982: 527). Humanity's development from the germinal state of the barbarous tribe has been a sequence of differentiations and innovations of social structure. A unilinear progression from savagery is proposed in which the rudiments of political institutions have developed from primitive god-king figures to the final fully developed form apparent in the separation of church and state. Propensities toward specialisation, leading to economic develop- ment, are nascent in the human mind and are seen concretely in the expression of aptitudes and demand for the resulting superior products. Our adaptation at this primary level overcomes the problems of subduing raw nature, but in so doing an increase in population levels provides further impetus for differentiation. At each level of progress the better adapted races are impelled – by pressure from their growing populations – to overcome the less adapted races.

The differences between Darwin's account of evolution by natural selection and Spencer's theory are marked. Under the terms of biological evolution, the variability of characteristics in a species population or the transmutation of species does not imply that the individuals who survive are the most fierce, fast or large. There is no 'improved model' of an organism that can be said to embody progress and nor can some animals of the same species, containing roughly similar individuals, be seen as competitors who are better adapted because of their strength and size and therefore more fit to supercede others.

Biologists such as Dobzhansky and Boesiger (1983) point out that the 'struggle for existence' is largely metaphorical and rarely takes the form of actual combat. Desert plants, for instance, adapt to arid conditions by various devices that protect them from excessive evaporation, not by extracting water from their neighbours. Animal combat is often ritualised and real harm is rare.

Evolution by natural selection means constant undirected adaptation to an environment that is itself changing, and such changes could favour more simple, smaller or weaker variants. The surviving fittest individuals are not super organisms: they are merely the parents of the largest number of offspring. It is therefore nonsensical to claim that evolution produces superior fitness or complexity or that any finite evolutionary stages can be demarcated: evolution is simply a continual adjustment of the organism to its ecological niche. Darwin's theory provides no grounds to justify the oppression of the weak by the strong.

Unless a metaphor is intended it is equally meaningless to refer to society as a 'social organism'. Societies are not biological organisms produced by genetic combination in the process of sexual reproduction nor do societies have a finite existence that is limited by birth and death. Neither can speciation through variation or mutation be said to apply to social forms. War or population pressure are not equivalents to natural selection. The resulting mortality, often from famine and disease, affects people indiscriminately. These disasters do not cause a surviving population to speciate by favouring variations in genotypes that lead to better adaptation. On the contrary they act to debilitate whole populations.

It is because human social organisation is an extra-somatic creation, a phenomenon which has definite biological prerequisites but is not in itself biological, that so many different environments have been claimed as human habitats. In all cases different societies have organised, transformed and been in contact with the environment in a *collective manner* rather than as individual members of a species. The cultural constructs, seen in the material and organisational dimensions of society, act as a powerful intermediary force between the environment and the human species. Culture thus acts as a buffer between humanity and nature. In this sense 'adaptation' by a human society to an environment is quite distinct from adaptation by individuals of an animal species.

Many highly diverse modes of human adaptation have existed to the same environment by different societies over time and each adaptation has constituted an accommodation with nature that has given a 'fitness' or viability. While it may be correctly asserted that one society exhibits qualities or potentialities that are not present in another, there are no meaningful criteria of relative or superior 'fitness' of a society to an environment that can be applied.

Hirst (1976) has argued that all theories of evolution which incorporate ideas of progress are teleological and arbitrary. Spencer's teleology lies in the inflexible, directional and deterministic presentation of history in which progress constitutes change from simple to complex. Why this increase in complexity should be inevitable in both the physical and human worlds cannot be directly known since it is an '. . . ultimate mystery which must ever transcend human intelligence' (Spencer 1982: 530). However, evidence for such a tendency is seen in empirical generalisation. The assumption that social life has the 'goal' of civilisation to work toward, that cosmic laws of progress provide the source for such a quest and that social advances are transmitted – having been refined and consolidated – from stage to stage, all point to a teleological theory that is based upon purposive change. Spencer's reading of history is not an innocent and faithful 'empirical generalisation' but an arbitrary selection of stages, marked by increased institutional complexity, which are deemed to be necessary preconditions for the stage which will follow. Empirical generalisations cannot prove that a singular path, that is routed via stages that are chosen arbitrarily, leads to an outcome that is already given. History does not move forward with a purpose or direction that is predetermined: a mass of other empirical examples could just as easily be cited to show that social complexity is often followed by social simplicity. Spencer's law states that forces and causes produce multiple changes and effects, this being the empirical reflection of an underlying cosmic process. But Spencer has not shown why the law operates in any particular direction without reference to a teleology which describes civilisation as a goal. Spencer is thus obliged to invoke cosmic law to use as a device to cover the teleology.

Natural selection does not carry with it any connotation of a purpose or direction and it is of no value in social analysis. Darwin did not endorse any social application of his theories by others and his collaborator and proponent, Thomas Huxley, repudiated such attempts. Huxley (1893) argues that evolution cannot be used to resolve moral questions – that are implicit in such issues as class hierarchies or imperial conquest – because evolution is a natural force which does not differentiate on the basis of moral criteria. The struggle for existence by plants and animals cannot be transposed to the human social world. While successful biological evolution consists of species adapting to whatever conditions are given by nature and perhaps becoming the 'fittest' or 'best' by developing the most simple characteristics, as are found in vegetable lichens and algae, human adaptation attempts to lessen the full impact of nature and overcome environmental forces. Socially organised cultural behaviour therefore mediates the effects of biological evolution in human populations and in this sense selection is for individuals who embody social behaviour patterns as much as biological fitness. In fact Huxley counterposes as opposites what is 'ethically' or socially best, and the forces of nature. Evolution by natural selection '. . . repudiates the gladiatorial theory of existence . . .' and the '. . . fanatical individualism' (Huxley 1893: 33) of social evolutionary theory: society depends not on imitating nature but upon ending its impositions. Huxley sees civilisation as a construction or '. . . an artificial world within the cosmos' (Huxley 1893: 35) which has changed the terms of evolution.

The complexity and contrived character of human society and culture, together with their considerable distance and independence from nature, which Huxley points to, should not be allowed to obscure the fact that the origin of human social life rests upon a biological basis that becomes comprehensible only in the context of evolution. In 1871 Darwin was to remark that in spite of an inordinate capacity for benevolence and a god-like intellect, '– Man still bears in his bodily frame the indelible stamp of his lowly origins' (Darwin 1981: ii: 405). In fact a variety of human behavioural patterns, as well as anatomical features, are now seen mirrored in some non-human primate species for whom culture is also of some significance. The social life and physical constitution of the primates that were directly ancestral to the human species provided a founding set of attributes for the human lineage. Yet, at the same time, the biological nature of modern humans and the direction taken by human evolution have been shaped by the practice of a way of life in which culture has been central. In this sense human evolution should be seen as a process in which human biology and culture have entered an interactive and mutually reinforcing relationship which goes well beyond that found among our ape relatives.

2

Culture and Evolution

Since the origin of life some 3.5 billion years ago, a series of significant evolutionary developments may be noted. The independent movement of organisms, the rise of animals with backbones, sexual reproduction, the emergence of organs for sight, flowering plants, the development of flight and the ability of animals to maintain a constant body temperature all mark important evolutionary milestones. While these developments did not imply progress in any directional sense – they were not cumulative, sequential or confined to a single lineage of animals – they were nonetheless biological revolutions. These innovations vastly increased the potentiality of life forms by opening and diversifying a range of niches.

Human evolution, featuring nearly unique abilities, is another such quantum leap in evolution. The significance of this leap lies in the appearance of an animal with a set of behaviour patterns that, though they remain attached to biology, are ultimately determined by culture. Culture in human and pre-human society represents a distinctly different means of regulating behaviour from that found among other animals. The emergence of culture and human characteristics is, as Dobzhansky and Boesiger (1983) argue, a unique moment in evolution. Human beings are descended from non-human ancestors who were the products of ancient environments. We will inevitably give rise to descendants whose form and behaviour will be 'post-human'. Human nature and morphology are not fixed: on the contrary our status represents only a conjuncture in the history of life and is as subject to change as any other part of the physical and animate environment.

To some extent it is possible to explain the emergence of culture by turning to evolution and natural history. An attempt will be made to present culture as an evolutionary product at the end of this chapter. However it is not possible to dismantle culture into a series of component parts which may be traced to separate forms of animal behaviour. Culture holds an independence from its biological origins since it is both self-generating and has a real existence only in social interaction. For this reason social scientists have usually discussed culture as a given entity without attention to origins and non-human forms of culture. In this chapter we shall look at culture both from the perspectives of social science and evolution.

THE NATURE OF CULTURE

For both humans and some social animals culture consists of all the behavioural patterns within a social group, that are passed on to descendants by learning. In any single species behaviour will reflect the idiosyncrasies of group members and will vary accordingly between groups but, if it persists for long, the behaviour will assist the whole group in coming to terms with its environment. Among animals a set of novel behaviour elements will distinguish a particular social tradition and in modern humans ideas and material objects are a part of such traditions. Biological mechanisms such as a large brain or a drive to gain information may provide each new generation with the capacity for cultural acquisition, but it is only by learning that actual cultural contents are gained.

Culture is therefore a superorganic or non-genetic form of adaptation which has a collective existence independent of single individuals. As such, culture is of inestimable adaptive value particularly to a human population which alone is able to make rapid creative changes and innovations. Genetic evolution is very lengthy, and may involve many thousands of generations to produce an extensively changed genotype. Culture, on the other hand, is highly flexible and is capable of diffusion to dispersed populations in less than a single generation. Bipedal human ancestors, with a cultural system that was probably not far removed from that of modern apes, developed a material culture in which at first tool-use and then tool-making became an essential facet of social life. Cultural behaviour now included a material element which came to grow in significance. Campbell comments:

Evolution is opportunistic, and any novel behaviour pattern with a selective advantage will, under appropriate circumstances, be incorporated into the behaviour of the evolving population. Such behaviour is necessary to develop and maintain material culture, for the use of external objects as tools increases the effective exploitation of the environment and decreases the danger of predators.   (Campbell 1975: 329)

Possession of a material culture at once implied environmental change which, however slight in its original extent, meant that an interactive process had begun in which human adaptation, in both the physical and cultural senses, was to be oriented towards an environment that was to a degree moulded and controlled by the very organism which was seeking an accommodation with it. Under the terms of this new compact with nature the emergent human species was to change the rules of evolution: natural selection was now to preserve those characteristics which best fitted an environment that had already been shaped by previous cultural life. Our socially determined activity had come to unconsciously influence future evolutionary trends in the human form and mentality. The original ancient environments that were natal to human ancestors have been transformed by cultural activities to the point that any attempt to relate modern humans to a 'natural' environment becomes meaningless. We have long since abandoned these habitats. Human adaptation is predominantly cultural: material artifacts and diverse modes of social organisation constitute an impermeable barrier between our species and the environment, that is essential for survival.

With the cumulative elaboration of culture the density of this barrier has progressively increased. Such elaboration consists of much more than an increase in the sophistication of material artifacts. Beliefs and values become crystallised within social behaviour and provide crucial landmarks for guiding action. The recognition of kin-relations or the virtues of sharing food are encapsulated in regular social practices or structures, and these structures, in turn, produce a collective social relationship with the environment which is a major influence on material development.

Each particular cultural form or type of society comes to accommodate itself and shape the environment in different ways. The environment has an influence on the formation of a culture but no absolute or determining role can be specified. Human culture is flexible and an innumerable variety of accommodations may be made to a single environment, or quite different environments may host cultures of great similarity. The use of culture has liberated human beings from determination by the environment alone: the necessary precise fitness of an organism to an ecological niche, that has in the end ruled the destiny of all other animals, has been escaped by a species that has been able to consistently reshape its niche to correspond with its social character.

The qualities which underlie this superior form of social organisation which now sustain a cultural way of life are conceptual thought and self-awareness, intelligence and the ability to use the symbolic forms of communication that are found in language. These characteristics, which are not totally exclusive to humans, are nonetheless a major part of our specialist adaptive pattern, and they should be seen in an evolutionary perspective that embodies an explanation for their origin and their connection with culture.

CULTURE AND THE SOCIAL SCIENCES

In the sociological tradition these qualities, together with our complex social organisation, have been taken as a confirmation that culture is an exclusively human property and that it is culture which demarcates us from the animal world. It is frequently argued, with justification, that the force of cultural determination in human society is unparalleled among animals. Culture is internalised in humans to the extent that even biological processes may be overridden. Thus the experience of pain may be mitigated or exacerbated by religious belief; fear of death counteracted by ideals, or death itself produced by a curse or by ostracism. Basic drives such as sex or aggression may be canalised by culture into a variety of socially approved forms of expression.

For Tylor (1832–1917), a founder of anthropology, culture and civilisation were synonymous. His definition of culture as, 'that complex whole which includes knowledge, belief, art, morals, laws, custom and any other capabilities and habits acquired by man as a member of society' (Tylor 1982: 18), excludes the possibility of culture in animal societies. Other anthropologists, like Kroeber (1982), recognise the evolutionary significance of culture but deny it to non-human animals. Kroeber maintains a formal distinction between the organic and the cultural which marks an essential difference between animals and humans.

Human society is marked by special forms of symbolic communication while animals are imbued with inbuilt behavioural repertoires. Learning in animals is restricted to the acquisition of an unchanging set of species characteristics, which does not constitute a social tradition.

In the work of Leslie White (1959A, 1959B, 1982) culture is seen as a concrete aspect of life open to scientific investigation. It is the ability to use symbols which makes culture possible and this ability is enjoyed by our species alone. White argues that human life is based upon 'symbolling' defined as the bestowal of meaning upon a thing or act and that cultural interaction is a process of trafficking in symbols which are conveyed by language. A veil of culture has been interposed between us and our environment which orients all human perception. Determining everything is language, which creates symbols that endow the environment with meanings and values that are not conveyed by the senses alone. Thus an object or act may acquire a variety of symbolic meanings. Water may be holy, a gesture rude, purple the colour for royalty, black or white the colour for mourning or sobriety or purity or whatever a cultural system may designate. Language, like culture, imposes meaning in an arbitrary fashion. It is this arbitrary property of language which allows conceptual thought and therefore culture itself.

White is insistent that symbols were the means by which our ancestors transformed themselves into human beings. The transformation has isolated us from animals who cannot use language to engage in symbolic transactions. Animals cannot manipulate a code of arbitrarily selected sounds to convey symbolic patterns and cannot participate in a social world whose basic structures include religious ideals, kinship relations and concepts of shame. Humans alone can 'arbitrarily impose signification' (White 1949: 36): animals cannot, and no intermediate stages are possible.

The ability of social scientists to maintain the rigidity of this distinction has diminished with the emergence of new forms of evidence. For one thing, pre-human ancestors made and used stone tools more than 2 million years ago. These small-brained hominids are thought not to have used language but, on the basis of their archaeological record, they cannot be denied culture or conceptual thought. This evidence alone should persuade us that culture, like other abilities, has a history in its appearance as a human speciality, and is not a unitary phenomenon that is either present or absent. Secondly, observation in natural habitats and laboratories has shown that animal communication is sufficient to allow social traditions and even a material culture. Many primates and animals, such as Cape hunting dogs, have been found to engage in practices that are peculiar to their social groups alone. This behaviour is acquired by learning and is retained by descendent generations. Chimpanzees are known to make simple tools and have been taught sign language. Even though chimpanzee 'language' competence is no greater than that of a two-year-old human child, some ability to use symbols and conceptual thought must be assumed. Thirdly, Hirst and Woolley (1982 and 1985) have argued that human infants at a pre-language stage construct elaborate fantasies and that their mental life includes conceptual thought well before the ability to speak occurs. Finally, these authors have also shown that modern linguistic theory will not support the notion that language consists of a series of sound signs that arbitrarily assign meaning. This evidence points to the fact that, in order to operate as a coherent system, sound signs are interdependent and take their value from each other. Further, sound signs are inextricably linked with the bundles of ideas and social organisation that comprise the everyday substance employed by language users to make sense of the world and are anything but arbitrary.

These arguments seriously undermine the proposal that language, culture, conceptual thought and symbol usage are interconnected in the way White assumes, and above all that they are exclusive human properties. It is important not to claim too much. Animal culture and communication is rudimentary by human standards but, if evolution is to be taken seriously, the rigid distinction of the social science tradition, which also mirrors many religious distinctions, cannot be upheld. There can only be continuities between ourselves and other animals with respect to human abilities rather than essential differences.

A harrowing illustration of these continuities is provided by work on rhesus monkeys done by Harlow and Harlow during the 1950s and 60s. In a series of controlled laboratory experiments infant monkeys were subjected to varying forms of deprivation. Groups of infants were raised in total isolation for different periods after early separation from their mothers. Some infants were allowed two surrogate mother figures, one made of wire and the other from cloth, in an otherwise bare cage. In all cases the 'cloth mothers' were preferred by the infants who clung to them for comfort even though the 'wire mothers' included feeder bottles. Other infants were raised with their natural mothers but denied contact with their peers. Further motherless groups were brought up only with other infants.

The experience of isolation was found to induce a range of abnormalities that are comparable to syndromes created in human infancy. The authors comment, 'Similar symptoms of emotional pathology are observed in deprived children in orphanages and in withdrawn adolescents and adults in mental hospitals' (Harlow and Harlow 1962: 4). Isolation had permanently damaged the ability to form real social bonds with other monkeys. Having matured, these experimental subjects were unable to play, engage in sexual behaviour or defend themselves in any normal fashion. Those who had been raised by mothers who were themselves deprived in infancy also exhibited abnormal behaviour. The mothering given here was clearly aberrant; it was often indifferent and even violent.

The Harlows have established important parallels between humans and other primates that are probably the results of a partially shared evolutionary journey. Many of the learning processes upon which social life depends are activated during the infancy of monkeys and humans. The crucial significance of this developmental phase for normal human psychic maturation leading to social integration and sexual identity has long been recognised. These and later experiments have also shown a connection between emotional security, environmental stimulation and the growth of intelligence in various primates including man. A desire for soft physical contact, warmth and gentle motion as a primary means of experiencing affection and security is found among primates in general. The need for social contact with other infants in a play context has also been found to be common to primates. There are occasions when humans and other primates seem to exhibit nearly identical forms of behaviour. Chimpanzee juveniles, for instance, have been observed in their natural habitat to express jealousy by regressive tantrums after the birth of a sibling.

There are many other aspects of primate social life which also serve to bridge essentialist distinctions between humans and animals. Attempts to demonstrate finite differences of this sort usually produce only ambiguities and fail to be convincing. The argument that certain abilities have become a human speciality, with powerful transforming effects, in no sense confirms this essential distinction. We cannot exclude animals from intelligent thought and behaviour, tool-making, self-awareness, social sensitivity or a range of feelings that would include joy and humour, empathy, grief and desolation merely because these are an indivisible part of the human condition. In fact there is no simple or singular criterion which distinguishes humans from other species.

CULTURE AND BEHAVIOUR

How then should human beings be seen in relation to other animals? Although we are a part of the natural world our behaviour has ensured that we have become detached from the conditions imposed by nature on other animals. The distinctive characteristics of humanity are marked by this separation. Culture dominates our existence by moulding environments and by imposing a barrier between individuals and their environment. Our survival rests upon the performance of a flexible range of learned behaviours that accord with a cultural repertoire rather than a closed genetic programme. Intuitive understanding and self-awareness are essential parts of cultural practice. These new dimensions of consciousness, which allow us to reflect upon our thoughts and interpret those of others, also enable us to have perception of time and an awareness of our existence in this context. There are other human characteristics such as tool-making and symbolic language which, as we have already noted, are found in a rudimentary form in some animals. However, taken as an interdependent complex whole, all these human abilities form a unique entity and are the basis of a novel mode of adaptation which seeks to control and change the environment. In practice this has meant intervention in both the environment and the process of evolution itself for humans and other species. But all this behaviour could not possibly exist without a cluster of physical attributes not least of which are our relatively large brain and upright posture.

Social scientists have usually avoided discussion of any physical aspect of social behaviour and have preferred instead to present human action as a cultural totality. In this approach, all forms of social organisation and behaviour are seen purely as cultural products. Both society and humans are 'blank pages' that are conditioned by cultural forces alone. No human nature or set of social structures that serve intrinsic needs is assumed. The extremes of cultural variability are often cited to demonstrate that, in theory, there seem to be almost no limits to such conditioning. Thus parental love or indulgence in one society can be contrasted with emotional austerity or even with sexually selective infanticide in another society. Similar contrasts are sometimes made between promiscuity and chastity or aggression and passivity in the discussion of gender differences, adolescence or criminality.

The use of such variation in social analysis tends to promote the idea that humanity is a formless species which gains a behavioural repertoire only during the experience of socialisation. This form of reasoning has the effect of denying humanity a place in the natural world: the significance of evolution is ignored and an artificial distinction is made between nature and culture.

While no purpose is served in demoting the role of culture in human society it is absurd not to recognise that cultural behaviour is both the product and cause of a range of biological and psychological mechanisms. Full human status – the possession of a collection of mental and behavioural properties – can only be realised as human biology unfolds in the presence of a cultural environment. In this sense no simple boundary may be drawn between nature and culture. Human development depends upon such an interaction. Irrespective of cultural variation our common evolution as a single species would suggest that, while we are a culture building animal, our cultures are built around a pattern of recognisable human qualities. Members of utterly dissimilar cultures have not the slightest doubt when encounters occur, that they are in contact with other humans. Whatever results from such meetings, a common species recognition occurs.

Midgley (1980) has argued that a human nature can be distinguished, and that nature and culture are not opposites but part of a coherent whole. Far from being infinitely plastic and structureless, human behaviour is a product of this whole. Cultures have to satisfy a natural pattern of motives and our species, like any animal, can be assumed to possess a range of drives and instincts. This claim does not imply fatalism nor is it allied to the determinism of sociobiology, which will be discussed in Chapter 3. A nature does not imply a destiny that is genetically fixed and cannot be compensated for or even superseded. Human instinctual drives are not a part of a closed programme serving in the way that birdsong, courtship rituals or web-making function in a range of animals. But it would be incredible if none of our behaviour contained genetic components that could be traced to our evolution. Midgley's claim takes the facts of human adaptation seriously while it also produces the understanding that culture and nature are bound together.

In simple acts like staring or smiling or the conservation of a personal body space, at least a part of everyday action may be inherited. Smiling is a part of the communicative repertoire of many primates, and a prolonged stare is interpreted as a threat by many animals. The smile, for instance, seems to be universally present in all societies while its use is of course culturally regulated. Since children who are born blind and deaf will smile and laugh in the same circumstances as normal children, the behaviour cannot be attributed to learning (Eibl-Eibesfeldt 1970). Apes and monkeys make facial expressions which appear to human observers as smiles and grins. These acts are used in play and greetings or to indicate submissiveness and reassurance. Human smiling may not be related to other primate facial gestures. There is no means as yet of isolating a common genetic mechanism. It is conceivable, as Passingham (1982) suggests, that similar forms of behaviour evolved independently to serve similar social functions in humans and other primates, but it would be foolish to dismiss the possibility of a common behavioural trait.

To propose that we possess an inherited emotional range that is evident from body and facial expressions, crying, sociability, aggressive feelings or sexual desire and practice, does not imply that such behaviours are independent of culture. There is no reason why such natural, inherited drives should not be activated and modified by culture. Many animals besides humans are also shaped by both social interaction and learning. Animal behaviour is not a set of standard reactions to stimuli: chimpanzees have to learn and interpret behaviour rather than produce an automatic response. Time, practice and an example are required to develop a full behavioural repertoire from an innate drive or capacity in most mammals. Indeed society and genetic programming are not alternatives either for us or other higher animals.

Objections to this form of 'open' genetic programme often focus on complex forms of human social behaviour. Social scientists have rightly objected that human *social institutions*, the building blocks of a society, through which people gain their daily sense of reality and integration, cannot be explained away by biology. Social institutions like the family or work groups or political parties are not mere cultural clothing which cover a pattern of innate urges. The family, in all its diverse forms, is the product of cultural factors even though it may encapsulate activities that are fundamental for biological survival and reproduction. Still less can complex human affairs be shown to result from a predetermined inner programme. The explanation of disputes which result in warfare or social hierarchies or the division of labour in society must be derived from social causes. These forms of conflict cannot be explained by reference to an individual aggressive drive or assumed to result from an aggregate psychology. They are the products of cultural forces: the collective manifestation of whole social structures that operate with a logic that is independent of biology. It is important that social science avoids such reductionist thinking, but no purpose is served by ignoring the fact of a common human nature and by insisting that human behaviour is directed only by culture.

There would seem to be an innate physical basis for language in the human brain, that is activated by society. Distinct developmental stages in language learning can be observed in children which include the introduction of original, unheard grammatical mistakes by children themselves. Similar forms of sequential development can be observed in spatial conceptualisation seen in the drawn shapes of children and in forms of play. Such development could not occur if humans were just 'blank pages' and there was no unfolding of an innate psychic mechanism in the presence of social interaction. We have evolved by using these skills and they are in a real sense natural human properties. Cultural 'conditioning' can only function along given pathways that cohere with human characteristics: conditioning cannot operate randomly or at the command of an arbitrary director.

It is equally important to stress that human evolution is a process of adaptation *through* culture and that the use of culture is natural to human beings; it is what we have evolved to use. We do not have culturally directed social behaviour imposed upon an otherwise savage or animal or even 'true' nature. Culture is not an alien affliction – the antithesis of freedom – it represents the only possible means of realising a human way of life: it is only through culture that human development can unfold and human needs can be fulfilled.

Culture serves to initiate processes of psychic development and maturation, provides a reservoir of examples for learning, and a predictable pattern through which to structure behaviour; culture conveys a social tradition together with rules, choices and symbols, and provides an understanding of why action is undertaken and a sense of identity. But irrespective of wide cultural divergences, a recognisable human way of life is reproduced that accords with human needs. Human social life stresses constant interaction: we are a gregarious species of primate with an urge to communicate quite unlike our solitary second cousin, the Orangutan. Such sociability everywhere includes humour and stories accompanied by some expressive movements and facial gestures that are universally understood. It would seem implausible that any society could exist without a basic currency of human interaction. Jokes, gifts, dietary rules and taboos, ceremonies, socially organised production and exchange, friendship, ritual, music, games, bodily decoration and art occur in a multiplicity of forms but recur throughout human societies.

Cultural practices vary in important ways yet recognisable patterns remain. Death and the dead may be variously perceived and disposed of, yet this rite of passage is always seriously marked and the deceased never treated with indifference. Despite our divisive marriage customs, marriage and bonding remain. Sexual behaviour is rule-bound – promiscuity has never been part of a human society for long – and mating has always been of social significance.

The most fundamental of all innate psychic drives, which develops by close interaction during the first few years of life and has such crucial importance in maturity, occurs in the bonding process between infant and parents. It is here that both natural and cultural forces come to interact in such a decisive manner. This first social relationship is closely coupled with innate drives involving both child and parent. The relationship will induce the formation of a cluster of human characteristics and it will structure a set of internal subjective meanings and sensibilities that are perceived through a specific culture.

The formation of this bonding process, the attachment of the child to its mother and later its father and the significance which this has for the development of adult personality is a basic concern of psychoanalysis. The process of bonding and the effects upon the child of its disruption by separation from parents are vividly described by Bowlby (1984). After the age of six months, prolonged separation from the mother without a surrogate parent figure produces a typical reactive pattern. A child experiences phases which Bowlby terms protest, despair and detachment. The sequence is one in which active anger at maternal loss gives way to dejection and withdrawal followed by an apparent recovery of normal well-being that masks what is in fact severe damage to the child's existing emotional bonds and an indifference and distrust of close personal relationships thereafter. These phases have been found to occur in a variety of cultures. They cannot be attributed to a new and strange environment since their onset rarely occurred when a child was accompanied by a parent during a stay in hospital for instance. A small child frequently construes the prolonged absence of a parent as an abandonment, and the traumatic effects of this separation have been observed in retrospective studies. The experience of pre-school children who were left in hospital for more than a week without a parent figure and were also denied visits has been found to be associated with long-term disturbances of behaviour and learning. In a sample of more than a thousand such children Douglas (1975) found that the number and length of hospital admissions significantly increased the likelihood of troublesome behaviour, poor reading and concentration and was positively correlated with delinquency.

Characteristic forms of instinctive parent–offspring behaviour can be noted among animals. Parents and young easily recognise each other, their behaviour together is distinctive, and constant attempts are made to be in close proximity. Observations of parental behaviour in primate species support the idea of evolutionary continuity with humans. Infant–mother bonding, seen in early clinging, grooming and continual close physical contact during infancy and a strong emotional tie which persists into adult life is fundamental to monkey and ape societies.

The quality of love and care-giving in human societies is strongly directed by the child-rearing norms of a culture. The historical record shows examples of indifference, the withholding of affection from children at specific ages and demands made upon children for stoical behaviour to repress any manifestation of liveliness. Rigorous forms of violent assault upon infants are even institutionalised in some cultures. There are examples of mutilation and restriction seen in male and female circumcision, scarification, foot-binding and swaddling which occur in sophisticated and complex societies. Culture then is in no sense a guarantor of 'natural' tenderness. However none of these cultural practices denies the facts of a parent–infant bonding process. A holistic interpretation of behaviour cannot be based upon culture alone. Attachment and bonding behaviour are present throughout the diversity of cultural regimes.

Bowlby's work shows that parent–child relations consist of far more than cultural conditioning. An important part of this behaviour lies in innate drives which are the products of a long chain of evolutionary development that connects humans with animals. Such basic parts of the human behavioural repertoire, along with other nascent skills, have been implanted to enable humans to acquire and live by culture.

CULTURE AND BIOLOGY

The evolution of human cultural behaviour is a novel and distinctive form of adaptation which departs radically from the regulating principles that govern even our closest animal relatives. Yet we were created by the same evolutionary process that shaped all other forms of life, and until the emergence of culture natural selection remained largely uninfluenced by our actions. It was argued above that human society and culture are possible only because of human biology. The converse of this statement is also a part of the argument being presented here. Our physical form, in all its complex developmental stages, requires interaction with social and cultural processes both to survive and to be realised as a biological entity. It is therefore necessary to pose two related problems. First, how and under what circumstances did culture arise as an evolutionary product? Second, what were the effects of our ancestors' practice of a culture-based life-style on subsequent human evolution? Both problems look at the question of how the set of standard human features, that have become the stamp of our species, developed from a pre-human form. These problems will be explored later but in the rest of this chapter the biological regulation of behaviour and evolutionary strategies that would favour the emergence of culture will be examined.

It can be shown that in some animals, behaviour is both inherited and adaptive. The breeding experiments of Dobzhansky (1972) provide evidence for a 'genetic architecture of behaviour'. In these experiments the abilities of fruit-flies to respond to gravity and to light was found to be controlled by genes. By selectively breeding flies in the laboratory it was possible to intensify and diminish these abilities. Two populations of flies with extreme opposite forms of behaviour pattern were created. A natural genetic response to fly upward towards light when disturbed, was accentuated in one group and lost in another after 20 to 30 generations of successive breeding. However when artificial selection was relaxed in both populations, the extremes of behaviour were lost in subsequent generations. Natural selection had produced an *optimum adaptability to the environment* as a whole. Dobzhansky concludes that the genetic diversity among whole populations provides the ability for some individuals to adapt by behaviour to potential environments. The supply of these variant individuals within a population enables species to adapt rather than remain irreversibly committed to any particular environment. This work shows that evolution and behaviour are closely interconnected.

However, the link between genes and behaviour is neither direct nor simple. Mayr (1974) distinguishes between two forms of genetic programme which organise behaviour. The *'closed'* programme, which is typical of short-lived lower animals such as insects, does not allow modification of behaviour by learning or experience. These fixed behavioural programmes provide for recognition of the animal's species, food sources, mating and courtship rituals, appropriate places to lay and secrete eggs and communication systems without any observation or practice of these behaviours. Such animals usually have little or no parental care. They must make the appropriate signals and responses to other animals of their species accurately and without ambiguity since, in mating for example, only one opportunity may be presented in a life-span numbered in only hours or days. Since there is neither time or an example for such animals to learn from experience, a full life sequence of behaviour is laid down in a closed genetic programme.

In more complex species, with prolonged periods of parental care, learning opportunities are provided. *'Open'* programmes allow new information, acquired through experience, to be translated into behaviour. A large part of behaviour is not determined by genes. While an ability to learn is inherited, it is parental tutelage which fills an open programme with environmental details, such as food, shelter and enemies, that become incorporated within an expanding behavioural repertoire. However this does not mean that any information or behaviour will be incorporated. Some information will be more easily acquired since the open programme favours certain types of learning. In humans, for instance, the highly complex task of learning speech occurs easily compared with learning simple arithmetic. In many species these two types of programme coexist and behaviour is determined by their interaction.

Human heredity determines whole developmental processes rather than a fixed character or set of traits. Human behaviour patterns are flexible, and therefore highly adaptive, because they are acquired during life and not transmitted through genes. The selective advantage enjoyed by humans lies in the fact that culture is acquired and that a great diversity of behaviour becomes possible. Many different social positions, sensitivities, skills or varieties of knowledge may be developed as a potential means of human adaptation. The human genetic endowment provides for an unrivalled range of behavioural reactions to a spectrum of real and potential environments. In this sense human behaviour can be seen as an evolutionary strategy resting upon highly effective problem-solving. Different or changing environments or new natural resources may be so readily utilised by human society because open behavioural programmes allow for the rapid assembly of cultural solutions.

Recent experimental work by biologists and psychologists confirms these arguments. Gould and Marler (1987) have shown that instinct and learning should not be considered as opposites. Many simple animals including some insects learn, and the learning process in more complex animals is guided by genetic programmes. Much learning is thus controlled by instinct: animals are programmed to learn a certain behaviour pattern in a certain manner at certain stages of their development. Experiments with bees have shown that this insect organises its knowledge of the environment according to a hierarchy of needs. Bees are inherently attracted to flowers but no closed programme could practically constitute a complete guide to the floral world. Learning to distinguish between species of flower and understanding which flower contains nectar and pollen occurs in a strict order. The bee favours scent as the most reliable cue, followed by colour and then shape in a descending sequence of priorities. This form of prioritised learning and memory can be seen to relate to the comparative reliability of cues in nature, and it can be assumed to be the product of an innate programme that induces certain forms of learned behaviour and communication (von Frisch 1962). Birds also function with genetic programming and genetically directed learning strategies. In activities such as identifying enemies and in song, both innate and learning processes are combined in order to achieve a complete behaviour pattern.

Important parallels with human behaviour are provided by this work. Learning to speak and to understand a language involves complex auditory discrimination, the acquisition of mental structures for syntax and meaning, the learning of large numbers of words and accurate pronunciation. As a complete package these language abilities present a learner with a formidable task and yet children appear to gain linguistic skills at a remarkable pace and with considerable ease and enjoyment. The magnitude of this achievement becomes clear when it is realised that much of this learning occurs unaided. By the age of three, children have mastered the basic structure of their language which, in addition to complex grammatical rules, includes an expanding vocabulary. Miller and Gildea (1987) have estimated that a child learns an average of 5000 words a year or about 13 words a day to attain the vocabulary of 80 000 words held by a young adult. Few parents can be expected to have explained more than a handful of these words to their children or to possess a formal understanding of the grammar which regulates their language and even less to instruct a toddler in its use.

There is evidence which suggests that children learn their language skills because they possess innate developmental programmes for these tasks. In early infancy a variety of perceptual mechanisms are present which serve to equip the child with primary language skills (Eimas 1985). Experiments show that babies learn to distinguish speech from other sound and that meaning begins to be explored as consonants are identified. This ability is provided by a perceptual mechanism that is inborn. Just as distinct stages in the process of language learning can be found in infants, such as babbling and a widening grammatical ability, that are now assumed to originate in a developmental programme, there can be little reason to doubt that a variety of other linguistic functions, such as those which provide the user with conceptual categories for meaning and grammatical logic, also have a similar cause.

How could evolution have produced such mechanisms? Can any form of evolutionary pathway that would favour the development of culture be detected? In a review of these problems Bonner (1980) suggests that a series of steps towards the evolution of culture in animals are present in natural history. In the earliest life-forms, as in modern bacteria, two strategies were present. A stationary, slow reacting, vegetative family of plant-like organisms with highly effective feeding modes, coexisted with a mobile, active group of organisms. Rudimentary social life is seen even in primitive organisms, but among certain insects a developed form of social organisation exists that includes a division of labour, cooperation to raise the young and castes who perform specialised tasks. These intricate social differences and their accompanying obligations are controlled entirely by chemicals and no claim to culture can be made by social insects. Despite some learning, the rigid stereotyped pattern of insect societies is induced by the controlled feeding of 'royal jelly' which determines social position and function.

Among vertebrate animals much social behaviour is also chemically regulated, but among mammals no morphological differences are associated with social divisions as they are among insects. More complex forms of communication, with a growing emphasis on sound among social mammals, has allowed the possibility of greater inventiveness that is denied to animals reliant on scent. Among mammals, learning increases as behaviour becomes more complex. A flexible adaptive strategy, enabling an animal to take decisions about its whole environment, food and predators, is much more effective when it is controlled by learning. When such a strategy has in addition the ability to make choices based upon past experience, the frontiers of cultural behaviour are approached.

Advanced behavioural flexibility is seen by Bonner to have the greatest evolutionary potential. Such behaviour will pioneer new habitats that may become future niches, and this in turn will produce selective pressures for new bodies and behaviours. The evolution of a flexible response in a variety of mammals directly correlates with the appearance of larger brains. The advantages provided by learned behaviour over genetic programming have led to selection for a more complex nervous system and a brain able to direct behaviour based upon multiple choice.

Observation in the field has revealed a wealth of novel learned behaviour patterns that have become transmitted to the young by a range of animals. The flexibility seen in the learning and teaching activities involved here constitutes a social tradition peculiar to a particular group of animals. Examples of tool-use in animals are found in birds and many mammals including primates. Animals have adjusted their behaviour patterns in accordance with the behaviour of predators. Some groups of African elephants, who have survived hunting, retained a highly aggressive and nocturnal tradition as much as 40 years after hunters had killed their forebears. An island-dwelling Welsh vole became non-aggressive when its predators became extinct. Among birds, dialects which facilitate group recognition and cohesion, and the destination of migrations are both learned social traditions. Inventive and innovative social behaviour, which quickly aids a group of animals in adapting to environmental change, has been widely observed. London blue tits have acquired a tradition of pecking through aluminium milk bottle tops, some troops of Japanese monkey now wash sweet potatoes before they are eaten and Welsh sheep have learned to roll across cattle grids.

Evolution has created the ability to transmit information and behaviour by learning because strong selective pressures for such adaptive behaviour were faced by a wide variety of species. For complex animals confronted by constant environmental change, social learning is a more efficient means of acquiring a behavioural repertoire than genetic programming. A genetic basis for a degree of cultural behaviour has been acquired by many animals. Learning the various whereabouts of locations for wintering or the use of a special call for contact with mates, for instance, will dramatically influence survival and reproductive success. Behaviour that can be learned within a social group could be changed with new needs. Information can be accumulated and new innovative behaviours tested without direct instruction from genes. Learning processes are massively in advance of gene mutation and physical evolution and assist rapid adaptation.

This view of natural history is not intended to imply that a directional or progressive force is at work within natural selection. No sustained pattern of transition from simple to complex form or behaviour can be demonstrated in the record. The development of cultural behaviour has occurred in a variety of unconnected species rather than a single lineage. The explanation advanced here suggests only that the development of a series of preconditions for cultural behaviour – mobility, social life, learning and a flexible response – can be seen in the history of life. Human cultural behaviour, which consists of much more than these characteristics, has an origin that derives from precultural ancestors. The full emergence of this complete behaviour pattern occurred during a long time-span and its genetic basis was due, as Dobzhansky argues, to a positive interaction between biological evolution and culture.

3

Biology and Culture

In the last chapter it was suggested that final distinctions between nature and culture are invalid. Nonetheless humanity has evolved a physical capacity for broadly based cultural behaviour that has become largely detached from biology and is in itself the primary force guiding social life. Human evolution has made human society possible, but society and culture hold an independence from biology. In the arguments presented by sociobiology, however, this position is reversed: culture emanates from biological mechanisms which ultimately determine social life. Sociobiologists claim that aspects of behaviour have been coded in human genes by natural selection. The most fundamental aspects of human society – social institutions, beliefs and practices – have been selected for by evolution because by acquiring these genes, and the behaviour which they give rise to, individuals will gain a higher rate of reproductive success. This chapter will be concerned with the ideas of human sociobiology as they relate to the social sciences. We shall also return to the problem of social evolution which was discussed in Chapter 1 in order that modern approaches to the question of culture and evolution may be considered.

THE SOCIOBIOLOGICAL THEORY OF CULTURE

Serious attempts have been made to understand the interrelationship between genes and culture in recent work by biologists. Boyd and Richerson (1985) have raised a series of vital issues concerning the impact of cultural practices on human phenotypes and the conditions under which new modes of social behaviour might convey culture. They propose a 'dual inheritance model' which, like the theory of Durham (1976 and 1978) who argues for the complementarity and coevolution of cultural and biological processes, fully respects both the role of culture in human social life and the distinction between genetic and cultural modes of transmission. These theories represent legitimate lines of inquiry and are distinct from the biological determinism of sociobiology. However the account of sociobiology presented here will be taken mainly from the work of the Harvard biologist E. O. Wilson, since it is this position that raises a series of controversial issues concerning the relationship between social sciences and biology.

Wilson asserts that developments in biology now make a comprehensive biologically based theory of human social organisation possible. 'Biology is the key to human nature and social scientists cannot afford to ignore its emerging principles' (Wilson 1977: 138). Advances in biology have widened this science so that it now embraces the concerns of the social sciences. In fact the expanding biological understanding, claimed by Wilson, threatens to engulf a host of once isolated disciplines. The new certainty provided by sociobiology will mean that the humanities and social sciences will be reduced to specialised branches of biology.

Wilson's justification for this audacious academic imperialism rests upon his conviction that evolution, comparative zoology and genetics together provide an explanation of human behaviour. By reconstructing the ecological constraints faced by the early hominids in their African Pleistocene environment; by placing humans among their primate relatives, and by understanding the biochemical logic of DNA, the innate rules which guide human cultural behaviour can be known. Social behaviour, which is assumed to be adaptive, cannot be explained by the 'cultural determinism' typical of the social sciences because whole behaviour programmes are assumed to be encoded in human genes. Tiger and Fox claim, 'We confidently assert that identifiable propensities for behaviour are in the wiring . . . Selection, then clearly favoured those creatures with a propensity to obey rules, to feel guilty about breaking them, and, generally, to control their sexual and aggressive tendencies . . .' (Tiger and Fox 1978: 57/60). Sociobiologists claim that far more than generalised developmental behaviour, like parent–child bonding, is a part of the human legacy. Such specific traits as homosexuality, incest avoidance, verbal and numerical abilities, xenophobia, entrepreneurship and ambition are a part of the DNA code. On this basis anthropology, economics, psychology and sociology will lose their independent standing and will become subsumed within the higher science of sociobiology.

The central problem around which Wilson has constructed his theory is that of altruistic behaviour. It has remained unclear to biologists since the acceptance of evolutionary explanation, why some animals engage in acts which appear to favour their fellows but place themselves in danger. Many animals act as sentinels and warn others of the approach of a predator. In doing so their own interests – their survival and successful reproduction – are in jeopardy. If the assumption that this trait has a genetic basis is correct, altruistic behaviour appears to be a contradiction in evolutionary theory. Animals performing such acts would be much less likely to breed and pass on the genes for this behaviour. In all probability, therefore, altruism along with its practitioners would become eliminated from the gene-pool of a population.

Wilson's task is to explain how natural selection has allowed altruism to have evolved and been retained by both animals and humans. If evolutionary adaptation is based on maximising the reproductive success of the individual, relative to other individuals, such unselfish behaviour would seem to be working against natural selection. Wilson's solution to the problem is firstly to assert that since relatives share significant amounts of genetic material with the individual who performs a self-sacrificing act, they will benefit accordingly. Secondly, sociobiology takes the individual and its small interbreeding group within a whole population as the subject of selection pressure. Within a related group, altruistic behaviour will increase the *'inclusive fitness'* of both an altruistic individual and his relatives. This term describes the 'fitness' of an individual plus the degree to which his genes are represented in the generations which follow, having been successfully transmitted by relatives because of altruistic behaviour by the individual. Even if an individual died in an attempt to protect the related group, and never reproduced, a significant proportion of his genes would become invested in future generations by close kin. Altruism, which reduces personal fitness, is therefore explained by kinship. If an altruistic act prevented an individual from reproducing but increased the chances of siblings having more offspring, the advantageous genes for this behaviour would be selected for and would spread through the population. Wilson refers to this process as *'kin selection'*. In the struggle to maximise reproductive advantage within a population, the selection of genes contributed by an individual is favoured if they promote the survival and highest reproductive output of relatives with a common ancestry.

Altruism is therefore explained in terms of a strategy by which successful gene transfer to subsequent generations is ensured. By solving this apparent anomaly in the theory of natural selection, sociobiology claims to have isolated an implanted genetic mechanism. A series of other behavioural traits are also assumed by Wilson on the basis that they too assist to maximise reproductive advantage. Genes contain a compulsion of their own which demands, unconsciously, that their owners successfully reproduce to perpetuate their genetic material. Wilson argues, 'The organism is only the DNA's way of making more DNA' (Wilson 1975: 3). Even acts of *'reciprocal altruism'*, which provide mutual benefit to unrelated individuals, are interpreted by Wilson as evidence of maximising reproductive advantage. Human sharing and exchange, the basis of all economic activity, are founded on reciprocal altruism and money has come to represent its quantification.

Wilson's belief that much of human social life is derived from genes is based upon the axioms that many universal aspects of behaviour, like altruism, are adaptive; that some traits are already recognised as inherited, and that study of closely related primate species reveals significant characteristics that are shared with humans. Since human societies are structured around varied social positions, this form of flexible behaviour directly aids adaptation to the environment. 'The ability to slip into such roles, shaping one's personality to fit, may be adaptive' (Wilson 1975: 549). Some forms of mental experience and illness, such as introversion/ extroversion, neurosis and depression, are assumed to have a genetic origin by some schools of psychiatry. A model of comparative ethology is proposed by Wilson to suggest that an evolutionary connection between the behaviour patterns of the apes and humans can be made. Such traits as aggressive dominance systems with males dominating females, matrilineal social organisation and bands of roving males living in cooperative groups, are taken as universal aspects of human society that are derived from our primate past.

Many social arrangements are seen by Wilson as ungainly additions to a primal human nature that evolved to pursue a foraging economy. Social alliances based upon kinship – part of an ancient bartering system that ensures male dominance and acts as a form of risk spreading – are seen as essentially human. Our lengthy habituation to life in a stone-age band of only fifty people has left us unequipped for mass society which is evident from our natural hostility to strangers and a preference for our own kin group. With the growth in social complexity these original bonds are now insufficient, yet we remain condemned to operate with the same internal rules as our hunting ancestors.

Wilson also presents the reverse of this position. Many crucial aspects of modern society, though alien to our evolved nature, may nonetheless rest upon a genetic basis. There may very well be genes for an upward shift in status which become concentrated in the upper reaches of society along with higher levels of intelligence. These superior mental traits would have the effect of raising the barriers between classes and make social stratification more rigid. Formal religious systems, though very distant from the magic and ritual from which they emerged, operate to promote fitness through a variety of functions. Religious dogma is clearly false yet doctrines and practices are so readily accepted that it is valid to suppose that an innate cause is responsible. Similarly, the high contribution of conformity, ethical behaviour and artistic expression to kin selection strongly suggests that these virtues are part of a biological programme.

It is for the questions of tribalism and conflict, however, that Wilson evokes the full record of prehistory. The fact of primate troop life within a home range allows us to extend territoriality to our earliest pre-human ancestors and assume that it has remained part of our inheritance that is now best exemplified by property ownership. The observation that modern hunter-gatherer peoples reserve their rights over game to themselves, suggests to Wilson that xenophobia is natural among human groups. Modern violence and rivalry between social, political, national or religious groups can be seen as an extension of our sense of common identity and natural hostility to outsiders. The hominid lineage began with a social life marked by small male dominated groups with a strong territorial identity to preserve hunting rights. Irrespective of elaborate social and technical change since these times, our original pattern of adaptation with its stress upon aggression and male leadership has produced an unreformed emotional heritage. Patriotism and war are biological urges replete with in-built cunning and a lust for conquest. The motivation to overrun neighbouring territory and increase genetic fitness is implanted.

The purpose of sociobiology is to show that culture remains tied to genes. An aberrant cultural change, unsupported by biology, could not persist for long: 'the genetic leash pulls culture back into line' (Lumsden and Wilson 1983: 65). Culture must be adaptive and seek out genetic approval for its multifariousness and in practice this means promoting genetic fitness.

The absence of any serious attention to social analysis, the cavalier reading of the cultural record and the tentative manner in which it is suggested that genes might or probably exist for given behaviours, provoked severe criticism obliging Wilson to modify his position. A new theory adopted by Wilson, and his collaborator Lumsden, proposes an interaction between genes and culture. Rather than genes being coded directly for behavioural traits, an argument reliant on biological determination alone, it is now proposed that genetic and cultural coevolution has occurred. Thus, 'people are neither genetically determined nor culturally determined. They are something in-between' (Lumsden and Wilson 1983: 84). It is important to be clear at the outset, however, that this new form taken by sociobiology in no sense relinquishes the claim that culture remains tied to biological mechanisms. No possibility for an independent, self-generating social realm is allowed once the biological foundations for culture had evolved. The contents of a culture and much social action are still seen to be directly linked to genes.

Lumsden and Wilson's gene–culture determination of behaviour postulates an interaction between genetic mental rules and culture to cause discrimination between types of behaviour. The mind is structured by genetic rules and modulated by culture. Cultural development produces new ideas, practices and artifacts, but the extent to which these become integrated with society depends upon the constraints and directions imposed by genes. Biological properties in the brain and some sense organs increase the probability that some choices in cultural behaviour will be preferred to others. Forms of behaviour which confer greater survival and reproductive ability will become genetically fixed. These new genetic patterns, favoured by natural selection, will themselves then become part of the rules governing mental development and will strongly influence further cultural change.

CRITICAL REVIEW

Wilson's intention to investigate the biological origins of human behaviour in evolutionary terms is legitimate. But the wholly naive idea that the content and meaning of behaviour are in any way directly generated by genes indicts sociobiology as a misuse of evolutionary theory. The eagerness of sociobiology to leap from genes to behaviour with only minimal regard for learning and adaptation or consideration of the historical and social records has produced a simplistic view of human society which misconceives the role of culture in human evolution. It is probable, though as yet unproven, that humans and animals have some specialist traits that have a genetic basis. But a genetic basis does not imply a genetic determination. Sociobiology's practice of citing traits that are apparently universal to human society as a proof of genetic causation, omits entirely the diverse meanings and motivations that are attached to these traits in their social context. Attention that should be directed to the interaction of the human organism and its environment, to adaptive learning and social analysis in fact, is dissipated in claiming genetic determination. Mary Midgley has warned that, 'sociobiology as a movement is a real menace because it provides simple-minded people who like the jargon of science with an exceptionally slick set of catchwords and formulae for universal explanation' (Midgley 1980A: 24).

GENES AND BEHAVIOUR

The first claim of sociobiology, that genes exist which are causally linked with behaviour, will now be examined. It is important to be quite clear at the beginning of this discussion that genetic research has not been able to show that any gene or set of genes is responsible for any aspect of human behaviour. A recent survey concluded that no identifiable genetic component has been isolated that can be positively linked with crime, depression, schizophrenia, alcoholism, intelligence or homosexuality (Horgan 1993). In the face of this complete lack of direct evidence, sociobiology has postulated a hypothetical gene–behaviour model. Genes for aggression or status or homosexuality are *assumed* to exist because those traits are said to have a direct influence upon reproductive success. Comparisons between animal and human society are used to suggest that corresponding genetic processes are at work in each case. This mode of reasoning is removed from the rigorous procedure used in biology to distinguish between genetic or environmental causes. But Lumsden and Wilson go further in assuming that there are genes which act as rules which decide the probability of changing from one form of behaviour to another. The geneticists, Alper and Lange, comment, 'There is absolutely no evidence that there are observable differences among people in the probabilities of their switching from one cultural trait to another, let alone that such differences might arise from differences in their genes' (Alper and Lange 1981: 3976).

While precise genetic differences of this kind cannot be shown to exist, the fact of genetic similarity can be demonstrated. We share 98.4 per cent of our DNA with our closest primate relative the chimpanzee and the human population shares an even larger proportion of genes in common. It is this genetic similarity, the product of a myriad of genes, which confers the inherited basis of human status. It is the general developmental programmes for, among other things, primary forms of interaction by infants, language learning, play and motor skills, which are the significant genetic contribution to behaviour (Bowlby 1984, Eibl-Eibesfeldt 1970).

The idea that single genes are coded for behaviour produces a distortion of social contexts. Wilson assumes that homosexuality is a genetically determined trait that is retained in populations because homosexuals provide important assistance to close relatives and thereby promote inclusive fitness. The homosexual is assumed not to reproduce but to help maximise the reproductive potential of siblings. This view is clearly a projection on the past of one type of contemporary homosexuality that fails to come to terms with the complex social location and meaning that attaches to this form of sexuality. The social context of homosexuality among New Guinea tribesmen consists of men and boys living permanently in men's houses while women, children and pigs have residences that exclude men (Washburn 1980). In this society, husbands and wives meet during the day for intercourse in forest clearings. Regular liaisons between married bisexual men have been a long-standing part of many European societies. It is also possible to cite cases of homosexuality where the male clients of male transsexual prostitutes consider their behaviour to have a heterosexual orientation. The social meaning attached to the behaviour in each of these examples is clearly different but no consequences for reproduction or genetic transmission can be specified.

Sociobiologists frequently cite the existence of apparently universal characteristics of human society as evidence for genetic determination. Religion, territoriality and entrepreneurship are advanced by Wilson as constant features of all societies. Evidence of universality is not, however, simultaneously evidence for genetic control. Under certain circumstances the characteristics Wilson proposes may arise and appear to recur in unrelated societies, but their existence in each case is supported by social factors. By the use of a sleight of hand Wilson infers that such 'universals' can be isolated in each society in an unequivocal fashion. On close examination, however, the religious institutions of various societies fail to correspond and agree with the homogenous label 'religion' which Wilson assigns to them. For the Greeks and for many other societies, no separate context or word for religion existed. It is a commonplace that religious ideas diverge markedly with respect to ethics and personal morality, but it is also the case that some religions, such as Confucianism, are unconcerned with metaphysical ideas. Yet other religious systems have no unified set of beliefs or practices and in many instances no priesthood exists. It seems curious then that such chemically precise and intricate entities as human genes can be responsible for such diversity. Wilson has accepted the sacred–secular dichotomy imposed by contemporary western societies as universal and naively transposed these cultural categories to the genetic record.

To make their use of evolutionary comparisons more convincing, sociobiologists present examples of animal social behaviour that is paralleled by humans in order that a genetic origin and connection may be presumed. But, as Rose et al. (1984) assert, this is a practice of confusing metaphors with real identity. A range of animals are claimed by Wilson to have social stratification: castes, slavery and even a division of labour exist among ants while dominance hierarchies characterise primates. None of these instances, however, have any bearing upon the class systems of human society which are the long-term products of unique histories and cultures in conjunction with large-scale economic organisation. The simple fact of such unchanging replication of animal social orders, compared with the infinite variety in which human society has been structured – from primitive communism to rigidly assigned class status reinforced by law – should be sufficient to dispel such crass comparisons. Similar claims of animal warfare, kinship, rape and even prostitution (among humming birds) are equally absurd.

The genetic arguments of sociobiology have generally failed to convince biologists. Wilson and his colleagues have been accused by one biologist, James King, of commitment to archaic forms of theory. Sociobiology is based upon, 'Old hat genetics that current research has rendered obsolete' (King 1980: 104). Wilson is working in an earlier tradition of genetics which considers that one genotype is best fitted to its environment and that all other variants will be eliminated. Natural selection will constantly purge mutations which do not conform to the ideal phenotype. This 'uniformist' position contrasts with the 'variabilist' tradition of Dobzhansky, Mayr and Lewontin, discussed in Chapter 2. An interaction of genes and environment is emphasised here, and the combined effects of genes on phenotypes is stressed rather than the aggregated contribution of single genes. Variabilist genetics proposes that a modal phenotype or an approximation of mean fitness is selected. There is no single or best genotype, only a diversity of genetically different individuals of similar fitness. Evolution works on a broad front with the aim that a variety of gene combinations are always present within a population, giving the probability that a modal phenotype will be produced. This prevailing tradition, which is well supported by experimental work, makes sociobiology's claim for single-gene control of behaviour appear highly doubtful. Wilson's idea of aggressive, indestructible genes which constantly impel their carriers to pass them on to the next generation is opposed by the variabilist conception of a population gene pool as a constantly varying, interacting complex in which individual genes are likely to disappear altogether.

The selfishness of genes is a recurring theme in the writings of sociobiologists. Altruism and the organism itself are merely devices to aid the survival of the genes. By claiming that genes ruthlessly strive for a place in future generations and impose their will on their human carriers, by formulating their behaviour, sociobiologists present an over determined and mystical view of social life. Genes are single bits of biochemical structure that cannot be credited with a capacity for conscious knowledge, calculation or single-mindedness. Nor is human behaviour an amalgam of separate actions which originate from single genes. Such minute units of life have no meaning except as a part of a human whole. Valid explanations of social behaviour cannot be based upon reductionist hypotheses. Selfless behaviour, like aggression, appears to be a constant part of the animal and human world and therefore not deterministically attached to genes. Mutual help and friendship between unrelated individuals are common among humans and are more likely to be a product of emotional bonding or an ethical code. In reality the social and biological worlds of humans are highly confused and acts of violence and even murder, as well as altruism, may be directed towards related individuals. Some animal species actively care for non-relatives and show altruism across blood-lines. Wolves for instance adopt the cubs of dead adults, giraffes have creche arrangements and elephants will come to the aid of distressed herd members (Zahn-Waxler 1986). Altruistic acts serve the well-being of individuals and whole social groups and are meaningful only in this context. For sociobiologists, though, the act of altruism is itself an expression of a genetic drive. This, as Mary Midgley remarks, is like attributing an autonomy to acts or behaviours; it is like asking what dances gain from people rather than why some people like dancing.

GENES AND CULTURE

Sociobiology's second claim is that, ultimately, human culture and primary social relationships are structured by genetic mechanisms. This is to deny any independent existence to social facts or social organisation. Under the theory of kin selection individual sacrifice will lead to a net genetic advantage provided that at least two siblings or eight cousins survive to reproduce. Kin groups are therefore synonymous with gene preservation and promotion. Kinsmen jealously guard the genetic integrity of their kin groups and pursue the well-being of relatives because of their vested genetic interest. Kin selection and reproductive success are seen as the principle motivation in human social life. However, if it can be shown that while kinship is fundamental to human social behaviour it is not based on individual reproductive success, the case for sociobiology is considerably weakened.

Sahlins (1977) has produced a valuable critique of sociobiology and kinship. Wilson's model of behaviour motivation by kin selection is reminiscent of cost–benefit analysis theories from economics. Any behaviour, whether altruistic, anti-social or spiteful, may be interpreted as a means of maximising an advantage in a system. However, in the real world, humanity presents a startling diversity of marriage and residence patterns, family organisation and means of reckoning descent. An even greater variety of common property ownership, mutual aid and cooperative production can be found. There are few cases where a simple correspondence between blood relatives and kinsmen exists, because kinship units comprise utterly diverse blood relatives. A person who discriminates in favour of blood relatives within a kin group will be forced at the same time to discriminate against other blood relatives, who are just as closely linked by genes, because they belong to another kin group. Sahlins gives an example of kin groups among Solomon Islanders where common residence occurred among people who share an ancestor as distant as nine generations. These kinsmen formed an economic unit by sharing their cooperative production, but their genetic similarity was often less than one in thirty-two thousand. In patrilineal societies, such as China and Japan, the practice of women marrying out of their natal families ensures that half the genes produced by a kin group are lost. Among aristocratic families in feudal Japan loyalty and sacrifice to a superior lord took precedence over family ties. The virtuous samurai would willingly forgo the welfare of his kin group in the interests of his lord. The Japanese practice of adopting an adult son to continue a lineage that had no male heirs, but could normally only be perpetuated by sons, shows even greater flexibility. In this case both female and male outsiders married into kin groups ensuring an even greater genetic dissimilarity between group members. Kinship organisation then is the product of cultural relationships that place a person in a social group which excludes many blood relatives.

Sahlins provides ethnographic examples of highly inventive kinship systems. In East African societies barren women become fathers and dead men marry. A childless woman among the Nuer is considered to be a man and therefore allowed to marry another woman. The 'husband' will choose a neighbour, friend or kinsman to consummate the marriage but any offspring are legally bound to the 'husband', and her rights include the disposal of her 'daughters' in marriage and the honour of being addressed as 'father' by her 'children'. The social, if not physical, survival of a man may be ensured through a ghost marriage whereby children are conceived on behalf of a deceased kinsman. In both of these examples the genetic connection between parents and children is either tenuous or non-existent. The Hawaiian practice of infanticide followed by the adoption of unrelated children among the noble families of this Polynesian society, and the high proportion of adoptees on other islands, suggests that whole lineages can be perpetuated without any necessary biological connection between

generations. Among Trobriand Islanders, children belong to the
mother and her brother's clan, the biological father contributes
nothing for their upkeep since no concept of paternity exists in this
society. In this case the father is discriminating against his own
offspring and, as in the other examples, the principles of kin
selection are violated. Human kinship is not organised in accord
with the genetic relationships proposed by sociobiology. Rules
governing marriage, residence, property and descent are socially
given and are therefore arbitrary with the result that biological and
social relatedness correspond poorly. 'Paradoxically, then, whenever we see people ordering their social life on the premises of
genealogy, it is good evidence that they are violating the dictates
of genetics' (Sahlins 1977: 60).

The application of sociobiological theories to the social sciences
has led authors to impose rigid parameters on human behaviour.
Symons' book, *The Evolution of Human Sexuality*, provides an
example of this form of sociobiological approach. The argument
presented here asserts that women and men have distinctly
different forms of sexuality because of natural selection. This
essentialist, or pre-social, distinction holds that the differences in
our sexual natures arose, 'because throughout the immensely long
hunting and gathering phase of human evolutionary history the
sexual desires and dispositions that were adaptive for either sex
were for the other tickets to reproductive oblivion' (Symons
1979: v). This difference consists of the assumption that men tend
to want multiple sexual partners while women tend not to,
because of the adaptive consequences which this had for either sex
during human prehistory. Successful reproductive strategies, the
product of significant evolutionary change in our nervous and
hormonal systems, are manifest in human biology and have the
effect of inducing different preferences in each sex. Men have a
relatively low physiological investment in reproduction and will
seek to impregnate as many women as possible to ensure
maximum genetic continuity. Women exhibit the reverse of this
position and will be concerned to mate with one fit male. It will be
adaptive therefore for a male to pretend fitness and for a female to
resist courtship to provoke a stronger display of commitment.
Symons considers men to be sexually competitive, polygynous
and jealous. Women on the other hand are preoccupied with
stability and the use of their eroticism as a means of control.

It would be facile to suggest that human biology, as it relates to
reproduction and sexual activity in each sex, has no effects upon

sexuality. The biology of pregnancy, birth and motherhood, regardless of a variety of socially constructed meanings, produces significant forms of psychic and bodily experience in women that have no equivalent in men (Freud 1977). There is every reason to believe that mental life, which is constructed from our perception of social life, develops in conjunction with our sexual biology. But there is no point in assuming that the adaptive consequence of our prolonged beginnings in prehistory has been a sexuality that is fixed. Natural selection has given us the reverse of this. Human sexuality has largely escaped the hormonal controls imposed upon the other primates and is, to an extent, free to form itself through social interaction. Beach (1978) has shown that sexual excitement and its results on the genitals are not caused by hormones. There is no evidence which shows human emotion in general to be essentially different in either sex. Similar emotional processes are observed regardless of sex. Sexual feelings have come under mental dominance and are the combined results of psychic development and life experience; they are not the manifestation of an innate behavioural programme that becomes mechanistically reproduced in each and every form of society. Hirst and Woolley (1982) point to Freud's discovery that, in its infantile state, the sexual drive has no given form. The object of sexual feelings is unspecified. Sexual drives are not directed towards reproduction or genetic survival and nor are they in any sense constrained by convention or even directed toward genital satisfaction. There are no genes for male jealousy, philandering or an obsession with paternity. Evolution has taken sexuality away from biology and allowed culture to be decisive.

Despite its excursions into the ethnographic record, Symons' book remains a projection of American dating culture onto prehistory. Patterns of sexual behaviour vary widely but in all cases they are subject to cultural regulation. There are a profusion of marriage types. Institutionalised promiscuity and tolerance of adulterous unions coexist with social arrangements that oppose these in other societies. It is important to be aware that the rules of human sexual behaviour, unlike those of animals, are both adhered to and transgressed. Little is known of actual patterns of sexual comportment in all societies, but it has become clear from the work of Kinsey and other researchers that extramarital sex is a reality for most American women and men and that as many as ten per cent of children born in the late 1940s were conceived by adultery (Diamond 1991).

Social change in the advanced industrial societies has included huge shifts in sexual behaviour which have occurred in the context of advances in contraceptive technique. In its simplest form this change consists of both greatly increased sexual activity for a longer part of the life span and an overwhelming willingness to use contraception. The connection between sexual behaviour and reproduction in these societies is slight as evidenced by their stable and even declining populations. The fact that sexual gratification has become largely separated from any biological consequence has been received with mass enthusiasm and it cannot be shown that men wish to sabotage contraceptive devices in accordance with a supposed genetic drive. Symons offers no evidence that sexually promiscuous men do in fact sire more children and there is no proof, beyond an adaptive story, that this behaviour is genetic in origin. Similarly he does not understand that where economic forces and biological consequences are not constraining for women, an independent sexual existence that inclines as much toward promiscuity as some male behaviour can be found.

GENES AND SOCIETY

Sociobiology's third claim is the insistence that biological processes can provide a comprehensive explanation of human society. Wilson's use of the term 'society' is consistent with biology in general. An animal society is seen as an aggregation of the behaviour performed by the individuals who compose a population. This usage when applied to human society results in a fundamental misconception: the organisation and structure of a society, together with the process of historical development and social change, which are distinctive and unique, are ignored. Human society is much more than the sum of the individuals who comprise a population. The idea of society as a historically created social whole is a product of the social sciences, with little precedent before the eighteenth century. This concept of sociology is an important intellectual advance which makes possible an explanation of the differences that exist between societies and allows a closer examination of the interrelationship of their elemental parts. Above all this insight provides a demystified approach to the problem of interpreting change, because of the understanding that human social organisation is a cultural product. This insight also opens up a major difference between human and animal societies which is based on the fact that human life consists of cumulative social change that has no connection with natural selection. Human social life possesses a unique historical dimension that is unparalleled by the social traditions of animal societies. It is the interaction of people in an organised and structured society which, in the most general terms, is the cause of historical change. Meaningful cultural and social experience in a historical context motivates human action rather than reproductive fitness. Sociobiology's failure to recognise that people are the products of their social and cultural settings, as these relate to time and place, condemns it as a set of ideas that is incapable of dealing with human affairs.

Sociobiology's attempts to utilise evolution in social analysis fail because natural selection cannot explain cultural variation and historical change. While natural selection must form the basis of any explanation for the origins of human society and human attributes, it cannot be usefully incorporated into social theory. Darwin did not provide a new perspective for the social sciences or the study of history. Irrespective of theories of social evolution used in anthropology before 1920, and earlier attempts to import biology into the social sciences, there has been no significant long-term impact on social theory caused by the Darwinian system. Social theory and biology pursue lines of inquiry that bypass one another. Despite a wealth of analogies between history and evolution during the nineteenth century, the sociological tradition has remained largely unaffected by *The Origin*.

Darwin's discoveries occurred in an era imbued with a philosophy of progress. While the theory of natural selection remains unideological Darwin was nonetheless influenced by such ideas. Kenneth Bock (1980) has argued that progress is implicit in *The Origin* in the sense that a tendency to, 'progress towards perfection' (Bock 1980: 40) could be detected in both organisms and civilisations. Darwin entertained no ideas of innate progress within organisms or human society but, in the struggle for existence, more efficient and superior forms that had overcome their competitors had emerged. Whatever the validity of this point it is clear from Darwin's writings that, like so many other nineteenth century thinkers, he conceived of history in a series of progressive stages. A hierarchy of stages reaching from savagery to civilisation is presented with Anglo-Saxon Europe occupying the zenith of cultural history. But Darwin had little understanding of social processes, particularly historical change, and he was inclined to attribute the diverse nature of human societies to the endowed qualities of their members. The peoples of the world and their societies possessed very different innate capacities. Darwin was a nineteenth century Englishman who was untroubled by ideas of natural superiority and a Eurocentric view of civilisation. Darwin had overcome theological and teleological modes of explanation in biology by proposing the theory of natural selection as we saw in Chapter 1. However the purpose of the theory was to explain change in the natural world not to promote ideas about human progress. Natural selection does not support any form of social theory and it may not be used to justify Darwin's own ideas of progress. Little if any consideration was given to social analysis by Darwin and the idea that cultural traditions are themselves a formative influence upon historical development escaped him.

Sociobiology is based upon an inappropriate intellectual foundation, and its attempts to explain society in biological terms raise a series of problems. Firstly, human societies are seen only as the statistical results of a population's behaviour. Society seen from the perspective of social science, as a complex interacting whole, has no existence. Secondly, models of early human society, taken from a reading of prehistory, are assumed to present a 'true' or natural state which has become fundamentally changed by 'cultural evolution'. But neither Wilson nor natural selection can offer any form of explanation that can account for historical change since the demise of hunting and gathering societies. Thirdly, the problem of explaining cultural differences is not addressed by sociobiologists. Cultural variety is frequently presented as a superficial variation on a theme: a universal human nature, with a genetic basis, which has diverse forms of expression. But cultural differences, the results of historical experience, are anything but superficial. The social life of hunting bands and industrial workers are profoundly different in almost all respects, including the organisation of their subjective mental life. The implicit denial of the significance of history and culture in determining this difference is a mask to hide an inability to explain social diversity.

ADAPTATION AND CULTURE

In what remains of this chapter we will return to the question of social evolution. Sahlins and Service's book, *Evolution and Culture*, is unconnected with sociobiology and represents a serious attempt to rescue the concept of evolution for the social sciences. The authors distinguish clearly between biological and cultural evolution, 'Culture and life have different properties, different means of transmission and change, and yet each has laws peculiar to itself'. Yet they insist that, 'both can be embraced within one total view of evolution' (1960: 8). Cultural evolution is seen to move simultaneously in two directions. 'Specific' evolution, the equivalent of biological speciation, modifies cultures by creating new forms. 'General' evolution generates progress as higher cultural forms arise from and surpass lower forms. A change in either a life-form or culture may be seen alternately as an adaptation or as an embodiment of progress. The criterion of progress advanced is that an organism or cultural adaptation uses energy more efficiently through superior structural organisation, as much as by technology. The higher, and more complex form, will contain more sub-parts and specialised adaptive means and it will in consequence experience greater freedom from environmental constraints and will possess a greater potential range for future adaptation.

Sahlins and Service argue that, 'culture continues the evolutionary process by new means' (1960: 23). Adaptive modification is the instigator of cultural differences. As societies adapt they also create, merge, adjust and diversify. A culture, being the sustaining vehicle of a society, will undergo adaptive development in accordance with varying problems of survival. Ironically, attempts to preserve existing social relations will demand the use of new methods which will have the effect of promoting change. The constant adjustment of a culture to changes in habitat, competing cultures and new needs will produce new modes of organisation, technology and ideas. The quest for stability becomes a source of change. Each cultural form with its own distinctive characteristics is, in the terms of 'specific' evolution, an adaptive entity which is no higher or better adapted than any other culture. Complex and sophisticated civilisations have arisen and passed to extinction while simpler, longer-lived cultures, persistently survive. This relativist position must be held in conjunction with the understanding that 'general' evolution has created progress which is, 'the total transformation of energy involved in the creation and perpetuation of a cultural organisation' (1960: 35). This principle allows the possibility of ordering various stages of social organisation in a ranked hierarchy. Hunter-gatherer bands without chiefs or clans are followed by larger-scale organisation, seen in tribes, which give way to societies based around chiefdoms and then to states. More effective energy capturing methods are implied by this transition from foraging to agriculture and then to urban industry. A hierarchy of progress in terms of improved adaptability in which higher forms dominate and replace lower ones is proposed. This is not a unilinear scheme of historical stages which supposes that a unique pathway must be followed to achieve the status of modern industrial society. The component sub-systems of a society will adapt at different rates in different directions, allowing a variety of routes to a particular form of society.

There is value in Sahlins and Service's book. Their description of the process of social change at work, the reasons proposed to explain the cultural dominance of one society over another, which are to be found ultimately in more sophisticated relations of production, and their specification of the conditions favourable for social change, though general, can be readily accepted. But can any of these processes be legitimately termed evolutionary? Some of the criticisms levelled against Spencer in Chapter 1 are also applicable here. Biological evolution consists of a succession of *individually* reproducing organic forms which have arisen through adaptation. Because evolutionary change is continuous and random the possibility of stages or improvements is precluded. The mechanism of natural selection, which governs the organic world, cannot be said to apply to human societies. There is no form of societal equivalent to random mutation within a single individual genotype and acceptance or rejection by the environment. Cultural adaptation is always a *collective* matter. The correspondence between the presentation of a novel genotype to the environment and social innovation must remain metaphorical. Cultures cannot be made to stand in the place of organisms. Elements of a culture – artifacts and ideas – may survive long after a social order has perished, and supply a potent influence on contemporary societies. The extinction of a species or variant, however, is a unique biological termination rarely seen in human society.

Sahlins and Service propose that history can be demarcated into a series of social forms that arise by cultural modification and adaptation. These successive forms change to produce a progressive sequence that is seen in movement from a lower to a higher position in a hierarchy. This organisation of history, irrespective of a variety of permutations offered, involves a directional form of change that in turn points to a particular end. But this generalisation of the historical record, to produce successive forms, has become confused with natural selection (Hirst 1976). Darwin's evolutionary mechanism does not imply a history or a series of stages. Natural selection specifies the manner in which organisms change without commitment to direction or quality.

Much of Sahlins and Service's argument rests upon analogy and metaphor. A society's adaptation can occur in a variety of ways and is not equivalent to natural selection. There is no built-in discriminating principle which serves to differentiate between social change that is to become incorporated by social institutions, and change which is determined unviable. Neither public opinion, market forces or economic performance can fill the void left by the mechanism of natural selection. Social change cannot be said to maintain or improve biological fitness or be caused by adaptation to a changing environment. Such change can be detrimental to human fitness or well being and is in itself the cause of environmental change. The equation of progress with energy efficiency can do little to explain historical change. 'Cultural dominance' cannot be seen to be governed by such a principle, as a few examples from key events in world history demonstrate. The fall of Rome to Barbarians, the defeat of the Sung dynasty by Mongol hordes, the invasion of Ottoman Turks into south-west Asia and the sacking of Constantinople by Crusaders exemplify domination by a simpler, less efficient society. In most historical societies the issue of energy efficiency was not on the agenda, even in an unconscious sense, and no consistent struggle to improve the forces of production can be detected.

Natural selection and social development should be considered as two quite distinct processes. Human ingenuity and social traditions allow highly diverse forms of adaptation to occur and no clear opposition between society and environment can be presented as in the case of organic evolution. The idea that cultural differences are the result of various adaptations to different environments is acceptable in a general sense but is unhelpful since none of the logic of biological evolution governed by natural selection is present. Adaptation may be used as a concept to explain all and any social arrangement with misleading and vacuous results. Both desert-living nomads and merchant bankers adapt and modify their actions in accordance with environmental change.

It has already been suggested that culture bestowed human ancestors with a highly adaptive new technique, but the reverse of this statement, that all aspects of culture serve to promote adaptation, must be false. Our social practices, ideas and social institutions can be frequently shown to be unadaptive. Humans do much to lessen fitness. Smoking, allowing free access to fire arms, the competitive consumption of alcohol, tattooing, flattening skulls or deforming feet by binding during infancy, ritual scarification, male and female circumcision and subincision or the denial of protein to pregnant and lactating women are all entrenched cultural traditions which various societies pursue regardless of their sometimes fatal consequences for the individuals involved. It is equally difficult to find any evolutionary logic in belief-systems which, in their extreme form, result in destruction and violent death. The powerful ability of religious ideals or nationalism to produce confrontation should not require comment. But concentration camps, warfare sanctioned by religion and ecological disaster are cultural products. Similarly the assumption that social institutions function to serve environmental pressures and seek 'adaptation' with a social environment is equally absurd. The family, for instance, which is itself a part of the environment, cannot be seen as an adaptive mechanism of reproduction without serious misunderstanding.

CULTURE AND ENVIRONMENTAL PRESSURE

The interpretation of culture as a mode of adaptation is often encountered in anthropology, particularly in relation to population pressure. In this form of argument mechanisms to control population are said to be employed to maintain optimum numbers that accord with local environmental resources. This form of materialistic explanation is employed by Harris (1978) and Harner (1977). Both attribute a variety of social practices directly to the constraints of over-population. Like sociobiology, this variety of argument uses reductionist and deterministic reasoning. Rather than gene determination, culture is seen as the product of environmental necessity.

Harner has adopted an evolutionary approach to explain the large-scale phenomenon of human sacrifice in the Aztec empire which flourished in pre-Columbian Mexico from the fourteenth to the sixteenth century. He argues that cultural forces alone, such as the Aztec religion, which required that the gods be given hearts and blood of victims to maintain equilibrium in the universe, could not explain the extraordinary lengths resorted to for the supply of captives destined to die by production line methods. Harner's thesis is based upon the interaction of population and ecology in Mesoamerica. Since this region lacked large herbivorous mammals such as cattle and pigs, animal protein was restricted to domesticated wild fowl, like the turkey, and the Chihuahua dog. Large game animals had been hunted to extinction and considerable environmental degradation had been perpetuated by civilisations before the rise of the Aztec empire. Faced by such nutritional stress Harner concludes that the Aztecs were forced to resort to, 'large scale cannibalism disguised as sacrifice' (1977: 119). The problem of gaining sufficient amino acids from a diet of maize and beans – which had to be eaten simultaneously to provide nourishment – coupled with frequent famine, produced an intense desire for the essential nutrients provided by meat. Any food source that would satisfy this craving was considered to be legitimate and the function of warfare was to take captives for sacrifice and consumption.

Spanish chroniclers learned that the scale of sacrifice could exceed one per cent of the population of Mexico each year. This estimate would yield an annual quarter of a million victims. In 1487, during a single ceremony, over 14 000 prisoners assembled in four lines, each two miles long, were sacrificed by a team of priests who worked continuously for four days and nights. Those intended for sacrifice were frequently fattened before being taken to the top of a temple pyramid where the heart was torn from the body and offered to the gods with blood. The corpse was then tumbled down the pyramid steps and dismembered, whereupon the victim's captor would take possession of the limbs which were served at a feast in a stew with peppers and tomatoes. Mortuary evidence of such large-scale carnage, in the form of skull racks on which heads were displayed after sacrifice, was observed by the Conquistadors who counted over 130 000 of these trophies on one occasion. Harner points to the rules which governed the apportionment of the various parts of the corpse to warriors, priests and rulers and to the fact that access to human flesh was probably reserved as a privilege for the upper classes. The weight of evidence pointing to a cannibalistic solution to over population and nutritional stress has been suppressed by a conspiracy between embarrassed anthropologists and modern Mexican nationalism.

How valid is the proposal that cannibalism is a cultural 'adaptation' to population pressure, and a nutritional supplement? Some anthropologists have denied the existence of cannibalism itself. Arens argues that it is very unlikely that cannibalism ever existed, 'as an accepted practice for any time' (1979: 9), except for survival in extreme circumstances. This denial is an iconoclastic assault on a central area of interest to anthropologists. Arens accuses his colleagues of perpetuating a myth to serve the needs of their discipline. Cannibalism is merely one of many exotic, primitive myths that anthropologists have a vested interest in sustaining. Irrespective of a host of academic works of analysis the lack of good ethnographic evidence for cannibalism in any contemporary society suggests that the practice exists only as a fantasy in the minds of observers. Constant references made by respondents during field work to the cannibalistic practices of other peoples, such as a neighbouring tribe, have been taken at face value. Mutually reinforcing accusations of barbarous practices, including cannibalism, by societies whose acquaintance with one another is restricted by notions of moral superiority have been recounted universally. Tribal and foraging societies frequently demarcate themselves from other societies through the categories of 'human' and 'non-human', and the name of many such societies is often the same as the word for 'man' or 'human'. Arens dismisses the existence of cannibalism in remote prehistory as ambiguous, resting on an inference from a supposed projection of primitive qualities onto the past. Historical accounts by European explorers of savages feasting on human flesh are re-examined in the light of contemporary ideology concerning the non-Christian world and also rejected.

Arens' charge that cannibalism can be a product of ethnocentricism should be taken seriously, but his total rejection of the practice as a social institution has failed to convince most authorities (Leach 1982), and by overstating his case he has become an extremist like Harner. The evidence which supports the existence of different types of cannibalism in a variety of cultures and times is vast and has certainly not been exhausted by Arens. Good first-hand evidence for many instances of cannibalism is to be found even in the cases Arens dismisses, according to Sanday (1986).

In a study of materials excavated from the Fontbregoua cave in southern France, Villa et al. (1986) conclude that dietary cannibalism was practised. The cultural remains of this neolithic community, dating from the fourth to fifth millennia BC, include a substantial amount of human and animal bones which appear to have been given similar treatment. Both animal and human bones bore similar cut-marks indicating that defleshing had occurred by similar butchering techniques. The long bones of all species had a similar pattern of breakage suggesting that marrow had been extracted. After processing and perhaps cooking, the bones were discarded without distinction between animal and human remains. Taken together this evidence strongly suggests that humans were used as food and that an alternative interpretation of secondary burial involving only dismemberment and defleshing is invalid.

Arens' dogged insistence that cannibalism never existed is unsupported by any reason which would explain its absence. Ritual sacrifice, murder, torture and a variety of mortuary practices have all been given public acceptance. Why should cannibalism not also be accommodated as an accepted social practice? Arens seems to be postulating an essentialist distinction by omission. The modern western taboo on eating human flesh is thought of as universal and its violation seen as a departure from human status. But in practice this supposed threshold between ultimate inhumanity and acceptable social behaviour has not always been drawn so starkly. Social indictments against cannibalism are not an equivalent of the incest taboo.

Brian Simpson's book *Cannibalism and the Common Law* reveals that among British seafarers during the nineteenth century, acts of survival cannibalism were common and well understood. An increase in shipwrecks during this era of commercial expansion led to a number of celebrated cases in which surviving sailors ate their shipmates. Cannibalism was an accepted element in Victorian nautical culture and, provided that lots were drawn to find a victim, no impropriety was thought to have occurred. In a case tried in 1884 a Master Mariner and his mate were accused of the murder of a cabin boy during 25 days in a open boat following the sinking of their vessel. Because both seamen had spoken openly of their ordeal the legal establishment decided to resolve its growing disquiet about this practice by a trial in which the state of the common law would be clarified in the light of this custom of the sea. Although both men were found to be guilty, their mandatory death sentences were overthrown and they were freed. This case, like many before it, had aroused considerable sympathy from a public that was notorious for its censoriousness. The point to be taken here is that there are no essential moral sentiments that are automatically outraged by the act of cannibalism.

But does acceptance of cannibalism as a common social practice confirm Harner's claim that Aztec human sacrifice was a cultural adaptation to protein deficiency? The facts of Aztec human sacrifice are well known and, apart from their magnitude, are not disputed. However there are a number of reasons that provoke immediate suspicion of this thesis. Why was cannibalism on this scale disguised by the Aztecs who dominated Mesoamerica? Since most sacrificial victims were male prisoners of war the carnage would have had no real effects upon population growth. Systematic butchery of corpses would lead to unmistakable cut-marks on bone yet no archaeological evidence is presented. But Harner's most significant failing is his inability to understand the full motivational force of culture. Aztec religion comprised a set of beliefs that were compelling, and there is no need to invoke an ecological explanation which assumes that cultures are only the practical results of rational self-interest worked out through the perception of environmental constraints. Aztec society believed quite literally that its religious practices were necessary for the world to continue. Both its own social structure and cosmic order could only be maintained through sacrifice. The symbolic meaning of human sacrifice and cannibalism was a part of a sacrament that brought about communion with the gods. The interdependence of humans and gods was central to Aztec religion. The gods gained the energy to maintain celestial motion from sacrificial blood and without it the sun would fail to rise and the return of an original state of chaos would follow throughout the universe. In the sacrificial ceremony the Aztecs mimicked the sun's progress across the sky. A victim would climb the eastern side of a pyramid and his corpse would be rolled down the western side. There is evidence which suggests that victims also embraced these ideas, accepted their own sacrifice, and cultivated a stoical attitude to death and a close personal relationship with their captors. Status was achieved in Aztec society through military prowess and the capture of prisoners. Membership of the aristocracy and high government office, which was non-heredi-tary, resulted in considerable privileges. This social group, and aspirants to it, had a vested interest in maintaining the military–religious complex. Only this top 25 per cent, who were the least likely to experience protein deficiency, ate human flesh. But the Aztec masses appear to have been equally enthusiastic in pursuing religious wars that led to sacrifice. The Aztecs were in no sense an aberration; there are many other examples in world history which show warfare and persecution to have been potent forms of mass motivation.

What may be said of Harner's ecological argument: were the Aztecs forced to gain necessary protein through cannibalism because of large numbers and a deficient environment? Ortiz de Montellano (1978) has reviewed Harner's argument against what is known of pre-Columbian diets. He concludes that while there is no doubt that ritual cannibalism was practised in pre-Columbian Mexico there is no evidence to support Harner's thesis that cannibalism was a dietary supplement. Compared with other large-scale pre-capitalist empires, such as China or India, the Aztec agrarian system did not appear to suffer from a peculiar ecological handicap or an inability to feed its moderate sized population. The absence of large domesticated animals in Mexico does not lead to the conclusion that the population faced protein deficiency. The Aztecs had a variety of good vegetable protein sources from grains, beans and tropical fruits, including the staple grain Amaranth which is exceptionally rich in protein.

Animal protein was available from deer, turkeys, armadillos, reptiles, including the iguana, fish and insects. A combination of tribute grain, received in the Aztec capital Tenochtitlan, and the products of the Chinampas, the rich lake-garden lands that could produce seven crops a year, was more than capable of feeding the city's population. Archaeological evidence shows that meat was widely eaten throughout the country and that malnutrition was not a constant threat as Harner supposes. However when famine did occur, the Aztec Chronicles lament that people who had starved to death were left unburied and fell prey to wild beasts. When the Spanish laid siege to Tenochtitlan the population resorted to eating bark. The reactions to these events seem odd if dietary cannibalism had been a usual practice.

The response of the Aztecs to population growth and famine in the mid-fifteenth century was an improvement of the agrarian economy and military expansion. Both of these measures appear to have been successful by the time of the fall of the Aztec empire in 1521, and ecological stress was not evident to the Conquistadors. Human flesh, though eaten for religious reasons, could not have provided the protein needs of the population. One calculation estimates that even as a supplement to a cereal and tuber based diet, cannibalism would mean the annual consumption of more than 86 per cent of the entire population. Even if cannibalism had been used as a five to ten per cent food supplement for a privileged quarter of the population the annual number of sacrificial victims would have been quite insufficient. Aztec sacrifice and cannibalism occurred at ceremonial occasions which correspond with times in the agricultural cycle when food was abundant. The most killing and eating occurred during the corn harvest and no killing occurred when food grains would have been almost consumed. Harner's thesis is flawed on all fronts. Social theories which rest upon social adaptation to ecological necessities ignore the significance of culture and fail because of simplistic reductionism.

Here and in the last chapter we have been concerned to establish the place of culture in social life. Human beings are animals, the products of natural selection with a non-human past, but this unique life-form has evolved an ability to live by culturally directed behaviour. Human genes provide the foundation for cultural behaviour but not for culture itself or its manifestation as social institutions. Evolution continues to modify the human form but culture has gained an independence from biology and is a powerful force in its own right. Cultural development and social change are driven by forces which originate within society and cannot be explained by an evolutionary mechanism. The emergence of a capacity to learn and the rise of cultural transmission are triumphs of evolution which bridge a gap between humans and animals but also allow the form of specialist characteristics seen in human abilities. These abilities have allowed information acquired through experience to become a part of social inheritance. It is of no consequence, for instance, that there are no surviving descendants of Marx or that Newton had no children at all. Socialist ideas and modern physics exist independently of genes. But the possibility of having cultural traditions as a part of a real history rests upon a lengthy interaction between the genetic and cultural processes which comprise human evolution.

4

Primate Societies

Modern humans and pre-human ancestors are now classified as members of the hominid family. Together with the apes, hominids are part of a superfamily known as the hominoids which, in turn, comprises the infraorder of primates known as catarrhines when Old World monkeys are included. This taxonomic relationship is presented in Figure 4.1. This chapter will be concerned with the evolutionary relationships among this group of animals. Similar forms of body construction are complemented by social behaviour and abilities. Some primate behaviour occurs only in a single species but other behaviour patterns are shared by close relatives. Many human characteristics are found in some degree among the apes and present us with an evolutionary echo of the earliest stages of pre-human society.

HUMAN EVOLUTION AND THE APES

Almost a century before the appearance of Darwin's *Origin*, Linnaeus (1707–1778), the founder of the modern system of biological classification, unequivocally assigned man, along with apes, monkeys and prosimians, to the order of primates. Humans were placed among the higher primates or *Anthropoidea* and given the title *H. sapiens*. The designation was not made because Linnaeus believed that all the species within the order were related through a common ancestor from whom they had all descended. Linnaeus held firmly to the eighteenth-century belief that all species should be regarded as separate acts of creation. By insisting that species were fixed, Linnaeus denied any possibility of evolution, yet his mode of classification had a powerful influence upon later evolutionary theories, including Darwin's, because it provided a conceptual framework which invited a unifying explanatory principle such as natural selection. However there were overwhelming morphological resemblances between humans and the apes: anatomical and physiological similarities were starkly apparent to Linnaeus and to later biologists who increasingly used the methods of comparative zoology.

Modern work has amplified and broadened these observations and has also introduced novel biochemical dimensions upon which further modes of comparison are based. Le Gros Clark (1978) outlines the resemblances between the skeletal structure of man and ape; both have the same construction and general plan, he continues:

> The muscular anatomy of men and apes is astonishingly alike even down to some of the smaller details of attachments of many of the individual muscles. The similarity in the structure and disposition of the visceral organs suggests that apes have a closer relationship to men than they have to the lower Primates. The human brain (though larger in relation to body size) in its morphology is little more than a magnified model of the brain of an anthropoid ape; indeed, there is no known element of the brain in the former that is not also to be found developed to some degree in the latter. All these facts, together with observations recording similar metabolic processes, serological reactions, chromosome patterns, blood groups, and so forth, are now well known . . . Anatomically, *H. sapiens* is unique among mammals only in the sense that every mammalian species is in some features unique among mammals. The unique character of *H. sapiens* lies primarily in his behaviour.
>
> (Le Gros Clark 1978: 3–4)

These biochemical parallels provide evidence for human and ape evolution that can be assessed by objective analysis. There are several areas of molecular biology where these parallels can be firmly established, and these findings confirm the zoological and fossil evidence that the closest living relatives to humans are the African apes. The closeness has been found to be quite marked: humans share with the chimpanzee a greater affinity than that which exists between sheep and goats or between the horse and the donkey.

Genetic material has been compared between species. Humans and chimpanzees have been found to have a closer genetic relationship than do the chimpanzee and gorilla. Washburn and

Moore (1974) report that the DNA distance between human and chimpanzee is small. The two DNA structures fit together with only a 1.6 per cent difference. By using a sophisticated biochemical technique, Sibley and Ahlquist (1984) have shown a receding order of similarity for humans with other primates: after the chimpanzee, the gorilla and then the orangutan follow. Other studies of the structures of amino acids, which comprise the protein present in blood, bone and skin, show a similar relationship between humans and the great apes; the human and chimpanzee proteins haemoglobin and fibrinopeptide are identical. Immunological studies have shown that the relative evolutionary distances between species can be measured. The work of Sarich and Wilson (1967) established our relationship with other primates in the same sequence as the methods already mentioned, but at first it caused a controversy.

It was proposed by Sarich and Wilson that because biochemical evolution has occurred as a regular process in different primates, that a 'molecular clock' may be derived from such measurements which would allow the calculation of a primate phylogeny or lineage that included the 'branching off' of the member species within the primate order together with the timing of these divergences. However the dates first proposed for these divergences were much later than those which came from studies of ape and human fossils. But with more reliable fossil dating and improvements in biochemical technique a new consensus has emerged. The timing of evolutionary divergence among the ancestors of modern primates is now generally agreed by biochemists and palaeontologists (Pilbeam 1984). Sibley and Ahlquist provide biochemical dates as follows: the divergence between chimpanzee and human occurred 6.3–7.7 million years ago; the gorilla divergence 8–10 million; the orangutan 13–16 million; the gibbon 18–22 million and Old World monkeys 27–33 million (see Figure 4.1).

This evidence shows that the human lineage is closely related to the chimpanzee and other hominoids, which suggests that study of ape ecology and social life may have value in reconstructing the early phases of hominid behaviour. Fossil remains that would establish a hominid–chimpanzee divergence 7 million years ago are meagre and ambiguous. However there is good fossil evidence which shows a sequence of human ancestors reaching back over 4 million years. There is direct evidence of bipedal walking 3.7 million years ago and fragmentary fossils that are 5 million years

94

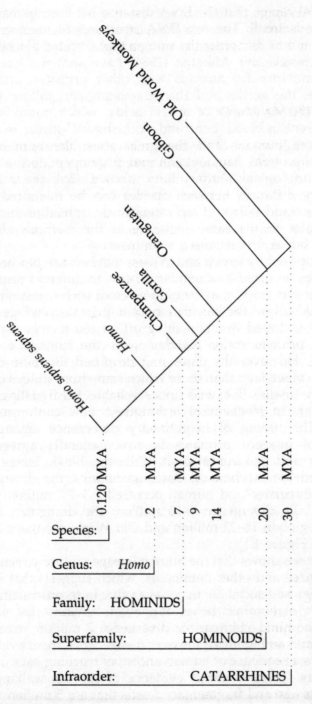

FIGURE 4.1 Taxonomic and Phylogenetic Relationships of Living Hominoids with Mean Divergence Dates in Millions of Years

MYA = Million Years Ago.
SOURCE: Based on Andrews (1988: 254)

old. These two-legged hominids possessed some 'human' features which are revealed in the archaeological record; in other respects they would not be recognised as human and their behaviour may have been ape-like.

MODERN PRIMATOLOGY

Darwin was fully aware of the anatomical resemblances of man to the great apes and Huxley compared primate and human locomotion and manipulative ability in his *Evidence as to Man's Place in Nature*. But neither scientist had access to either a fossil sequence or knowledge of ape behaviour gained from field studies. The African forest environments, which contain chimpanzee and gorilla troops, were, with only a few rare exceptions, considered to be impenetrable, disease-ridden barriers to academic zoology until the 1950s. In lieu of any real knowledge of ape society within its natural habitat, the tales of white hunters were believed and these contributed to a mythology that could not be challenged by the observation of captive animals. The 'ape-as-savage-monster' conception embodying the antithesis of all that was human, was taken as a popular article of faith. But elementary facts about primates in their environment were unknown. A number of basic questions had yet to be answered by ethological research. Was the gorilla arboreal or terrestrial? Did chimpanzee behaviour differ between troops? Did primate diets include any non-vegetable items? Could new patterns of behaviour be learned by monkeys and apes and transmitted to succeeding generations? Was the mating behaviour of apes based upon the promiscuous behaviour of females?

Long-term studies of primate ecology based on prolonged observation have produced a wealth of new knowledge about a range of primate species. In general many of the conclusions reached by these studies indicate that there are a cluster of characteristics that are common to both humans and other primates. The social behaviour of apes, monkeys and humans is to varying degrees limited in scope and content by bodily structures and functions which have a similar evolutionary origin. A broadly coherent sensory apparatus and reproductive experience, together with the dietary versatility of many primates, have acted as constraints on social behaviour. Without these physical endowments the various patterns of human culture which have come to serve production, birth, kinship and sustenance would have emerged in other forms. If, for instance, multiple births were the norm for primates, the human family would have become organised in distinctly different ways. If the apes could digest cellulose, like some monkey species who are ruminants, hunting might not have become a human speciality and culture might now designate time for cud chewing (White 1959A, 1959B).

Humans possess an extension of some features found in other primates. Both apes and humans share a biological heritage that originated in an arboreal niche which has favoured the evolution of colour stereoscopic vision. Visual acuity has been a major goal of natural selection in primates. Precise limb coordination, prehensile hands and a propensity to manipulate objects are trans-specific traits. The primate brain gives prominence to a special linkage between eyes and hands. In humans the tactile sense is reduced to the hands with the muzzle playing a minor part. Smell and hearing are correspondingly less significant sensory sources for primates. For both, pregnancy usually results in only one offspring which requires extensive maternal care during infant and juvenile dependence, which is long in relation to life span. Bonding between mother and infant is crucial among primates. Besides acting to set the pattern of bonds made later in life, bonding usually marks the beginning of an enduring social unit which is basic to society. Such slow maturation is accompanied by learning, and much of the behavioural repertoire of monkeys and apes is learned rather than reflexive. The social life of primates therefore contains far more complex patterns of responses and flexibility than are found among other gregarious species (Washburn and Avis 1958).

A strong link exists between the environment of a primate species and a characteristic form of social organisation. Yet this does not prevent monkeys and apes from actively seeking out and adapting to quite novel environments with an associated change in social behaviour. Washburn et al. (1968) report that rhesus monkeys living close to humans in Indian cities behave quite differently from troops who live in forest habitats. In one instance of tool-use by a chimpanzee, a young medium-ranking male clashed large paraffin cans together as part of a charging display. This inventiveness intimidated other older males and led to a dramatic rise in rank for the noisy challenger (Goodall 1971). Japanese macaques have developed a local culture associated with food processing in response to provisioning by scientists (Jolly

1985). Another innovation used by this monkey is group bathing in hot volcanic springs during winter when thick snow covers the Japanese Alps.

The complexity of non-human primate societies has perhaps been the major finding of field studies. No single underlying principle of social organisation can be specified: neither sex nor dominance, as was once thought, provide the cohesive element which holds primate society together (Reynolds 1980). An equally complex dietary spectrum has also been revealed in recent research. The idea that primates may be conveniently assigned dietary labels such as 'frugivorous' or 'folivorous' is exploded by Harding and Teleki (1981). Most non-human primates are in fact omnivorous in the same way that most human societies have always consumed meat and animal proteins as a supplement to vegetable food. The chimpanzee is now known to eat 200 species of plant and 40 species of fauna. This, however, should not obscure the fact that primate diets are often specialised and that dietary adaptations seem to correlate with body size (Richard 1985). The smallest primates, weighing less than 1 kg, can afford to specialise in an insect, gum and sap diet with the addition of a little fruit. Such foods are available in sufficient quantities to provide for the high energy requirements of small bodies. For larger monkeys and apes, with less energy requirements per unit of body weight, but much higher total nutritional needs, a more abundant food source such as fruit and leaves is necessary.

THE ECOLOGY AND SOCIAL LIFE OF PRIMATES

The presentation of primate social life provided here is a generalisation drawn from the observation of different species of monkeys and apes. Because of the evolutionary closeness of humans to the African apes, particular attention will be given to chimpanzee and gorilla societies. However the principles which govern primate societies cannot be taken as a foundation for human social behaviour. Comparisons between human and animal societies must be made with caution and the significance of culture to our species fully recognised.

Modern primates are themselves the products of evolution, rather than the preserved remnants of earlier stages of hominid development. It cannot be argued that because culture is of less significance to primates than to humans that a 'failure' or termination of evolutionary development has occurred, since over 100 primate species have successfully adapted to a variety of specialised niches. Yet it is also true that a knowledge of primate society can be used in reconstructing the stage in hominid development which came immediately before the emergence of culture. Elements of primate social life may be used to construct a model which attempts to envisage how a slow but cumulative process of cultural development began among the early hominids. The social life of early hominids, like that of any other animal, was determined and limited by their zoological status and '. . . must have evolved from within the range of social variations possible to this mammalian group' (Schultz 1962: 58).

The discovery and understanding of a great diversity of social regimes among different primate species has been a major achievement of modern ethology. Observers have spent prolonged periods close to animals who have become habituated to their presence. An open but passive mode of observation has been adopted which involved moving with the primate troop through its range and in one case regularly sleeping a few yards away from a gorilla community (Schaller 1972). Acceptance of a human as another animal in the habitat has allowed scientists to witness intimate aspects of ape and monkey society. Complex patterns of social behaviour which relate to friendship, socialisation, violence, social status, sex and feeding have been revealed. The writings of these scientists contain a respect and affection for their subjects who they have known in some cases for a life-span of several decades.

The ecology of a primate species is one of the principle determinants of its social life. As a simple general rule the more a species shares its environment with potential predators the more close knit is its social organisation. Thus the highly acrobatic gibbon, which lives almost exclusively in tree tops, is easily able to avoid its enemies and has little need for life within a highly structured troop. Gibbon society consists of no more than a monogamous couple and up to three of their dependent offspring. For baboons, however, matters are quite otherwise. This plains-living species shares its open country habitat with a variety of carnivores and has large highly ordered troops of up to 100 animals. The bonding of an individual to its troop is an adaptive advantage in this form of social organisation.

A more precise understanding is provided by Clutton-Brock and Harvey (1977) who have shown that an association exists between the social behaviour of a primate species and its particular form of ecological adaptation. Nocturnal primates are lighter and live in smaller social groups than diurnal species. Small-bodied tree-top living species feed alone or in pairs at night within a very small home range to avoid predators. Body weight tends to increase the less a species relies upon trees for feeding or as sleeping places. Ground living primates tend therefore to be large bodied, to range over large territories and to live in larger social groups. This is an effective adaptation for species who range through dangerous open country to feed at dispersed points.

Chimpanzee society consists of anything from 15 to 80 individuals who live in a loosely coordinated community within a lightly forested home range. Unlike most primate societies, which have a surplus of females, chimpanzees live with a balanced sex ratio. In this open society individual animals join and leave groups that are formed for particular purposes. The membership of temporary parties which gather for feeding, patrolling the community's territory, infant care and play, mating and companionship is constantly changing. Goodall (1986) found that the chimpanzees in Gombe had a far more varied daily social experience than that found in a baboon troop. The sexual composition of groups and their activities constantly changed. A chimpanzee could be part of a noisy excited gathering one day and alone the next. Individuals rarely saw all of their community on a single day and patterns of interaction were governed by personal feelings rather than group pressures. This high degree of personal autonomy and flexibility of social organisation is unique among non-human primates. Adult chimpanzees have a large freedom of choice in matters such as travel, activity and associates that contrasts with other species who are either troop-bound or solitary.

This mode of social organisation has evolved to facilitate a ranging pattern in which chimpanzees travel from 2 to 5 km each day within territorial boundaries of 10–30 sq km. The chimpanzee niche involves a series of slow seasonal migrations to fruiting trees which, together with leaves, provide the bulk of food needs. A fluid group structure which permits a diversity of ranging in order to discover new food sources, that are then reported to other community members, is an effective adaptation.

The social organisation of the mountain gorilla embraces some 2 to 20 individuals but typically consists of a dominant silver-backed male leader, who is twice the size of a female, a younger male and three or four sexually mature females together with their immature offspring (Fossey 1985, Schaller 1972). Although individuals join and leave the group, observers have reported considerable social unity as all members sleep, range and feed in close proximity. Dominant males can weigh 400 lb, yet despite this overwhelming strength gorilla society is one of a close knit group of animals bound together by affection. Gorillas are ground feeders having a specialisation for eating large quantities of abundant, relatively low-grade food, found in leaves, stems and shoots of plants such as wild celery, thistles and nettles. This primate is entirely terrestrial and has no predators but man because of its size. The gorilla's bulk does, however, restrict movement to about four to seven hundred yards a day within a territory of less than 8 sq km and a cohesive social structure is a response to this restriction.

Most of the dilemmas of life that among other animals are met individually, are given a collective solution by primates. Within the typically mixed sex group that contains adults and young, the individual animal will experience the meeting of an entire range of life needs from its earliest infant dependency to maturity and death. The daily necessities of feeding, sleeping, protection from predators and mating are met by social organisation.

DOMINANCE AND VIOLENCE

Apes and monkeys live within a system of rank-ordering that has been termed a 'dominance hierarchy'. Animals are assigned a position within the troop hierarchy according to age, strength, health and alliances with other troop members. Both sexes are ranked but males tend to control the dominance structure and engage in the most social mobility. The status of females is usually taken from their mother and this is often retained in later life.

This form of ranking should not be confused with divisions in human societies caused by social class or with male dominance (Leibowitz 1979) even though our species has a clear propensity to form hierarchies within the small informal groups that frequently develop in an educational or work context, for instance. In primate societies, social organisation by dominance probably has a genetic basis since a standard pattern of hierarchy is associated with particular species. However the social position of young monkeys and apes is not genetically predetermined. Each generation adapts to an unchanging social system by learning, and gains rank by individual performance. The rise to power of a young challenger may occur with the failing health of a leading alpha male who is no longer able to resist the threats and aggressive behaviour of a subordinate. But high status will be arduous to maintain and few material benefits, apart from a priority for some food and greater sexual access to females, will accompany the position. Top rank will be retained only while strength and alliances last and cannot be inherited by the holder's offspring.

In human societies, class systems have varied widely between cultures and historical periods. The patterns of domination and subjection associated with hierarchical societies originate in particular sets of human social relations that are unique to particular cultures. The material inequalities and distinctive life styles, that are found in the ranking systems of some societies, derive from a division of labour, property and inheritance rights which have no counterpart among animals. The absence of social class divisions in hunter-gatherer societies, which have persisted for over 90 per cent of our existence, argues strongly that there is no biological basis for inequalities in human society.

The dominance hierarchy functions to reduce conflict within the primate troop. Unrestrained conflict would be wasteful in terms of lost energy, increased vulnerability to predators and might result in infected wounds or death. By prescribing a social position for each animal, troop life is usually relaxed. An individual's behaviour will result in a predictable outcome, especially when it is oriented to a particular animal in the hierarchy or if group competition in feeding or mating are involved. Real conflict ending in wounding or even prolonged aggressive behaviour is rare. Certainly threat displays of mock biting, noisy charging and the exhibition of canine teeth occur as ritual performances and act as a means of reminding the troop of the positions which they originally occupied. Such displays can lead to a rise of an individual's status within the troop but usually serve only to demonstrate – by stamping and throwing branches – the feelings of the individual.

Infants learn the social position of their mother during play with contemporaries and take on this rank with independence. As they mature, a rise or fall in status will depend upon prowess, but although positions are established by fighting, social organisation does not rest upon aggression. Once relative positions have become clear only slight threat reminders and submissive gestures are necessary to maintain peace.

The hierarchy is usually monopolised by mature males who, in order of age, dominate the rest of the troop. A linear ranking system, in which descending subordinates are placed in subjection to those above, appears at first sight to be the sole structure of the hierarchy. But in fact the troop contains a series of alliances which have the effect of cross-cutting the authority of high-ranking individuals. It is therefore possible for low-ranking individuals to affect decisions through alliances or to gain first access to food or to take a disproportionate part in mating within the troop. Similarly the most dominant male animal may not enjoy all the privileges of sovereignty because groups formed by alliances are beyond his manipulation. The degree to which the dominance hierarchy is activated varies proportionately with an increase in danger or competition for food and space. Thus troops moving from an exposed savannah to a forest habitat will rely less on hierarchy.

Gorilla males go to extremes to avoid physical clashes and will provide reassurance to a challenger that has been intimidated. The alpha silverback dominates by virtue of strength and size all subordinate statuses. But the females, juveniles and infants who submit to the dominant male live in a social group which, as observers have remarked, contains strong ties of affection for their leader.

In chimpanzee society larger numbers and the complex fusion–fission pattern of groups lead to a more dynamic dominance hierarchy. Top status is usually the preserve of a small coalition of males who have combined their fighting skills with an intelligent series of alliances. Once a rank order has become stable and accepted, chimpanzee society is usually calm and aggression less common. Dominant males exhibit considerable psychological power. In one instance Goodall observed an alpha male confidently intimidating four angry males united against him. High rank may be gained after a male is fourteen years old and retained until the late twenties. However, desire for social mobility appears to vary with personality. Some chimpanzees are highly competitive while others, although fully aggressive, seemed indifferent to status. Male chimpanzees will normally be able to dominate all females in a community, but a female ranking system based on aggressive interaction and similarly intricate alliances is also present. Social position here will crucially affect a female's ability to appropriate desirable food and thereby provide a reproductive advantage. The material advantages of high status for males are relatively small in relation to the risks and energy required to sustain the position. Goodall suggests that chimpanzees may be one of the few animals who wish to maintain status for purely psychological benefits.

The idea of a 'war of all against all' as the basis of primate society is unfounded in observation: violence is not the cause of social cohesion and peace normally prevails. Nonetheless the occurrence of violent acts, some of which mirror pathological behaviour in human societies, has ended the conception of exclusively gentle, peace-loving apes which was current in primatology during the 1960s. While real violence is used infrequently it is still a real part of chimpanzee society (Goodall 1986). Only about 15 per cent of all attacks contained potentially wounding assaults and lasted for more than 30 seconds and, of these, only a quarter produced wounds. Attacks upon members of neighbouring chimpanzee communities, though rare, could last for 20 minutes and result in severe wounding and death. Female strangers who were not sexually receptive, and their infants, were in particular danger from males when attempting to integrate with a new troop. Extreme brutality by females was witnessed in a series of attacks which caused death. Injured or sick animals, in one case a polio victim, were attacked or callously ignored. The unexplained disappearance of infants in the first few days of life and the discovery that cannibalism was occasionally practised by chimpanzees came as a shock to many scientists. Six infants were known to have been killed and eaten by a group of related females, having been forced away from their mothers in the community studied by Goodall.

Gorillas behave with much greater tolerance towards neighbouring communities with whom they may share females and territory. Yet despite normally gentle and amiable behaviour, evidence of aggression is present in gorilla society. Large healed wounds and broken canine teeth suggest that lone males attempting to entice females away to establish a new community may have been responsible. There are also reports of infanticide and cannibalism among gorillas (Fossey 1985).

EARLY LIFE EXPERIENCE

The lasting emotional bonds developed in infancy between mother and offspring form another important element of primate social organisation. The quality of an adult's social relations, seen in the ability to form friendship groups and alliances, is known to be related to experience in infancy. A number of 'matrifocal subunits' exist within primate troops which are organised around mature females. Besides providing emotional satisfaction these social networks constitute centres for learning (Lancaster 1975). Natural selection has endowed an animal with a long period of dependence between infancy and mating with a set of emotional attachments which enable the formation of stable and continuous relationships which will, in turn, facilitate social learning. Laboratory and field studies of maternal deprivation among primates were referred to in Chapter 2. These studies show the vital importance of the mother–infant bond. Deprived monkeys and apes developed as dejected and withdrawn adult animals, if indeed they were able to survive at all. Lancaster comments that '. . . the emotional content of the bond between mother and infant is very similar in monkeys, apes and humans; it is deep and enduring' (Lancaster 1975: 22). Infants and mothers will recognise each other even after six months of enforced separation, and many primatologists report that simian mothers will devote themselves to sick and even to dead infants who they may carry with them until decomposition occurs.

In a number of ways the matrifocal unit is the true hub of the primate troop: males may enter or leave the troop or change ranks within the hierarchy yet the groups centred on mature females remain as the principal focal points for all troop members. Collections of animals of both sexes, who are related through the mother, often come to determine troop affairs as whole genealogies. Related female animals interact with males in the dominance hierarchy to determine when grooming, feeding, sleeping or movement will occur. The matrifocal units usually have a more lengthy attachment to the troop and hence a better knowledge of its territory than dominant males. Old females in dominant genealogies will know of the location of food and water and will often determine the direction of troop movement, while high ranking males may travel at the front of the troop and appear to be giving leadership.

Although parallels with the human family have been drawn with this aspect of primate society, the matrifocal subunit differs in several crucial respects. Primate societies lack any concept of paternity and, among the apes, there is no special bond between adult male and female that is marked by mating with the exception of the monogamous gibbon. The matrifocal subunit is limited to mothers and their offspring and is not a part of a kinship system which links relatives of both sexes and different generations. Nor are there any socially defined rights and obligations with attached roles for brother, sister, father or aunt in primate societies. The subgroup functions as a series of affectionate ties rather than as a social group containing prescribed positions.

Infancy in a chimpanzee society is described by Goodall as intimate and relaxed. At birth an infant is totally helpless and must be carried or cling to the mother for the first five months of life. Chimpanzees mature more rapidly than humans but infantile dependence remains long, lasting until at least age six. Styles of infant care vary with the personality of mothers; however prolonged close maternal contact with extensive cuddling, grooming and play, alternating with bursts of aggressive chastisement is typical of the experience of a growing chimpanzee. Sibling rivalries and quarrels followed by romping and play are a major part of early socialisation. The learning of social positions, sexual differences, mothering skills and familiarity with the environment occurs in the company of adults who normally remain tolerant to infants and juveniles. Young chimpanzees frequently attempt to intervene when their mother copulates, yet normally ferocious males will usually tolerate this interference benignly.

The same close bonded affectionate ties have been observed between gorilla mother and infant as well as considerable individuality in maternal care. Dominant silver-backed males are exceptionally gentle with helpless newborn infants which are smaller than human babies. The polygamous group provides a young gorilla with an environmental setting which is as decisively formative for later development as the human family.

The relatively long period of infant and juvenile dependence for primates, relative to their life span, necessitates a recognised status for the young. Between 20 to 25 per cent of a primate's life is spent in such a status. Protection and guidance are marked by a tolerant emotional attitude of all members of the adult troop and an appropriate context in which learning for social, foraging and physical skills is given. But this separation of roles on the basis of age does not imply any form of economic division in which adults provide the young with food. After weaning, primates must forage for themselves, though within the troop's protective embrace, and no assistance will be given to other adult animals even if they are sick or wounded. Washburn and DeVore (1961) describe the abandonment of handicapped baboons who are no longer able to keep up with the troop.

Similarly no other form of division of labour, such as one based upon sex, is present among primates. Sexual dimorphism – the characteristic differences between the sexes of the same species – varies markedly between primate species. Female baboons, orangutans and gorillas are usually only half the body weight of males, while among chimpanzees and humans the difference narrows down to about 90 per cent. Female roles will typically revolve around the matrifocal unit. Male chimpanzee and baboon groups may range away from the main body of the troop for reconnaissance or hunting. Males also have a major defensive role to play in troop life but their predominance here will vary directly with the degree of sexual dimorphism of the species.

SEXUAL BEHAVIOUR

The mating behaviour of primates varies between species but it is usually tied to the female's menstrual cycle and is characteristically unemotional and promiscuous. Mating is initiated by behavioural and scent signals indicating that a female has commenced estrus and is sexually receptive. In some monkey species it is possible to make a female attractive to males artificially by swabbing the genital area with appropriately scented chemicals. The period of mating is usually keyed to the ecological pattern of the species. There is a tendency for primates to allow birth, primary nursing and weaning to occur before the beginning of the rainy season. But seasonality remains complex and varies from one population to another. Females have been known to enhance and suppress their own sexual cycles and those of other females as a response to the politics of the female hierarchy as much as to an annual change in food supply.

For most primates, other than humans, sexual activity is infrequent and even rare. Irrespective of differences in the menstrual cycle of species the facts of pregnancy, lactation and the post-birth period of infertility mean that the presence of a sexually receptive female in a primate community is uncommon. If pregnancy does not occur during the mating period, females generally cease to cycle. The estrus period itself often lasts only a few days and since this is usually followed by pregnancy, years may elapse before further sexual activity. Among gorillas, the opportunity for a male to copulate may occur less than once a year. Ringtail lemurs mate only during a two-week season each year and for gibbons monogamy means that sex may not occur for several years. This relative lack of sexual activity means that social stability in primate societies cannot be ascribed to a reproductive urge.

The manner in which mating pairs are formed varies widely in different primate societies. Among Capuchin monkeys and gorillas it is the female alone who normally initiates sex. The desert living hamadryas baboon mates by dominant males monopolising a number of females in a harem. Some females are captured in infancy and are mated at maturity. Female fidelity is only maintained by constant male wariness and this social pattern is most probably the result of an adaptation to an arid environment which necessitates troop dispersal for foraging. Exposed females and young would be at a great disadvantage if they used the promiscuous organisation of savannah living baboon species and therefore the harem functions primarily for economic reasons rather than sexual ones. Gibbons mate as monogamous couples with an emphasis on sexual bonding for both sexes for life. When this bond ends in death, the surviving member of the couple is frequently unable to attract a new mate and remains alone.

Among most monkeys and apes, however, sexual interaction is promiscuous and consort pairs form only for a brief period. Ground living species such as baboons, Vervet monkeys and gorillas engage in intercourse which involves the whole community. While dominant males may enjoy the greatest access to an estrus female, and perhaps form a consort relationship, the female will still mate with other males of lower rank. Despite the existence of a female scale of preferences for males – savannah baboons, for instance, form long lasting male–female friendships – females will willingly mate widely. This absence of fidelity shown by females suggests that while sexual behaviour is influenced by the power relations of the hierarchy it is not controlled by dominant males alone. Dominant baboon males have in fact been known to mate with females closer to the time of ovulation than lower-ranking males. However because sperm can survive within fallopian tubes for several days it is impossible to determine which male causes a female to conceive (Martin and May 1981). It is therefore not possible to demonstrate that mating behaviour involves the selection of the 'fittest' or most dominant male as the most prolific breeder.

A chimpanzee female becomes sexually mature at ten or eleven and may leave her natal troop for a neighbouring community. This pattern of female transfer has been widely interpreted as a means of avoiding incest. But it has also been observed that among Japanese macaques, rhesus monkeys and olive baboons, as well as chimpanzees, mating between close relatives who remain together as adults is rare. The promiscuous mating of chimpanzees means that the relationship of father and daughter remains unknown, and incest remains a possibility for females who have not transferred. However the age difference and a female's visits to other communities during estrus may help to minimise incest.

Estrus in chimpanzees becomes apparent when a large rear swelling and scent signals appear that last about ten days. The presence of a receptive female in a community produces what Goodall refers to as a 'sexual jamboree' in which individuals, who may not usually see each other for weeks at a time, flock together for a week of unrestricted mating. A female may copulate with as many as nine males within an hour without any sign of possessiveness or jealousy provided that the dominance hierarchy is well established. Males will wait patiently without fighting for an intromission of about 15 seconds.

The sexual apparatus of male chimpanzees appears to be the result of selection for reproductive success in a promiscuous society where males outnumber females. The chimpanzee has a large penis relative to the gorilla, but not to man – 8 cm, 3 cm and 13 cm respectively (Goodall 1986) – and large testicles for high sperm production. Rapid intercourse of less than ten pelvic thrusts is a functional means of ensuring multiple male access to a single female during infrequent periods of estrus. Males are capable of ejaculating three times within five minutes and a single alpha male was observed to have had 21 copulations in a single day.

These varied examples of primate mating patterns are cited to indicate that simplistic analogies between primate sexual behaviour and the formation of the human family on what is assumed to be a homogeneous primate model should be avoided. Early hominid societies cannot be assumed to have been organised around either universal promiscuity or the hoarding of females by a few dominant males. Primate mating systems tend to relate directly to environmental circumstances and their use in comparative analysis with the human family only serves to show how great is both the gap between the two systems and the problems posed in attempting to comprehend an evolutionary connection.

SOCIAL LEARNING

Much primate behaviour is learned and some behaviour can be traced to a cultural tradition. Baboon infants that have been isolated in laboratories fail to develop 'baboon' behaviour beyond a few reflex actions such as clinging. Evolution has selected for an ability to learn; to acquire a knowledge of the environment and a behavioural pattern by social interaction. Wild baboon troops with similar territories will choose different diets; the plants, roots, grasses and animal proteins that are eagerly taken by one troop are avoided by another. The range and variety of diet can be considerable. By one year Japanese macaques have learned to recognise up to 120 items of diet. There are, however, limitations on the extent of this learning. Both baboons and chimpanzees relish live termites but only chimpanzees strip branches for insertion into a termite nest to extract insects. Chimpanzee tool-use is frequently observed by baboons, yet despite sufficient dexterity baboons have never been known to attempt to unearth termites with any form of tool.

It has been shown experimentally, and through training in circuses, that apes can perform quite complex tasks. Wright (1972) was able to teach a captive orangutan to make stone tools and to use them to open a box containing fruit. There appeared to be no difficulties of manipulation or coordination in striking stone to make tools but the orangutan seemed unable to make any connection between the tool and its potential until Wright had repeatedly demonstrated one particular use. Successful learning, signified by the ape's eventual performance of the tasks, occurred only after a long series of exposures had given a model of behaviour that could be copied.

This form of imitative or observational learning is the usual means of gaining knowledge. Learning usually occurs between close kin, and subordinates are more likely to copy the behaviour of dominant troop members. It had been thought that active teaching was not present in primate societies, but this distinction is now more difficult to maintain. Primates generally are known to place their young in situations which aid learning; considerable non-verbal communication has been observed during the learning process and, in a laboratory, a chimpanzee using sign language has taught an infant relevant signs and the objects which they denote.

New learned behaviour patterns have been observed to accumulate within troops to form social traditions. There are now a large number of examples of cultural variation between primate societies. Different forms of tool-use are found in the termite-fishing methods of chimpanzees. All chimpanzees form social grooming clusters but only in one society do participants hold hands above each others' heads for this purpose. Chimpanzees in West Africa open nuts with hammer stones which is unknown among troops in East Africa. Certain baboon troops also use stone hammers but their neighbours do not. Chimpanzees both in nature and in captivity have shown that their social group is capable of inventing new techniques for food gathering which are applied for a shared benefit. Japanese macaques learned to remove earth from sweet potatoes and to separate grit from wheat by washing these foods in the sea. The innovations began among young monkeys and spread to the rest of the troop through the matrifocal units, but rarely reached the adults at the top of the dominance hierarchy until several generations had elapsed. Learning potential here seems to be concentrated in the period before puberty and adolescence; thereafter a conservatism prevents adults from taking up new learned patterns. However experimental work with other primate species suggests that while adults are slower in solving new tasks, they possess more environmental knowledge with which to integrate novel facts. Dominant males especially are much less likely to adopt new patterns of behaviour, but Jolly (1985) argues that this conservatism is often functional for the troop. Males would often prevent younger troop members meddling with strange objects that were potentially dangerous.

PRIMATE SOCIAL ORGANISATION

This chapter has pointed to the complex nature of primate societies which stems from a high degree of social interaction involving recognised roles and statuses. Primate societies are maintained by a learned behavioural repertoire that may be varied with greater flexibility and innovation than is found among any other gregarious species. There are however severe limitations here. Primate abilities fall far short of the characteristics upon which human society rests. The close resemblances between chimpanzees and humans are astonishing at many levels, but this ape cannot be seen merely as a 'person without language': two distinctive evolutionary histories divide these species. Despite its elasticity primate society seems to constitute a configuration of behaviour and roles which hold together as a matrix from which comes an endlessly duplicated form of organisation.

One attempt to explain this consistency and to distinguish the vital differences between animal and human society is provided by Reynolds (1980). He argues that primate social behaviour derives from innate physiological mechanisms for self maintenance which are functionally related to the environment. He cites an example taken from the observation of a colony of captive rhesus monkeys to illustrate this mechanistic process. With the death of a top-ranking female her dominant male consort sought out and promoted a low-ranking female as a new mate. This female underwent a change of personality; she rose from utter subjection to regal dominance by taking over, in total, the behavioural performances of her predecessor. The colony found the disruption caused by the loss of a leading member intolerable and induced the same behaviour pattern to fill the void. A range of latent behaviour could be drawn upon by the colony from any member in order to reconstruct a social system with very definite boundaries. Reynolds' view does not amount to a simple formulation of primate behaviour as one which is genetically preprogrammed with the addition of possible learning. He presents a theory of interaction in which behavioural signals of troop members are responded to with matching signals that are linked to an ability to respond in new ways as the behaviour of others changes. Social life operates to maintain an unyielding and predetermined matrix of social relations.

What distances this mode of social life from a human one is its absence of conceptual thought. Only humans think conceptually, which means for Reynolds that ordinary consciousness – present to an extent among animals – is for humans encoded in the symbolic forms of language. Only via the symbolic properties of language can the cognitive means for concept formation occur. In humans the unique endowment of conceptual thought allows the naming and classifying of objects according to their qualities and the arrangement of concepts ordered as a system of knowledge.

Conceptual thought then is a primary factor which allows human social behaviour to be possible. Reynolds' argument implies that the varied and changing social positions and social rules which confront us in human society are the results of an infinity of diverse conceptual formulations which have been formed during interaction with environmental circumstances. As there are no real limits to the permutation of the symbols from which concepts derive, innumerable variations in human behaviour can occur. Language thus allows humans to manipulate symbols and have an open social life based on culture, but animal signalling systems function within an interminably circumscribed pattern of social action.

However this distinction cannot be made absolute. Animals with a close evolutionary affinity are usually found to possess characteristics that are held in common, though not in equal parts. The human–chimpanzee affinity is overwhelming and will not support a sharp animal–human dichotomy. It must be presumed from the archaeological evidence that pre-human hominids possessed at least rudimentary consciousness and concept formation. On this basis it would be unjust to deny at least a semblance of these qualities to the apes. Studies of the communicative abilities of chimpanzees confirm that at least a nascent ability to form concepts and use symbols exists here.

But no arguments in support of an overlap or of shared features can hide the enormous differences that exist between humans and closely related animal species. These of course are modern differences, the results of some millions of years of independent evolution; but at one time a propensity to acquire a way of life based upon culture must have been present among the earliest hominids and a knowledge of primate society greatly assists in the reconstruction of this phase.

5

The First Hominids

What were the circumstances in which culture became a significant mode of adaptation among human ancestors and then began to produce a sequence of related changes in both the environment and the biological constitution of the early hominids? Henceforth this new mode of non-genetic adaptation would nurture forms of behaviour and technology, which is the material expression of culture, that were transmitted across generations. A constant accumulation and modification of cultural practices has allowed a variety of separate adaptations or traditions with distinctive social identities to arise, since the use of culture implies a whole range of different possible relationships with the environment. The cultural way of life interposed itself between the individual pre-human hominid and the environment, and acted as both a protective barrier and as a prism which mediated and defined a collective response to the forces of nature.

Culture has come to constitute an enveloping 'social environment' which has changed the terms of natural selection. Dependence on cultural solutions has considerably extended human fitness. But in the course of evolution the human *capacity* for culture became biologically implanted. The practice of a cultural way of life influenced the direction of physical evolution. The overwhelming propensity to live by culture has become part of the human genetic code and therefore a species characteristic of *Homo sapiens*. The human genotype has interacted with culture and in this reciprocal feedback relationship natural selection has directed that a capacity for culture should be developed. A capacity for culture, but not its contents, is therefore as much a part of the human genetic programme as a body temperature of 37°C or a nine month pregnancy term. There are of course no genes for particular languages or for most forms of social

113

behaviour, but the capacity to learn a language, and to learn in general, is a function of human biology that is realised in social interaction (Dobzhansky 1963).

This chapter will be concerned with the evolutionary steps which led to the emergence of the first bipedal human ancestors. An examination of the psychological and social properties which may have accompanied this transition are presented first. An account of the earliest stages of hominid evolution and a discussion of the possible causes of bipedal walking complete the chapter.

MENTAL ABILITIES

In the last chapter it was claimed that while no human qualities could be isolated that were not to some degree shared by other primates, modern humans nonetheless possessed a collection of mental characteristics that allowed culture to be maintained. A common assertion in the social sciences is that such characteristics are proof of an essential and unbridgeable distinction between humans and the animal world. What is at stake here is the rigidity of this distinction and the possibility that continuities may span this rift. Some authorities have suggested that particular mental abilities not only underpin human status but may also be responsible for its emergence. Although elusive and unamenable to positive analysis, Dobzhansky nevertheless insists that:

> . . . mind is scientifically an unknown, though existentially the basic entity. Manuals of psychology and neurophysiology often do not even mention the slippery words 'mind' or 'self-awareness'. An evolutionist, however, is obligated to deal with them, since the emergence of mankind is incomprehensible without them. . . Human self-awareness obviously differs greatly from any rudiments of mind that may be present in non-human animals. The magnitude of the difference makes it a difference in kind and not one of degree.
>
> (Dobzhansky 1977: 453)

The need to explain human mental powers in evolutionary terms is inescapable. But which stage of human evolution is being proposed as synonymous with human status: the emergence of bipedal hominids, the first use or making of tools, the rise of novel human patterns of social organisation or the use of language? These are elements of a human way of life which were accumulated by different hominid species over several millions of years.

Reynolds' proposal that conceptual thought and self-awareness are *now* part of the mental foundations for culture was discussed at the end of the last chapter. Reynolds argues that language allows symbolic conceptualisation; classification and order allow a social life based on abstract conceptions seen in rules and statuses. Language therefore appears to be a prerequisite for culture. Reynolds continues:

> . . . conceptual thought needs to be considered as a primary datum causally underlying the development of the big brain in our genus, *Homo* [and is] . . . of equal or greater magnitude than the development of tool using . . . The emergence of man was essentially a coming into self-awareness of what was already a highly complex social being, and . . . *this coming into conceptual awareness was the crucial and fundamental adaptation that enabled man to master first the savanna environment and later the rest of the world.* By it he was enabled to achieve a quite rigid and regular pattern of social relations . . . This was man's key social adaptation to the open country: the conscious ordering of social relations by conceptualisation of the self and of the group, its structure and roles. (Reynolds 1980: 63/74, original emphasis).

There is much to agree with here. These modern forms of consciousness are a significant part of the foundations of human behaviour. However Reynolds' categories, conceptual thought and conceptual awareness, cannot be clearly delineated or assigned to humans and non-humans with any precision. Nor did these qualities of mind act alone as privileged salients which preceded and spearheaded the emergence of full human status. Such qualities must be seen in the context of a cluster of interacting characteristics that were chosen by natural selection.

There is no reason to suggest that some form of altered mentality and communication was a necessary part of the transition to the bipedal open-country niche made by the first hominids. Nor is there any necessity to postulate any special mental qualities, that are now unique to modern humans, to explain either new modes of social organisation or the emergence of tool-use and tool-making. Nothing so dramatic or spectacular was necessary for the emergence of a culture-based way of life since all the prerequisites of a cultural adaptation can be found in primate society. Higher forms of conscious awareness and conceptual thought gained prominence in human minds because of the historical experience of pre-human life and probably emerged only in the later stages of human evolution as little as 40 000 years ago.

However Dobzhansky's claim that the differences between human and animal minds are ones of kind and not degree would find broad acceptance in the social science tradition. Only humans would be able to project a mental image, which was the product of past abstractions and insights, and realise this at some future time because of conscious awareness. An animal might 'know' by possessing knowledge in the form of conceptual categories, but at the same time be unaware that it 'knew'. The deficiency of consciousness prevents the imaginative juxtaposition of concepts, called forth either at will or spontaneously – but in either case through the agency of language – to comprise an image or idea. In humans, conceptual awareness enables social relationships and behaviour to be classified together with norms and notions of deviance, since these abstractions depend upon and act in concert with symbolic language.

What can be said of consciousness itself and how significant is conscious awareness for human behaviour? This most intractable of mental qualities, which positivistic psychology has usually avoided and even refused to recognise, has no clear set of boundaries. Humphrey (1986) proposes that consciousness, like all organisms and aptitudes, is a product of natural selection and arose to facilitate complex social interaction. He sees consciousness as the presence of an inner picture, which we call 'I', and through which we gain an awareness of self in the context of time and environment. This definition is close to that of Crook (1980) for whom consciousness is also an evolutionary product. Consciousness is an integrated whole made up of information processed by the senses, which produces an image of the external world like an internal visual panorama. Consciousness can be called upon directly to review immediate problems or used to focus upon something in another time or place while current activity is somehow directed by a mental autopilot.

However for Jaynes (1976) consciousness arose only in historical times when language, and even writing, had long been human accomplishments. Jaynes sees consciousness as a unique human quality which vastly extends perception and understanding because it grants the facility to make connections by metaphors that are independent of the constraints imposed by time, space or self. Consciousness can project meaning by analogy since consciousness is located in the gap between the thing to be described and the thing, or relation, used to elucidate it. If I claim therefore that 'rain fell like stair-rods' my consciousness is at work.

Jaynes argues that in modern humans the use of concepts, learning, thinking and reasoning do not require consciousness. Even complex skills where coordination of the limbs occur, or in reading or music-making, consciousness is not present. A self-conscious awareness here would endanger the activity itself as every performer, cyclist and water-skier well knows. Like animals we have no need to be aware that we are using concepts even though our concepts are more general and more abstract. The act of learning occurs without our awareness of it. Experiments show also that learners can be induced to learn without awareness being present. Nor does thought or reason occur in company with consciousness. Acts of judgement seem to occur spontaneously, as a property of the nervous system, and like complex skills these tend to bypass consciousness. We are also apt to make acts of imaginative insight that reach correct conclusions that only latterly are justified by conscious logic; our reasoning processes work faster and ahead of consciousness.

In Jaynes' terms, many distinctive human abilities can thus occur without conscious awareness. But in certain basic functions which we share with animals, such as perception, consciousness – as described by Humphrey and Crook – seems to be essential. Patients who had suffered the removal of their visual cortex were technically 'blind' because they were unaware of what they 'saw'. Yet the brain and eyes of these patients were otherwise undamaged and continued to process visual information. When asked to guess the nature and location of objects it became clear that sight was present without consciousness of what was seen (Passingham 1982). This phenomenon of 'blind sight', which has also been found in a monkey that learned to 'see' again (Humphrey 1986), occurred because of the mechanical disassociation of conscious and unconscious perception. Much human behaviour then, including complex mental operations, can occur independently of consciousness. In other brain activities though, which are shared with some animals, conscious awareness seems to be a property necessary for survival.

Can there be any bridge between the mental activity of humans and that of other primates? Do only humans use concepts, think, and have a sense of self-awareness? Admittedly human imaginative insight, as it is observed now, is without rival from the animal world. It is certainly possible to demonstrate by experimentation that animal mental abilities are of a lesser degree than those of humans. However captive apes have shown a range of imaginative solutions to technical problems suggesting an evolutionary continuity of mental qualities.

Some mental processes are thought to be primate traits which preceded human evolution. These include conceptual thought, incorporating a perception of the external environment, an analysis of this sensory input and the production of an abstract formulation (Campbell 1975). The prominence of visual perception and its integration with other sensory data has given primates a particularly clear sensitivity to spatial relationships. This tendency has been reinforced by the habitual manipulation of objects and has allowed the higher primates to extract individual objects from their place in the environment. This analytical form of perception, derived from primate heritage but reinforced by protracted social learning, provides the basis for conceptual abstraction. Concept formation occurs when the perception of a particular object is transcended and the object thereafter ceases to be unique by being subsumed within a class. The concept holds an independence from any single instance in time or experience. Categorising is not a conscious process and it is likely that animals other than primates form concepts, although their degree of abstraction and their range in time are much more limited than those of humans.

Donald Griffin's book, *The Question of Animal Awareness*, pursues the theme that complex processes appear to occur in animal brains which have much in common with human conscious mental experience. During the past few decades research in animal behaviour has revealed a wealth of unsuspected abilities and has invalidated many criteria used to establish human uniqueness. Animals have been found to recognise themselves and their fellows as individuals, have the ability to use tools and use symbolic forms of communication and to learn from past experience.

A further challenge to human pride may be presented by the possibility that some animals possess conscious awareness. Griffin argues that animals must have cognitive maps in the form of a representation of the outside world in the brain, in order to have proper orientation. Similarly, animals cannot be denied a semblance of mental experience in the form of a series of interrelated mental images. In as much as animals combine the different types of information that are produced by their senses into a unified representation, which is integrated with memories and needs, a form of consciousness is present. Even though this consciousness may be restricted to a single sensory channel, such as vision, unlike the multi-dimensional form in humans.

Many animals have been shown to be aware of what they are doing. In one experiment rats pressed one of a series of levers according to the activity which they were currently engaged in, showing that they were conscious of their own actions and able to report on them (Passingham 1982). Animals are also capable of learning. Numerous observations and experiments have shown a change or modification of behaviour according to expected consequences even in simple animals. Animals also use concepts in the sense of classes of 'equivalent things', such as food or trees to escape up, and in as much as animals also engage in acts of judgement which concern the use of concepts, they can be said to engage in thinking.

At one time in hominid evolution the differences in mental ability between ape and human ancestors must have been small, and evidence of this stage and of an evolutionary continuity is still present among modern apes. Chimpanzee studies show that conceptual thought, conscious awareness, reasoning ability, symbolic communication and culture all exist though in rudimentary forms in this animal (Calvin 1994). Chimpanzees have a concept of self since experimental subjects were found to react unequivocally to their own images in mirrors (Gallup 1975). In another experiment a chimpanzee put an arm through a wall and searched for food by looking at a television screen.

Chimpanzees have used sign language, or arranged plastic symbols which stand for units of language, in original ways. A brazil nut was signed as a 'rock berry', celery as 'pipe-food', watermelon as 'drink fruit' and Alka Seltzer as 'listen drink'. New signs have been invented by chimpanzees and, in at least one case, a younger chimp has been taught to make signs by an experienced older animal with a vocabulary of several hundred signs. These language studies have shown that chimps have an ability to associate, categorise and use concepts and abstractions. On an occasion when the sign for 'plate' had been forgotten, the signs for 'bowl' and 'cup' were used as an alternative. The sign for 'open', normally used for the door, was applied by the chimp to a water tap and a drink can. The ability to use signs as symbols is clearly present here. This ability was found to extend to a recognition and understanding of photographs. Chimpanzees put their ears to pictures of watches, salivated over food advertisements and, in one case, a female became sexually aroused as she looked at erotic material featuring nude male humans.

Chimpanzees have good memories and cannot be said to live permanently in the present. Spatial memory for solitary trees, termite mounds, the site of a past incident or the placing of hammer stones near a nut tree has been noted by observers in the field (Goodall 1986). In one community living in an arid part of Senegal, chimps memorised an exceptionally large range of over three hundred square kilometres. Tool-making and using by this ape, as well as what appeared to observers as pre-planned group foraging, constitute evidence of an ability to anticipate the future. However crude the adaptation of a natural object into a tool may be, the very act of modification prior to its use implies intention and foresight.

Experiments showing that chimpanzees can impute the mental state of others, that they are capable of deception and lying and that social traditions are a significant part of their societies, surely point to the presence of an evolutionary continuity between ourselves and animals. Television scenes of a human actor trying to escape from a room, and an actor shivering in front of an unconnected electric heater, were shown to a chimpanzee who then correctly chose pictures for either situation showing a key and a plugged-in heater. On many occasions wild chimpanzees have been observed to lead rivals away from food or adopt a strategy of giving false information to others for an advantage. A chimp fluent in sign language had a conversation with her keeper in which she disclaimed knowledge and then blamed others for her own misdemeanour before being shamed to confess and apologise. Examples of primate culture were cited in Chapter 4, but we should add the early discovery of Kortlandt (1962) that in chimpanzee troops respect for individuals who had once held high social rank continued even though they were no longer able to defend themselves.

Why should natural selection have elicited such advanced mental processes? For many primate species consciousness and intelligence would appear to be unnecessary for the pursuit of a simple foraging life style. But while primate societies may be ecologically simple, relative to human hunter-gatherers, they are socially complex. Even lemurs who are only distant relatives of apes and hominids, as well as many monkey species, acquire much of their foraging and social behaviour by participating in troop life. This suggests that the adaptive use of intellect as a means of social learning was a long-term feature of the primate order (Jolly 1966). The comparatively large brain and high intelligence of the gorilla and chimpanzee is thought by Humphrey (1988) to be functional for social skills rather than practical invention. An endless series of small disputes revolving around ranking, access to favoured food, sleeping sites or estrus females, the membership of foraging parties, grooming or the acceptance of outsiders into the community demands a high level of social awareness. Pure intelligence, for both apes and humans, would be insufficient for social survival. What is required is an introspective ability which can understand, respond and manipulate social situations. But it is only through an internal understanding of the behaviour of the self that an intuitive process can be used in social interaction. Having access to one's own thoughts allows intuition to be added to perception, but it also allows an individual to become identified with the goals of another and to make common cause or oppose these. The underlying fears, motives, intentions, opinions and hopes which imbue social contact can be decoded. This ability would enhance fitness for a primate leading a complex social life and would also elicit new traits such as compassion, empathy, deception and treachery. In this argument the emergence of these new mental qualities is the result of selection for creative intelligence which underlies successful social relations. Byrne and Whiten (1988) have called this collection of qualities 'Machiavellian Intelligence', a form of thinking evident in the social strategies of higher primates.

Regardless of continuities in evolution, the mental capacities that underlie habitual tool-use and tool-making and advances on animal forms of communication were distinctive for human ancestors. The ability of ancestral hominids to mentally extract from raw material the form of a manufactured tool must have far surpassed the abilities of any contemporary animal species. The later ability to monitor and reflect on thought and to think in word symbols had momentous consequences. A broad span of experience over a great depth in time could now be incorporated and become a source for classification of a widening environment. These mental processes, which comprise a significant part of the human mentality, are together with other anatomical features the result of a new adaptive pattern which arose cumulatively over millennia. They should not be seen merely as an insertion of novel features within an ape psychology.

However it is interesting to note that in the early twentieth century, human evolution was erroneously seen as a 'brain-first' or 'brain-led' process (Reader 1988). An intelligent ape was thought to have adopted bipedalism, and a human-like way of life followed as intelligence illuminated the way forward. It has now been well established that brain expansion occurred after a successful line of evolving hominids had embraced new foraging strategies.

To sum up: it seems reasonable to conjecture that although consciousness and intelligence, as it now resides in humans, may not be reduced to elements of a primate mentality – it is something quite apart from the experience of apes and monkeys – it nevertheless evolved from earlier forms of animal conscious awareness. On this basis it can be claimed, firstly that no *essential* distinction between humans and other animals can be maintained – differences are of degree, and not of kind, and cannot be specified with accuracy – and, secondly, that no special mental qualities need to be postulated as causes of a cultural tradition. The primatologist Robert Yerkes wrote in 1943 that 'The ape is at the beginning of a road on which man has advanced far' (Goodall 1986: 40). It is now clear that chimpanzees at least have remained in step with humans for some distance along this road.

THE EMERGENCE OF HOMINID SOCIAL LIFE

What is implied by the transition to hominid and then full human status? The time-scale of human evolution covers several million years, and the primacy that may be assigned to any single feature, even to bipedalism, is difficult to establish unequivocally in the fossil and archaeological record. In broad outline the interlinked changes can be presented as follows. Proto-human hominids developed bipedal movement. Hands freed from locomotion allowed tool-use to assume great significance and in consequence laid less emphasis on teeth and jaws as a means of manipulation and defence. The loss of estrus and the continual sexual receptivity of females probably accompanied these changes. These features may have coincided with what was the beginning of a mixed economy of foraging for both animal protein and vegetable matter. Neurological reorganisation, which produced new mental abilities, occurred with a change in the size and capacity of the brain which was to give prominence to new manipulative and communicative skills. All of these changes both encouraged and relied upon advances in culture which must be seen simultaneously as both new modes of social organisation and an enhancement of technology seen in tool-making. Cultural patterns with no precedent among social primates developed and became crucial for hominids. These included the carrying of food and habitual food-sharing at a home base, which enabled food storage and further division of labour, care of adult dependents, an affiliation with other groups particularly for the exchange of mates – leading to the recognition of kinship as a basis for reciprocity and the use of rules and norms of conduct. This process is shown in Figure 5.1.

Almost nothing is known of the form of social life of the first hominid species or their immediate precursors. A preadaptive stage in which novel forms of behaviour became adaptive and gave the potentialities for further change has been suggested by some authorities (Hallowell 1962). This 'proto-cultural foundation' constituted a behavioural complex which differentiated the earliest hominids from other primates. It was seen in Chapter 4 that the wide diversity of social behaviour and mating patterns among primates presents an obstacle to the reconstruction of the early hominid pattern. But common to almost all primates is a social structure consisting of several generations in a mixed-sex group which operates to provide all adult members with a social context in which to forage and satisfy needs for procreation, nurture and protection. It is from within this social system that a cultural system could emerge. The particular species from which hominids diverged no longer exists, but nevertheless the life styles of surviving primates may provide an insight into the earliest stage in human evolution.

Baboon society consists of large hierarchically organised troops of some 40 to 80 members. The troop incorporates individual animals as much by its male dominance system, which usually precludes any contact with other troops, as it does by determining behaviour which, in turn, governs every aspect of survival. Foraging occurs as a whole troop moves in an organised convoy through dangerous open country with appointed defensive

FIGURE 5.1 *Diagrammatic Representation of the Reciprocal Feedback Process involved in Hominid Evolution*

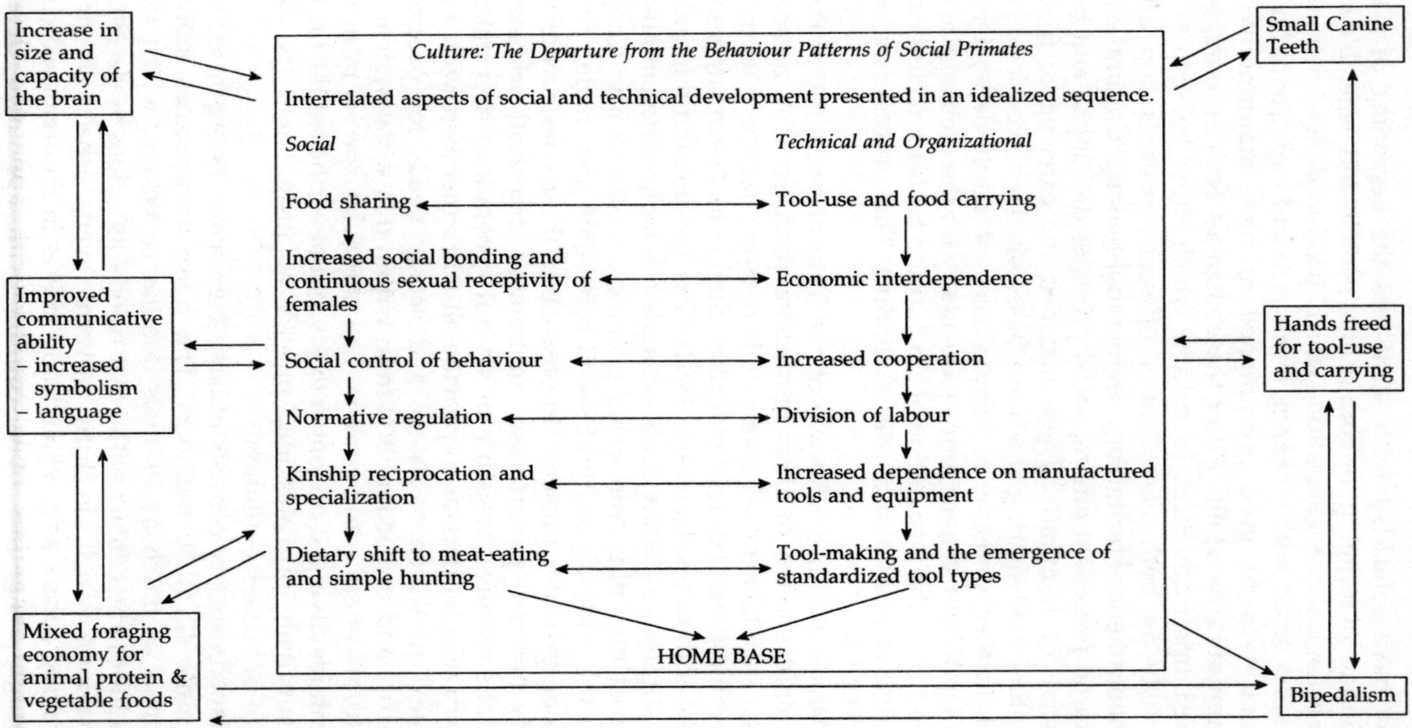

SOURCE: Adapted from Leakey and Lewin (1979) and Isaac (1976B)

positions for each member (Washburn and DeVore 1961). Most food is vegetable in origin with roots, fruit, seeds and nuts predominating with occasional feeding on invertebrates and exceptionally on meat taken from larger animals (Harding 1981). This somewhat rigid social organisation contrasts acutely with that expected of the early hominids in which a broadly-based subsistence strategy was pursued by open groups. Strum (1987) has since modified this totalitarian conception of baboons, particularly with regard to male dominance, yet apart from the savanna adaptation few of the behavioural traits of this monkey agree with the hominid pattern.

The flexible foraging patterns of chimpanzees best approximate the ecology of the early hominids. Observation in the field has disproved the long-held assumption that humans are the only hunters among the primates. Goodall (1971) and Teleki (1973) have shown that chimpanzee hunting followed a predictable pattern which suggests that it is an established tradition. Teleki observed members of an all-male chimpanzee group pursue a variety of game including young bush pigs and bush bucks and even other primates such as baboons and colobus monkeys. The division of meat was a social event that appeared not to coincide with the troop's rank structure. Requests for food addressed to the carcass holder were met or denied according to bonds of affection rather than social position.

Considerably more chimpanzee behaviour has been observed to be the result of learning than that of any other primate except humans: the individual animal is moulded by maternal education and the troop's social traditions. Chimpanzees make tools and share food: two characteristics once thought of as uniquely human. Chimpanzee communication is complex and cannot be reduced to stereotyped or repetitive call signs. There are a number of elaborate, vocal and silent, features in the chimpanzee's messages, including the coded use of gesture and facial expressions, which anticipate language.

In terms of troop structure, chimpanzee society consists of a series of interacting open groups without permanent ties of membership. The spartan corporate group with its mechanistic forms of control in baboon troops is not seen among chimpanzees. Goodall (1986) found a loose confederation of groups: chimpanzees roamed the forest in single-sex and mixed groups, and nursery groups of infants and mothers which sometimes included male kin. 'Stranger' female chimpanzees from distant troops occasionally joined the observed troop. Groups merged or split as individuals freely joined or left any group.

A basis for the emergence of hominid society is provided by this nomadic and exploratory pattern. The open groups of chimpanzees are consolidated by ties of affection, and in this respect as in mating, where individual preference governs pairing, action is guided by choice rather than by a dominance hierarchy. Chimpanzee 'out-mating', which contrasts with baboon 'in-mating', allows an individual animal to mate with any other chimpanzee regardless of rank or original troop membership. The action of adult male bands, sometimes with the addition of childless females, in foraging ahead of other groups and signalling the presence of food sources to the whole troop constitutes a rudimentary sexual division of labour. This model of social organisation agrees with the foraging needs of the first hominids living at the forest edge. Flexible group structures with cooperation and food-sharing could effectively develop on this social matrix around an ecological adaptation of gathering and scavenging (Reynolds 1966, 1968, 1980).

The emergence of hominids was based upon new ecological relations which, in turn, began a sequence of changes in social structure, communication and psychology. Bipedality appears to have given the initial impetus to hominid evolution rather than growth of the brain which, as we have noted, is known to have occurred long after this new form of locomotion had been perfected. But it was the interaction of these factors with social life which makes hominid evolution comprehensible as Hallowell comments:

> While no one would wish to minimise the importance of the later expansion of the brain in behavioural evolution, cause-and-effect relations are over simplified if we do not take into account the continuing social context of behaviour, the potentialities for change in the pattern of inter-individual relations and, consequently, in the attributes of the social order considered as an evolving system.   (Hallowell 1962: 240)

Fundamental to the emergence of hominids was a shift in environmental relations which revolved around bipedalism and terrestrial adaptations. Future capacities as a tool-maker or fire-user are unthinkable in the context of an arboreal existence. It was ground living which rendered a new ecological niche available since new food resources and a change towards a partially carnivorous diet opened up potentialities denied to tree dwellers. Bipedalism accompanied this shift, but under the new ecological regime powerful 'pre-adaptations' that had been fashioned for a life in the trees were to be crucial for a cultural way of life. Dextrous hands allied to colour stereoscopic vision and a complex brain capable of considerable environmental discrimination were coupled with a gregarious behaviour pattern.

At some point in hominid evolution the usual primate pattern of closed groups with exclusive territories was overcome in favour of flexible bands with shared territories. An internal differentiation established within a group for a division of labour, perhaps based on sex, would have implied a reciprocal exchange of services. Open groups, with changing size and composition, that migrated with game and seasonal vegetation would benefit from unimpeded entry to territories that were far larger than those of any other primate.

Using the chimpanzee model it is possible to conceive of this development through an alliance of male roving groups and females with infants which was based upon sharing the products of gathering and hunting. This new formation would have constituted the first historical social unit of pre-human society, the 'cooperative band'. Mate exchanges between bands, as with cooperation within the band, may not have been an outcome of purposeful choice but the result of an inherited ape-like aspect of social organisation. Bipedalism, with its attachment to tool-using practices, allowed wider and richer omnivorous foraging. The ape tendency to nomadism became stronger and earlier associations with a particular territory weakened until the distribution of hominids became governed by food supply alone. Once embarked upon this course, future selective pressures for tool-making, enhanced forms of communication and increased mental capacities to negotiate a subsistence pattern of growing complexity would have been marked.

Human language consists of more than an extension and perfection of primate traits. But nonetheless a complex series of expressive gestures and sounds, that are a part of primate communication, were available for early hominids. For spoken language to have become a hominid ability, an evolutionary package involving modification and control of the vocal tract and a major change in the capacity and size of the brain was necessary (Laitman 1988). While it is unclear when this process began, it is not until the emergence of *H. sapiens* about 400 to 300 thousand years ago that fully articulate speech would have been possible. But the beginnings of a cultural record over 2 million years ago lead to the assumption that at least an enhancement, if not a departure, from the most elaborate form of primate communication had occurred. Some persistence of primate communication and its mixture with symbolic forms would have accompanied hominid life on the open plains where tool-use and tool-making had become an adjunct to foraging. The cultural patterns of the first hominids would have been transmitted with minimal symbolic elements during a proto-language stage.

This pattern of development was eminently suited to elicit a cultural way of life: hominids had behind them several epochs of intelligent, variable responses which could generate new behaviour patterns. As a cultural adaptation became increasingly important, in terms of simple tools and division of labour, selection was accented towards an increased propensity to learn and communicate. This long-term process was marked by an integration of organic and behavioural changes: full bipedalism, delayed sexual maturity and a longer life-span, and a reorganised enlarged brain, all developed with habitual sharing, food transport and tool-making.

AN APPROPRIATE EVOLUTIONARY MODEL

This presentation of human evolution as an interacting cluster of social and biological factors (Figure 5.1) should not surprise a sociologist or psychologist. The methodology of the social sciences has always emphasised the need for holistic forms of explanation. A particular element in a set of social relations, it is argued, can be properly understood only in relation to a whole context. This whole is seen as more than just the sum of its individual parts. Separate elements are thought to be dynamically linked and it is only by analysis of their interactions that a real understanding, which would include causal connections, is gained. Even though single elements within a complex whole may be imperfectly understood, they must nonetheless be retained within the bounds of a general theory. Such elements must not be excluded from consideration on the grounds that they cannot be specified with the normal precision of science or that their existence can only be known through intuition or that their effects upon other variables remain speculative. It has therefore become an axiom among social scientists that desegregated forms of explanation are inherently misleading.

This approach, however, has become unacceptable to many anthropologists who explicitly choose a reductionist methodology which excludes cultural behaviour from evolutionary models. Foley (1987) argues that the evolution of our distinguishing characteristics can be explained by ecological pressures alone. He proposes that hominid behavioural development is best understood by reference to models which reconstruct the environmental forces faced by our ancestors. This procedure is illuminating in itself, but the consistent rejection of culture and social life as elements with at least equal determining force to the other variables presented in ecological models results in inchoate theories. Reductionist explanation requires that a minimum of variables are used to explain a complex whole. Foley has therefore pared down his variables to include only aspects of natural selection and adaptation that are revealed by fossils and environmental change. Cultural behaviour is excluded on the grounds that culture is a composite term for a cumulative set of emerging human properties, and because the effects of culture on other variables are speculative.

Foley argues that the principles of evolutionary ecology apply to hominids as much as to any other group of animals. Ordinary biological processes common to all life forms can produce complex and unique behaviours or biological structures such as the eye or human brain. Nor should human evolution be envisaged as an ascent to a predetermined form, since the appearance of human characteristics was not inevitable. All ancestral species, including varieties of hominid, had to adapt to their immediate environment and cannot be seen as part of a teleological progression of forms which make sense only because they culminate in organisms such as humans. Evolutionary change makes sense only as a means of solving contemporary problems, not as part of a final state. Hominid emergence occurred as part of a wave of coevolving African mammals whose evolution was a response to changes in local environments. The development of bipedality, rather than intelligence, initiated our career as hominids. Thereafter, as our activities encountered a complex changing environment, a variety of hominid species evolved in response. In short, we became human because our mammalian and primate heritage met ecological conditions which made each stage of our evolutionary pathway the best adaptive solution.

These principles must of course be incorporated in any modern theory of human origins. However, in practice Foley's apparently comprehensive view of evolution comprises a set of explanatory models which give attention to ecological factors alone. Hominid activity is confined to foraging and is seen as comparable with that of any other animal while incipient forms of cultural behaviour are ignored. Major changes in the behaviour and form of human ancestors are explained without reference to what must have been emerging cultural traditions and social institutions. Hunting, stone tool-making and food-sharing are seen only as the products of an adaptation to savanna ecology. The migration of hominids out of Africa at the *Homo erectus* stage is explained as part of a 'biographic event' in which a new niche for large carnivores induced hominids, along with the lion, leopard, wolf and hyena, to exploit a new abundance of grazing animals in Eurasia. The rapid increase in hominid brain size occurred as an energy strategy for new environments demanded a larger body size. To be sure Foley asserts that brain growth was positively selected for by an animal with complex social relations, but the driving force here was the increase in body size. In each case cultural practices are seen as a consequence of ecologically-led evolution. Stone technology, superior hunting organisation or greater cognitive abilities are induced by ecology while the power of socially based behaviour to direct evolution is denied. The question is then, whether human evolution can be explained by a conjunction of ecological circumstances which alone formed hominid abilities as the external environment interacted with hominid biology? The argument pursued here is that the use of hands and brains for new forms of elaborate learned behaviour produced an *internally generated environment* of rules, technology and organisation which was of evolutionary significance. The point at which culture joined with ecology to become a new force in human evolution is hard to determine. Natural selection would not have been influenced by culture in creating hominid status with bipedalism at *c.* 5 to 7 million years ago. However once an elementary cultural system had become established among hominids, social and ecological factors became fused, and attempts to disentangle the two are artificial.

This is not to claim that natural selection is inadequate in shaping hominid characteristics, but rather that in human evolution culture was an important new part of the environment and had itself become a selective pressure. It is not teleological to suggest that human evolution presents us with the gradual acquisition of new behavioural features such as rules and norms, that were socially variable, which point to a pattern of growing social complexity. Stone tool-making, language and art, like the other elements of material and intellectual culture, formed an internal environment which was interposed between the individual and the external environment. For these reasons it was argued in Chapter 2 that culture came to change the terms of natural selection. Our ancestors were not just like any other evolving animal: hominid cultural practices became essential for survival, and the use of culture as an adaptive strategy was a breakthrough in the history of life.

It would be facile to deny that reductionism has produced significant discoveries and advances in a range of sciences, but the inappropriateness of this form of methodology for human evolution needs to be underscored. Reductionism can successfully explain unrelated aspects of an organism but is rarely able to do this for a whole entity, for the obvious reason that the whole is more than the sum of its parts. If, for example, a steam locomotive was to be 'explained' by reductionism alone we would learn, by scrutinising increasingly smaller units of this machine ever more closely, only that its real existence as an articulated system of mechanical processes lay in the physics, chemistry and metallurgy of its component parts. But this procedure is only partially helpful because the full reality of the locomotive is revealed only when its mechanical properties are seen in the context of the machine as a socially produced artifact. Now obsolete, but undreamed of 300 years ago, the steam engine is much more than a furnace, pressure vessel and set of energy transmitting pistons and connecting rods. The locomotive was brought into existence in a particular historical era by specific social forces to serve definite ends which are not amenable to positivistic explanation. Similarly, the evolving human organism cannot be understood as a whole without an examination of the interactive social context in which it arose.

It is now common to downgrade or deny the evolutionary significance of culture by resorting to a hyper-empiricist methodology. Here what cannot be easily measured and observed among hominid remains is assumed to be unimportant or entirely absent. Such an approach is artificial and misleading but worst of all it bars the possibility of gaining real understanding. As a wealth of new 'hard' evidence has become available since the 1960s, specialists have often become preoccupied with small-scale studies relating to fossils, artifacts, objective dating methods, comparative zoology or the data which reconstruct ancient environments. With the burgeoning of these new fields, comprehensive views of evolution have been rare and the reality of human ancestors as members of complex social groups who adapted culturally, as much as biologically, has been frequently overlooked.

HOMINID EVOLUTION

Bipedalism can be claimed to be the most fundamental of all hominid adaptations since it began a series of complex changes which led to the emergence of *Homo* over 2 million years ago. Modern radiometric dating methods and the wealth of archaeological and fossil evidence that has been collected in the past few decades have confirmed that hominids with brains of around 400 ml, which is little larger than that of a chimpanzee, were walking erect as much as 3.75 to 4 million years ago. There is also fragmentary evidence which suggests that these hominids, known as the australopithecines, could have been active 6 to 5 million years ago. This is well before the appearance of a large brain and manufactured stone tools. Brain size is significant in relation to body weight. Modern humans have a brain which is over three times the size of a primate of equivalent body weight. Deliberately fashioned stone tools and primitive hunting have also been found to precede the development of a large brain. In fact it is only after tool traditions had been in use for several hundred millennia, indicating the success of a new way of life, that the large brain characteristics of *Homo* began to emerge. These conclusions have been taken by Washburn (1960, 1978) to mean that it was the change in the terms of natural selection, that came with the tool-using way of life, which resulted in the structure of modern humans.

The study of human evolution is pursued by a number of specialised sciences that have produced a series of complex and rapidly changing evolutionary scenarios on the basis of current evidence. Only a simplified and incomplete version of such findings can be provided here, and in what remains of this chapter the context of hominid emergence and some of the theories which attempt to explain bipedalism as an evolutionary strategy will be examined.

The beginnings of primate evolution occurred some 55 million years ago when insect-eating mammals first pioneered an arboreal habitat after the spread of tropical forests across large parts of the Old World. It is now known that finding sufficient food in these conditions is difficult and that selection pressure for enhanced foraging abilities would have been severe (Milton 1993). Any physical or mental advantage that allowed dietary needs to be more easily satisfied would have become incorporated as a part of primate heritage. The development of clasping hands, improved visual perception exemplified by colour stereoscopic vision, an unusually large brain and a considerable capacity for flexible behaviour and learning were preadaptations that were gained in the forest well before the emergence of hominids.

While habitual locomotion on two legs is unique to hominids among the primate order, its use by other mammals as well as by birds and dinosaurs reaches back over a 100 million years into the Mesozoic era. But even among primates bipedalism is not an entirely radical innovation. Apes and monkeys frequently stand erect to raise their level of sight, but unlike other animals who also do this, primates possess *prehensile hands* that are systematically used for catching and picking out food, for fighting, grooming and nest building. Primates do not feed directly by mouth, and the use of hands for eating and a range of other purposes can be seen as an inducement for bipedal walking. The first hominids are assumed to have been adapted to an open-country environment in which a high level of vision and free hands would have been advantageous. An evolutionary tendency towards bipedalism can be seen in primate behaviour patterns which include, among some apes, the defensive throwing of objects and the use of sticks as weapons while standing on two legs.

The ecological context for hominid emergence begins in the Miocene epoch which lasted from 22.5 to 5 million years ago. During Miocene times an extensive series of environmental changes caused dramatic readjustment to the ecological niches available to flora and fauna. Widespread continental drift, which was associated with mountain building, produced a realignment of Africa and Asia and a further widening of the Atlantic. A marked climatic change was seen in falls of both temperature and rainfall. In the Eocene and Oligocene epochs, immediately before the Miocene, warm tropical and subtropical climates had prevailed as far north as Greenland and Spitzbergen. Lush tropical forests stretched across Eurasia and Africa well into Miocene times. These enormous tracts of forests contained the Sivapithe-cine apes, the undifferentiated primate ancestor of both the modern apes and hominid lineages, known collectively as the hominoids.

The long-term drying of the Miocene climate had removed five million square miles of forest by the middle of the epoch; the Asian and African forest belt became dissected by areas of grasslands that were much less suitable as arboreal primate habitats. Shrinking forests yielded to a new woodland-savanna environment which presented primates with a number of novel niches. This transitional ecotype acted to promote a variety of adaptations. Some apes pursued an arboreal specialisation in the forests that remained. Ancestral gibbons adapted to movement in the upper canopy of the forest and orangutans to a mid-tree life. Ground living developed among some monkey species, in particular the baboon and gelada groups. Terrestrial life held open new food opportunities in the form of fruit, roots and insects while it also exposed primates to carnivorous predators never encountered in a forest niche. Bipedal movement is slow and would have offered little advantage as a means of flight before dangerous animals. An occasional upright stance used to fight or threaten, by giving the appearance of greater size, or to carry over short distances or provide surveillance would all have constituted a selective advantage and yet still be incorporated with fast quadrupedal locomotion.

Miocene ecological change provided the setting for new adaptive differentiations among a whole range of animals. With the unparalleled creation of grasslands, *Hipparion*, the small ancestor of the modern horse, spread rapidly over the world. Other plains-living animals evolved under the selection pressures imposed by the Miocene climatic shift to produce the immediate ancestors of the modern savanna and steppe fauna. Selection pressures also caused the Miocene hominoids, the Sivapithecine apes, to undergo differentiation. A number of separate species evolved from this hominoid stock 10 to 15 million years ago in what can be seen to be an adaptation to new ecological opportunities (Simons 1977). The later evolution of the hominids 6 to 7 million years ago should be seen within the context of a general speciation of the hominoids, and other mammals, with the creation of grasslands at the end of the Miocene epoch. New species which were derived from the Sivapithecine line in mid-Miocene times, such as *Gigantopithecus*, *Ouranopithecus* and *Ramapithecus*, exhibited characteristics which resemble modern apes in terms of their dental morphology and post-cranial structure, although there is much less specialisation present here than among the modern apes. These species appear to have lived in a mixed open-woodland and forest environment close to water. Tooth wear patterns and the presence of thick cheek-tooth enamel for all of these species indicates a specialisation for a diet of fruit and nuts which required the cracking of rinds and removal of husks and shells (Simons 1981). No indication is given here of a meat or grass-eating diet.

At the end of the Miocene and the beginning of the following brief Pliocene epoch from 5.3 to 1.6 million years ago, a further shrinkage of tropical forests occurred with lower temperatures and rainfall. The onset of glacial conditions in the northern hemisphere produced global climatic changes which transformed habitats even in middle latitudes. This was a period of extinction for many Eurasian primate species deprived of a tropical habitat but, as grasslands spread in Africa, a new range of animals evolved to life on open plains.

What circumstances produced the first bipedal hominids 6 to 7 million years ago? With encroaching deforestation, primates would have been confronted with a major adaptive problem. In place of continuous forest belts, a mosaic of environments developed in Africa which ranged from semi-desert and savanna through open woodland to remaining tropical rainforest. Rift Valley development and volcanic incidents would only have exacerbated this process. As this variegated pattern of environments spread across the continent, isolated habitats held small populations who faced severe selection pressures. These new conditions were quite unlike the more homogeneous early Miocene environments where primate populations formed large gene pools that changed rarely. By Pliocene times, novel adaptations which spread rapidly in small populations could be expected as products of an unstable environment, because isolation had reduced gene flow. A convincing case for environmental transformation leading to speciation has been made by Vrba (1985) who has shown that in Southern Africa, between two and three million years ago, a gross vegetational change occurred as predominantly bush and tree cover gave way to open grassland.

Different bovid species are known to be closely related to particular habitats and their fossilised remains leave a record of ancient environments. Both bovid and hominid lineages experienced an 'evolutionary pulse' as new species arose at this time. It seems very likely that another period of extensive habitat change occurred at the time of the ape–hominid divergence which also witnessed similar evolutionary changes among bovids.

There is almost no direct evidence from fossils which can illustrate the original speciation which led to ape and hominid forms or of the circumstances in which the forest habitat was abandoned and the earliest form of bipedal locomotion adopted. Nor can the Sivapithecine apes be traced directly through fossil evidence to the earliest hominids in the australopithecine stage. But for most of the last 4 million years the fossil and archaeological records allow us to trace the enlargement of the brain and the emergence of cultural practices among a sequence of related hominid species from a point in time when the term 'human' is quite inappropriate.

It is in the australopithecine genus that bipedal walking is first encountered in the fossil record. These hominids have an assumed chronological range of over 4 to 1.3 million years ago. The earliest australopithecine fossil is *A. ramidus* from Ethiopia. These recently described remains, which are 4.4 million years old, are claimed to represent the most ape-like and primitive hominid known (White et al. 1994). In fact the fossils of this new australopithecine species, which is presumed to be ancestral to later hominids, were formed in an era which is near to the ape–hominid divergence. Preliminary work on the environment and ecology of this earliest known hominid indicates that *A. ramidus* occupied a woodland niche along with monkey species. This finding suggests that, though bipedal, *A. ramidus* remained committed to an arboreal existence which, it is increasingly thought, hominids did not relinquish much before 2 million years ago (Wood 1994). However there is as yet no detailed account of this hominid's form of bipedality and this, plus the fact that the status of *A. ramidus* within the hominid lineage has yet to be fully established, obliges us to turn to later but more firmly authenticated evidence.

The Tanzanian site of Laetoli has yielded a set of fossil hominid tracks made in volcanic ash that can be dated to 3.7 to 3.5 million years ago. Mary Leakey (1981) has described the track makers as three individual australopithecines who crossed the 27 metres of ash together, with two of the hominids walking beside each other and the third treading in the steps of the leader. The walkers must have held on to each other since they all take the same course and have the same stride length despite having different sized feet; this could have been a family group, as Leakey suggests. There is a gap of over a million years between this securely dated bipedal activity and the earliest well-documented stone tools that have been dated to 2.4 million years ago. Leakey has suggested that during this gap:

> . . . the early hominids must have progressed slowly through the stages of selecting naturally shaped objects to assist in their activities, adapting objects by means of their hands or teeth, haphazardly breaking stones and finally shaping stone tools to a recognisable pattern, an achievement made possible by the development of the precision grip and intellectual ability for conceptual thought. (Mary Leakey 1981: 102)

The earliest fossil bone remains which provide unequivocal evidence of bipedalism come from the Afar depression of Ethiopia and are dated from at least 3 to 3.7 million years ago. These finds include about 40 per cent of an adult female skeleton known as 'Lucy', dated at 3.5 to 2.8 million years ago, which is now named as *Australopithecus afarensis* (Johanson 1980: 64). *A. afarensis* is remarkably close to a missing link between later hominids and living African apes. Lucy's limb proportions are intermediate between humans and chimpanzees while her teeth and head are ape-like. The pelvis reveals that full bipedal locomotion was present while it is clear from the skeletal structure that some arboreal features had also been retained. Lucy's long curved fingers and toes are similar to those of apes yet her hands are almost as dextrous as those of humans and would have possessed many of the manipulative abilities needed for tool-use. The importance of climbing and tree-life, perhaps for sleeping, is strongly indicated (Fleagle 1988).

The consequences of bipedalism were far-reaching as a sequence of evolutionary changes transformed hominid anatomy to facilitate this highly specialised mode of locomotion. In order to support more weight the pelvis changed from the elongated ape form to become rounded and bowl shaped. The spine thickened and formed an S-shaped curve to improve weight distribution. Modification of the feet for improved body support was achieved by realignment of the toes and, as a result, the grasping ability of the foot was lost. Limb proportions came to reverse the ape configuration as legs lengthened in relation to arms. Free hands were available for grasping and lessened the need for long sharp teeth. As carrying shifted from the teeth to the hands, reduction in tooth size and disposition occurred with the creation of the semi-circular dental pattern that is characteristic of the hominid lineage. The large canine teeth, that are seen in all ground living apes and monkeys, are not present in hominids.

Reduction of canine teeth, and later the grinding surfaces of molars, was accompanied by smaller jaws and a snout that had a decreasing tendency to protrude beyond the face. This recession of the muzzle began a flattening and hollowing of the face itself. A reorientation of the head to accord with bipedal movement required a change of balance on the spine and this, together with smaller teeth, produced a remodelling of the neck muscles and skull. Later a change in the size of the head occurred with the enlargement of the brain. While this cerebral increase is not a direct consequence of bipedalism – it is usually associated with selection for a cultural way of life and a larger body size – these two features have had important interactive consequences for human evolution. The maximum cranial size that a viable foetus may attain is set by the dimension of the mother's pelvis because it is through this bony structure that birth must occur. Hominid bipedalism has selected a pelvic structure which has limited the size of the birth-canal while cultural life has simultaneously selected for a larger brain. These two evolutionary processes are thus working in opposition and the resolution of this problem has had a profound effect upon human social life. The compromise reached between these opposed tendencies has been a slowing-down in the rate of maturation of the human foetus and a change in the stage at which birth occurs.

Among many mammals the newborn must possess sufficient bodily control to cling to the mother or even move independently to avoid predators. Ape babies are born having attained a far more advanced stage of neurological development than human babies, after a gestation period that is close to the human 38 week term. In the orangutan, gorilla and chimpanzee, pregnancy lasts 39, 36 and 33 weeks respectively and yet the life-span of these species is only about a half of the 75 year expectation of humans. These apes give birth to a baby that has 60 to 65 per cent of the adult brain size. This would be equivalent to a woman giving birth to a baby with a brain volume which is over twice the normal foetus head size (Tobias 1981). To overcome this obstetrical problem the bipedal hominids have given birth to children whose brains are a quarter of their adult size.

The consequence of belated expansion of the brain has been a human infant that is helpless and utterly dependent at birth. However this fact has important further consequences that should be seen in terms of the positive feedback system in which human evolution is situated. A long period of infantile dependency necessitates bipedal parents able to carry their offspring and a bonding process that ensures that the child's needs will be satisfied. The evolutionary advantages here consist of a protracted learning period which begins in humans at an earlier developmental stage than in other primates. Long infant experience of the social and physical environment coincides with major changes in brain function as growth occurs and it is this process which allows both great versatility of adaptation and determines the structure of personality. The helpless hominid baby is exposed to adult behaviour and has of necessity to form deep affectional bonds with its parents at a time after birth when other primate offspring have attained a considerable degree of independence.

HOMINIDS IN EVOLUTION

In spite of increasing amounts of evidence from a range of sciences, the precise causes of the bipedal adaptation remain unknown. None of the theories which attempt to explain this primary and definitive characteristic of the hominid line have gained universal acceptance. Current opinion broadly favours the adoption of a new foraging and reproductive strategy as the source of bipedality, but neither of these causes can be advanced with final certainty.

Having made this radical adaptation the first hominids cannot be assumed to have embarked upon a way of life which resembled that of modern humans beyond their common form of locomotion. *Australopithecus afarensis* is in many instances more ape-like than human and suggests a direct link with our hominoid ancestors. The teeth and jaws of this hominid suggest the processing of large amounts of tough vegetable food. Hunting and meat-eating seem to be unlikely activities. The development of teeth here also indicate that *A. afarensis* grew to maturity at roughly the same rate as the African apes and that the prolonged growth period used by modern humans for extensive learning was not present. These early hominids had brains which, when relative body size is considered, were hardly larger than those of apes indicating that language ability had not emerged. In many ways the first hominids can be described as walking apes with superior manipulative abilities who lived and foraged both on the ground and in trees. Anthropomorphic interpretations of this stage in human evolution are to be avoided. Despite their ancestral relationship to modern humans, any relict austropithecines found today would be consigned to a zoo.

The tendency to envisage all hominids before *H. sapiens sapiens* as diminutive versions of modern humans has been a part of earlier accounts of hominid evolution. Landau (1984) detects a narrative or heroic epic form used as a structure for the theories of scientists in the early twentieth century. These authorities had unconsciously adopted this literary device because of their adherence to a historicist interpretation of evolution. A teleological or linear view of prehistory was implied here. An ultimate goal of evolution – modern humans – was seen to be preceded by a set of stages which led progressively to this destination. Each step in a chain of events was seen as preparation for the next stage. The stages themselves were synonymous with progress and humans were the inevitable outcome of the whole process.

The use of a narrative structure by these writers is evident in their presentation of stages running from the trees to civilisation. A humble hero is expelled from a stable environment and then overcomes a series of obstacles by virtue of his developing intelligence, reason and technology. The hero constantly strains to realise his destiny as a distinctly human character. This triumph over animality is contrasted with the indolence shown by the apes whose abilities remain static in a world of plenty. All the gifts of a larger brain, tools and a moral sense are seen as the products of a struggle by human ancestors to transform themselves.

The organisation of human evolution as a sequence of events – terrestriality, tool-use, brain development – is in no sense an explanation for any of these steps. Evolution is a random process without a route, goal or destiny. Interaction between organism and environment includes an element of chance which means that the current pattern of life on earth is only one of many possible configurations. Without Miocene climatic change, hominid evolution might not have happened at all (Lewin 1987). Accounts of human origins which envisage the early hominids as in some sense 'transitional' or failures, because they fell short of human status, are teleological and invalid because they imply that a definite direction and end has been set for human evolution. The persistence of the early hominids is impressive; *A. afarensis*, for instance, had a successful career at least ten times as long as modern *H. sapiens*, without full human abilities. The first bipeds make sense only in terms of their adaptation to the open landscape of southern Africa during the Pliocene epoch.

Teleological narratives are misleading for both social scientists and biologists. Bipedality was not inevitable but its preservation in descendant hominid populations was due to the flexibility it gave to later adaptations. Had this radical means of movement foreclosed the possibility of following new evolutionary routes it would probably have disappeared. Hominid evolution was circumscribed by a primate and mammalian heritage which strongly influenced later adaptive pathways. Without being teleological one may legitimately speak of 'preadaptations' or current characteristics which evolution may elaborate upon in a new species. Bipedality may owe its origins to a feeding adaptation, but it also served as a powerful preadaptation for the enlargement of cultural life.

THE ORIGINS OF BIPEDALISM

The particular adaptive advantage given by bipedalism and the circumstances in which it arose are disputed but there is now general agreement that the first hominids emerged in Africa during the boundary between the Miocene and Pliocene epochs some 7.5 to 4.5 million years ago (Pilbeam 1984). It is also assumed that the immediate ancestor here was a tree-adapted ape with a way of life, but not body, similar to the chimpanzee. This period of environmental transformation, which involved extensive change in a wide variety of animals and habitats, was described earlier in this chapter.

It has recently been proposed that the beginnings of the hominid lineage and the final divergence from our last ape relative occurred either side of the East African Rift Valley, as this major land form developed 8 million years ago to produce extensive climatic change (Coppens 1994). On the west of the Rift Valley humid conditions prevailed allowing the preservation of forest habitats that contained the ancestors of modern chimpanzees. To the east, however, drier conditions caused the loss of woodlands and demanded a new adaptive repertoire for an open landscape. But these times were also synonymous with a frustrating gap in the hominid fossil record from 8 to 4 million years ago (EHEP 1988). For this reason molecular dates, which estimate the emergence of bipedalism at about 6 million years ago, are commonly accepted. However work by Lovejoy (1988) on the pelvis of the *A. afarensis* specimen 'Lucy' shows that in this hominid, which was active 3 to 4 million years ago, upright walking was not a novelty. Since bipedalism was so well established in these times Lovejoy concludes that the original divergence with other primates may have occurred as much as some 8 to 10 million years ago. All of these dates, however, contrast with the prevailing views of the 1970s which envisaged the beginnings of bipedalism 15 million years ago in mid-Miocene times. *Ramapithecus*, a primate now considered to be a Sivapithecine ape, and perhaps an ancestor of the orangutan, had been proposed as the first hominid. But new molecular dating methods and a revision of the fossil evidence have ended this candidate's claim.

The identity of the first biped is therefore mysterious, as are the driving forces for this adaptation. The thesis that tool-use and even tool-making were a cause of bipedalism, while attractive in terms of the arguments pursued in this book, suffers from lack of evidence and cannot be proposed with confidence. Recognisable stone tools do not appear until at least 3 million years of bipedal walking had elapsed and any earlier vegetable tools have not been preserved. In 1960 Washburn argued that tool-using man-ape creatures had evolved a form of bipedal running, but not walking, and that full bipedality arose with continued use of this cultural practice. But new dating methods and the discovery of earlier hominid species produced the long gap between full bipedalism and the earliest tools. Louis Leakey's proposal in 1968 that *Ramapithecus* was using tools 14 million years ago on the grounds that fossil remains of this primate were found in association with battered lava cobble stones and fractured bones has been disproved. Certainly no stones that had been altered for tool-use have been found with *A. afarensis*. But, as Pilbeam (1984) speculates, these hominids would probably have used tools more frequently than chimpanzees. In foraging, food-processing and display behaviour these hominids could have casually adapted wood and stone to their needs without their appearance in the archaeological record. This form of tool-use, he continues, 'might

have been an important component of those behaviours that
stimulated walking on two legs and the reduction of the canine
teeth' (Pilbeam 1984: 67).

A combination of bipedal walking and tool-use would have
been highly advantageous. The additional nutritive value that can
be gained from the most rudimentary pebble or wood tool used to
cut, dig, pound and scrape is enormous. Compared with other
savanna living primates, such as baboons, who dig for roots, tool-
using hominids might easily have doubled their food resources.
Tool-use, even by partially bipedal hominids, would have opened
up a wider range of foods dispersed over a larger area. Two-
legged walking is an adaptation which can outdistance all other
primates, whose ranges remain tiny when compared with all
known hunting and gathering peoples. However, these advantages may have been realised *after* bipedalism had arisen and they
may not be the cause of the adaptation. Without further evidence
this hypothesis cannot be proven.

Rather than engage in discussion of any causal connection
between hominid cultural behaviour and bipedalism, the prevailing tendency among physical anthropologists now is to concentrate upon specific answerable questions concerning human
origins (Lewin 1984). Explanations for bipedalism are proposed
in terms of the selection pressures that would have been
experienced in the context of the emerging savanna ecology.
Among these are the form of diet and the energy expenditure of
proto-hominids.

A method long in use by comparative zoology for reconstructing diets of extinct animals is to note the correspondence between
ancient and modern tooth and jaw specialisations. While this
method has produced broadly coherent classification of animals
into general categories, as for instance carnivores or browsers, it
suffers from being unable to detect any dietary shift or to
determine what range of food sources may have been present in
an omnivorous diet. Nevertheless recent work has provided a
comprehensive review of diet in hominid evolution (Walker 1981).
The mechanical properties of australopithecine teeth and jaws
reveal a much larger cross-sectional area of teeth and yet the same
chewing muscle pattern as modern humans. On this basis one may
ask what form of diet required four or five times the quantity of
tooth surface and muscle force to create the same jaw pressures as
those of modern humans? The australopithecine diet must have
included large amounts of indigestible material since the teeth are

capable of processing large quantities of food, but whatever this food was it could also have been part of a human diet. The microwear patterns of fossil teeth, revealed by an electron microscope, enable a fine determination of ancient diets to be obtained and can differentiate between even browsing and grazing herbivorous animals. The dietary pattern that is most evident on australopithecine teeth is one of fruit eating. The frequency of minute scratches and pits on the teeth of this hominid indicate the consumption of tree gathered fruits which had thick indigestible shells, rinds, pods and husks. Microwear analysis can certainly rule out feeding on small objects such as seeds or a carnivorous diet.

The tooth wear patterns of both the late hominoids, the Sivapithecine apes, and the earliest known hominids, the australopithecines, indicate a frugivorous diet. It is not until the rise of *Homo erectus*, some 1.6 million years ago, that evidence for a meat and vegetable diet can be shown. It would seem then that if the forest primate fruit-eating dietary pattern had not been abandoned by the australopithecines, after several million years of full bipedality, the original hominid adoption of bipedal locomotion cannot be explained by a shift in diet. It would appear that the first hominids brought their taste for forest fruit with them onto the savannas. But, even in this new open environment, the early hominids remained closely associated with trees (Isaac 1983). This is seen both in bodily adaptations which favour climbing and from the fact that the earliest archaeological sites often occur where groves of trees would have grown. The forest may have been abandoned at the point that we became hominids yet we appear to have retained our desire to live in trees until much more recently.

Savanna fruit, which is much less abundant than in the forest habitat, required foraging hominids to cover longer distances and bipedalism could be seen as the means which allowed the preservation of the forest diet in a new environment. The proposal that evolution allows the continuation of an old lifestyle in new circumstances, which was introduced in Chapter 1, makes sense in this context (Romer 1958). The new form of locomotion would have been used initially to move between widely dispersed clumps of trees. But extensive ranging on two legs also allowed the carrying of scarce food.

Later food-sharing practices combined with enhanced forms of communication meant that food resources could be optimised. At least one factor in the later growth of the hominid brain must have been the need to store information of the spatial distribution and available times of fruit sources. Accompanying and related to these later developments was the probable use of wood and stone tools as probes and levers and as processing equipment for the outer covering of nuts and fruit. This is not meant to imply an absence of meat or animal products in the diet of the early hominids. The omnivorous primate stock gave rise to a line of hominids whose taste was even more general. Hominid evolution was a process in which a broadening range of foods became important as a widening pattern of subsistence developed (Isaac 1980). Opportunistic feeding on big game was probably rare. But animal protein from eggs, nesting birds, fish and molluscs, insects and small mammals would have been usually available to a hominid with no more than a digging stick (Mann 1981). The archaeological record begins with bone refuse bearing the marks of stone tools, and stone tools themselves. But the fossilisation of bone and the persistence of stone has given a highly biased cultural record which has excluded all vegetable matter from preservation. A wider range of vegetable foods such as buds, shoots, roots, tubers and fungi would also have been available to a tool-using hominid. The tool-making ability that is required to produce a digging stick is no greater than that of the chimpanzee yet the rewards and further incentives for a habitual bipedal tool-user are enormous.

The development of bipedalism can be seen to be part of a general evolutionary trend. Many primates, beside hominids, are ground based. Over 40 per cent of Old World monkeys, apes and hominids are terrestrial (Foley 1987) and a tendency toward ground-living has been present among this group of animals – the catarrhines – for some 20 million years. The 18 million year old remains of *Proconsul*, which have been taken to be the common ancestor of all the great apes and humans, are those of a generalised arboreal frugivore, but even here there are indications of ground life (Walker and Teaford 1989). As Miocene deforestation proceeded, a series of specialised apes suited to a variety of habitats are seen in the fossil record. But the necessity to adapt to life on the ground produced a variety of solutions. Why did the proto-hominids not adopt the quadrupedal movement of baboons or the great ape mode of knuckle walking when no other savanna mammal is bipedal? Given that the new environment was now an open mosaic of diverse and dispersed food sources, answers to this question are best posed in terms of two factors. These relate to the efficiency of bipedal locomotion and to the evolutionary history of the common ancestor which gave rise to both the hominids and African apes.

Because fast human bipedal running is energetically inefficient when compared with most quadrupedal animals, anthropologists had assumed that this rare form of locomotion must have evolved in the presence of special factors such as carrying or tool-use. However, Rodman and McHenry (1980) argue that this form of comparison is not valid. They find that at normal walking speeds human bipedalism is as efficient as quadrupedal movement in animals. Even more striking is the fact that when chimpanzee quadrupedal knuckle-walking is compared with human bipedalism the latter is found to be considerably more efficient. They conclude that chimpanzee locomotion is the result of an evolutionary compromise between feeding in trees and travel on the ground, whereas the hominid adaptation arises from the need to forage more widely for the same quantity of food. Hominid walking, at speeds less than three miles per hour, could have been selected for because it provided improved energetic efficiency for long-distance terrestrial movement.

If bipedalism arose to facilitate endurance foraging, unlike most quadrupedal locomotion which allows rapid but short bursts of travel, it must be remembered that this adaptation occurred in a tropical climate which posed particular problems of heat stress. The normal activity pattern of ground-adapted monkeys and apes is for movement and foraging during the warmest period of the day while carnivores and grazing animals are most active early and late. However, the great apes rarely spend prolonged periods at midday out of the shade. For the immediate ancestor of the proto-hominids, slow movement between scattered groves of vegetation in high temperatures would have produced hyperthermia. But hominids have evolved the most effective cooling system among living mammals (Wheeler 1984). Bipedalism itself was one means of achieving this since under the midday sun a hominid presents only about 40 per cent of the body area which would be heated in a quadrupedal position. The other mechanisms, which are also unique among primates, were nakedness, which occurred with the loss of body hair, and sweating. Bipedal movement would have allowed intense periods of activity, including walking and running in high temperatures to isolated food sources, and this was probably a principal selection pressure which led to its evolution. This theory also explains why we have a layer of subcutaneous fat which serves as insulation at night.

Primates have adapted to the same basic problem of ground living in different ways. Natural selection has elicited a variety of solutions each of which have been strongly influenced by the evolutionary heritage of particular species. The idea of an 'evolutionary pathway' is useful here as this concept presents both the possible route which may be taken and the existing characteristics or preadaptations which will comprise the material available to natural selection for creating a solution (Foley 1987). Natural selection can preserve or alter only existing bodily structures and adaptations. The direction of evolutionary change is constrained by the historical experience of a species. A pig, for instance, cannot undergo immediate selection for bipedalism: its adaptations will be porcine. Colour stereoscopic vision and superior hand-eye coordination, which derive from life in the trees, were general primate preadaptations that have been incorporated in the human genotype. These preadapations now allow complex mechanical processes using all limbs simultaneously.

Anthropologists have frequently searched for significant forms of locomotive preadaptations which may have led to hominid bipedalism (Tuttle 1981). A range of preadaptations can in fact be detected among the Miocene hominoids (Aiello 1981). While these apes were generalised arboreal quadrupeds they can be categorised as above-branch and below-branch feeders depending upon body size. The possibility that these different forms of feeding were related to types of locomotion was investigated. The body and limb proportions of some larger below-branch feeding fossil hominoids show specialisations similar to those of the modern apes. Among the medium sized hominoids, which include *Proconsul*, who fed mostly below branches but were also able to engage in some above-branch feeding, a preadaptation for bipedalism is found. The last common ancestor of apes and hominids was not chimpanzee-like, it was probably similar in body proportions to the *Alouatta* monkey of Central America. When deforestation forced adaptive change a number of routes were open to primates. For the heavier hominoids the rare knuckle-walking adaptation developed. Some ability to move in trees was preserved for the chimpanzee but not for the gorilla. Small bodied above-branch locomotion in some monkeys led to baboon-like quadrupedalism. On the other hand, fully fledged bipedalism and terrestrial life were selected for in a medium sized primate with a versatile pattern of arboreal movement.

THEORIES OF BIPEDALISM

A host of explanations for bipedalism have been proposed since the nineteenth century, and a comprehensive review of these ideas would require a volume in itself. One way of extracting an understanding from what has become an intricate labyrinth of closely interwoven theories is to follow Isaac (1983) who distinguishes between two forms of explanation. In one group is a type of theory, which is generally favoured by physical anthropologists, where the causes of bipedalism are not associated with their consequences. Here the adaptation has arisen only as a direct result of its advantages as a form of locomotion. In the other group are theories which propose that the causes of bipedalism are virtually the same as their consequences. This form of explanation sees the adaptation as a product of the advantages it provides. In particular, bipedalism is seen to have arisen to facilitate new forms of behaviour. Among the first group of theories the 'aquatic phase' and the 'seed-eating' hypotheses can be singled out for attention.

The idea that hominids owe their origins to a marine adaptation has few proponents and yet, because the evidence for this thesis, indirect though it is, has not been fully explained by physical anthropology, the theory cannot be eliminated. In 1960 a few inchoate remarks made by Hardy that humans had an aquatic past – a seaside life in which the search for shellfish had produced bipedalism, a hairless body and tool-making – has been taken up with enthusiasm by others (Morgan 1972, Richards 1987). Life close to the shallow coastal waters of tropical seas would have induced wading, diving and swimming which are all unknown among the apes. Our nakedness, existing hair pattern and subcutaneous fat are best explained as aquatic adaptations to those who favour this idea, while tool-making is seen to have arisen from the need to open shells. Although there is no fossil evidence to support these propositions La Lumiere (1981) has suggested that a geological survey of the ancient north-east African littoral might reveal an isolated location where the ecological pressures of Miocene deforestation had pushed hominids into a semi-marine existence.

In 1970 Jolly presented a theory which gained wide acceptance among physical anthropologists. The theory became popular at a time when growing doubts had been raised about the validity of the culture-led reciprocal feedback model, current since Darwin, which had championed tools and behaviour as the cause of hominid beginnings. Jolly led a wave of change in proposing that bipedalism and a range of features including remodelled teeth and jaws had arisen before and independently of culture. Jolly proposed the gelada, a ground-living quadrupedal baboon, as an analogue of the first hominids. He claimed that comparative anatomical study of the two species shows considerable functional similarity. Of all the primates the gelada exhibits a mode of feeding that is most likely to lead to a bipedal specialism. The gelada is a 'small object' feeder which takes scattered foods that are round and hard such as nuts, seeds and berries from a semi-arid savanna habitat. For the proto-hominids a dietary shift and similar form of foraging adaptation had to be acquired before all the advantages of quadrupedal movement were abandoned. Such a dietary pattern would have favoured a bipedal adaptation because the collection of small hard spheres, such as seeds, would require careful use of two hands and movement between dispersed groves of vegetation. The gelada foraging position is a squatting or crouched stance in which the trunk is fully erect allowing full use of both hands. Geladas engage in a form of proto-bipedal shuffling between food sources.

Several arguments can be brought to bear against Jolly's theory. Firstly, it has already been noted that the microscopic analysis of australopithecine teeth shows no sign of small-object feeding. Secondly, Szalay (1975) argues that quite distinct dental processes were involved in the proto-hominids and modern baboons and that therefore functional similarities cannot be made to imply particular evolutionary trends. The teeth of the two species show quite different mechanical functions which point to separate biological roles. Proto-hominids would need to carefully separate berries and seeds from indigestible leaves and grass blades, which would have been inefficient and wasteful. Finally, in Jolly's argument that the proto-hominids were in fact seed eaters, with a feeding position like the gelada, parallel evolution is assumed. Parallel evolution may, in fact, produce remarkably similar forms or behaviour patterns among totally distinct lineages of animals. The Permian 'fish-lizard' ichthyosaur and the modern dolphin both exhibit a remarkable similarity in their body structures, and both occupied shark-type niches despite having reptilian and mammalian origins which are 200 million years apart. Both animals have converged on a fish-like form which has optimised their roles as predators. But despite this similarity the animal lines have converged only in one respect and significant differences will be present as a result of separate evolutionary histories which produce diverging adaptive routes.

The second group of theories, which accounts for bipedalism in terms of the beneficial behaviour which it made possible, presents a rich tradition of thought reaching back to Darwin. Arguments that tools and technology induced this primal adaptation have already been mentioned. However theories which propose new forms of reproductive behaviour, hunting and food-sharing will be presented here.

Lovejoy (1981) considers that bipedalism began in the Miocene forests as the result of a new reproductive strategy. His theory has been given wide attention and has been endorsed by Lucy's discoverer Johanson (Johanson and Edey 1982). Using well established zoological models Lovejoy argues that in evolutionary history two extreme types of reproductive strategy, known in ecological theory as 'r' and 'k', can be detected, which in general relate to the number of offspring that are normally produced by a species. Selection for an r strategy is common among short-lived small animals living in unstable environments, and will involve a high production of fertilised eggs which mature rapidly. The r strategy usually leads to phenomenally high overproduction and mortality. An oyster, for instance, may release half a billion eggs in a year, of which only a tiny fraction become fertilised and reach maturity. A single frog may give rise to 200 tadpoles a year but parental care is not given to these offspring and few survive. Consequently, drastic changes in population level accompany this strategy. At the other extreme, k selection occurs among longer-living animals who produce few young, like the gorilla or chimpanzee whose single births would usually be followed by periods of four or five years. In k-type species, offspring develop slowly and remain highly dependent upon parents. Population sizes are more stable – as are environments – as long as the high levels of infant care succeed in preventing early deaths.

The Miocene hominids had opted for k selection but, with such infrequent births and long infant dependency, they had pushed this strategy to its limits. Without new behavioural mechanisms evolving to facilitate this reproductive system, extinction was a possibility. Lovejoy proposes that in order to ensure more frequent births and thereby enable a mother to care for two dependent young simultaneously, males were obliged to assist in infant care and particularly in foraging. Provisioning of females with young by males greatly increased survival. Mothers and infants, who could now remain in a group, were less prone to attack and accident. Pair-bonding, monogamous mating and an end of estrus developed. But it was the innovation of bipedal walking which underlay these changes. Bipedalism provided the most efficient means of holding and carrying food. Allied with a new form of cooperative social behaviour, bipedality made sense as a means of avoiding demographic oblivion. Lovejoy's ideas are considered to be well situated in evolutionary theory, but nonetheless they have been criticised for failing to explain why modern apes with low reproductive rates, like chimpanzees whose environment is unstable, have not also evolved compensating forms of behaviour. Doubt has also been cast on the evidence for food-sharing and monogamous pair-bonding.

Food carrying as a cause of bipedalism is a well established line of argument. Hewes (1961) also proposes that females and their young may have been provisioned by males, but the food envisaged here was meat since it is among social carnivores that food-sharing commonly arises. In this theory the proto-hominids developed a taste for scavenged meat and began bipedal walking as a means of removing carcasses from dangerous kill-sites. The idea that hunting behaviour was a cause of bipedalism has frequently been canvassed (Washburn and Lancaster 1968). Merker (1984) points out that Lovejoy's thesis does not exclude meat in the diet of the first hominids. Lovejoy had considered that meat was probably not hunted because primitive bipedality would have prevented the tracking and capture of large game. However, Merker suggests that human legs and feet are well designed for slow, silent stalking and watching while bipedal semi-crouching allows a hominid to closely approach and grab an animal. This posture could explain intermediary forms of locomotion and Merker claims that evidence for its use can be found in the earliest fossil hominids. Meat might have been bartered with females for sexual favours in a society with polygynous mating, which would lead to the evolution of continual receptivity.

Modern human abilities seen in endurance running and hunting are not in doubt. In one study Carrier (1984) shows that while the energetic costs of running are high, humans have exceptional stamina. We appear to have evolved mechanisms for long-distance pursuit which, despite our slowness, allow us to run-down much faster animals. A variety of recent examples gathered by anthropologists show that hunter-gatherers can outrun fast animals such as antelope, kangaroo and zebra in a chase which may last many hours or even days. Important as hunting and meat-eating are in human evolution these new forms of behaviour are more likely to have evolved long after the emergence of bipedality. The first hominids were small and without weapons, tools or the intelligence required to organise this form of hunting. In any case the type of teeth and tooth wear patterns found in the earliest fossil hominids are not those which show a predilection for meat.

Finally, Lancaster and Whitten (1987) propose that unique forms of social behaviour associated with sharing, distinguished our forebears from other primates. Our proto-hominid ancestors were cooperative apes whose bipedality arose to facilitate carrying food back to others for communal meals. Evidence for this thesis is found in the parenting practices of modern primates and the economic behaviour of hunting and gathering peoples. Among Old World apes and monkeys, infants receive considerate maternal protection and care but the responsibility for maintaining this relationship rests with the infant who must be able to cling to the mother. After weaning, an infant will remain close to the mother for some years but will always be obliged to feed itself. A mother will continue to provide affection, grooming and protection but not solid food. Sick and injured offspring have been observed to languish and die in the presence of a distraught mother who was unable to comprehend the need to provide foraged food. However the possibility of behavioural change is seen among apes. While chimpanzees discard their tools they do make frequent but ineffectual attempts to carry food. They are also known to hunt small animals and share meat in social gatherings. Modern hunter-gatherer diet is based on a mixture of plant and animal food produced by a sexual division of labour which would be impossible without carrying and sharing. At some time in our evolution the basic form of primate society was transcended by a hominid social organisation which gave prominence to cooperation and mutual exchange. The authors claim that archaeological evidence of this behaviour occurs in East Africa where the sites of home bases have been discovered. The formation of this behaviour pattern involved a unique change in ecological and social relations. In addition to feeding the young, new reciprocal forms of interdependence emerged based upon a subsistence of animal and plant food.

The emphasis on social behaviour in the context of adaptive change is valuable and distances this hypothesis from empirical models, but the theory remains vulnerable on a number of fronts. This line of argument is in fact based upon extrapolation from either side of a huge time scale. Evidence for bipedalism and cooperation are separated by at least 4 million years and causal connections here are not easily established. Modern hunter-gatherer societies are composed of thoroughly modern people with mentalities and normative behaviour as sophisticated as those found in complex societies. This theory is perhaps too comprehensive and works with an over compacted time scale; however, many of the issues raised here lead us directly to the next chapter.

6

Tools and Culture

Here and in the following chapter we shall be concerned with the adoption of stone tool-making by human ancestors. The purpose of this chapter will be to explore the nature and significance of tools in the context of the known pattern of hominid evolution. In particular we shall consider who the tool-makers may have been and what processes may have initiated this novel practice. Finally the new mode of adaptation initiated by *Homo*, one of foraging for meat and plant food which relied upon stone tools, will be examined.

ANIMALS AND TOOLS

It was noted in Chapter 4 that chimpanzees learned to make and use tools. Until observers in the field revealed this, these activities were considered to be an exclusive and defining characteristic of our own species (Goodall 1971, 1986). Chimpanzees are now known to use and modify more objects for more purposes than any animal except humans. Twig probes are used to fish for termites, ants and honey. Leaf sponges are made for drinking water, soaking up juices in the carcasses of prey species and for cleaning the body. Modified branches, leaves and sticks are used as toothpicks, for brushes during grooming, as fly whisks, toys and as containers. Sticks, stones and foliage are used as missiles for defensive and aggressive purposes and during play. Stone hammers are used to open nuts and to remove shells from fruit. Objects may be used as a means of making the user appear larger or more dangerous in a charging display which intimidates rivals and re-establishes supremacy in the dominance hierarchy. Baboons and pigs are often attacked by chimpanzees using sticks and rocks. However, in fighting, these tools are ancillary and are not true weapons, which means that selection pressures favouring large canine teeth remain. But the degree to which tools are relied upon in foraging may be greater than was at first thought. Up to 20 per cent of feeding time is spent on termite fishing and this food, like nuts and fruit with thick husks, would be unavailable without tools (Goodall 1986). The adaptation and modification of objects to make them suitable for a particular purpose or to solve a new problem shows once again that chimpanzees possess cognitive processes which echo human abilities. Clearly the ability to abstract from raw material the pattern and form of a tool which will be used at a future time and in a distant location is found in this ape.

The full extent to which primates use and manufacture tools has been catalogued by Beck (1975). He shows that all the apes and even some monkey species can be included here. But this behaviour is a far more developed part of the chimpanzee repertoire and provides us with a model for hominids.

There are also extensive examples of tool-use, but not tool-making, by non-primate animals. The burrowing wasp *Ammophila urnaria* uses small pebbles as a hammer to pound soil over its nest; the Galapagos woodpecker finch uses a cactus spine to dig for grubs in tree bark; certain crabs use living sea anemones as a defensive aid; sea otters are thought to use stones to open mollusc shells, and polar bears to hurl blocks of ice on the heads of sleeping walruses (Wilson 1975). In all these cases tool-use occurs because extraneous, extra-corporeal equipment is applied to the environment to extend the user's efficiency. The tool-user category cannot be applied when an animal uses or modifies natural materials as an end in itself, in for instance nest construction or in the breaking of shells by gulls and thrushes by dropping them onto rocks. Nor can the indiscriminate manipulation of objects, such as the pulling down of a vine, be seen as tool-use. But dropping rocks onto eggs or pounding fruit with another object does count as tool-use. Tool-use occurs therefore when an animal directly controls objects which it has separated from the environment, by holding or carrying, and employs these for specific results for which it is responsible (Hall 1963).

However in non-primate tool-using animals no indication of specially intelligent behavioural adaptability is present. These animals have evolved what is mostly a stereotyped, inflexible behaviour pattern that is usually directed towards feeding or reproduction which is part of the species' inheritance. It is, of course, a matter of great interest that tool-using has evolved here as an inbuilt behavioural trait rather than as a physical adaptation to deal with a particular ecological circumstance, but these examples in no sense rival primate tool-use and tool-making.

Chimpanzee tool-use is for one thing the product of learning, albeit of observational rather than active learning. For this reason the tools made and used by this primate do not occur as part of a stereotyped sequence of movements; rather the innate primate propensity to manipulate objects and to explore the environment is allied with tools as a voluntary and multi-purpose activity. The long period of chimpanzee immaturity gives an intensive basis for learning which occurs not as mimicry but with an understanding of cause and effect. Anatomy too has an important role here since the chimpanzee has a brain twice as large as a monkey, dextrous hands and a shoulder construction that allows throwing. Finally, non-primate tool-use merely allows species to occupy a niche but not to broaden it beyond a fraction, or to subsume and incorporate the niches of innumerable other animals as the early hominids did.

Early hominid tool-use and tool-making before the appearance of stone tools can therefore be inferred from the defensive and food-acquiring behaviour of modern primates. Hominid behaviour was, however, more advanced and represents the primary phase of human skill that is marked by an absence of any recognisable tool tradition that would be evident from standardised tool types (Oakley 1972). The evidence for human manual skills, which argues that hand and tool were decisive in the evolution of the hominids, is revealed in comparative studies of brain function. Since the nineteenth century it has become possible to map the surfaces of the human brain and to determine the purposes served by different regions. Particularly large brain areas are devoted to mental functions that have become human specialisms such as language, speech and memory. The two hemispheres of the brain are by no means identical in structure or function. An asymmetrical relation exists between the hemispheres which is an evolved species characteristic (Geschwind 1979). Both language and 'handedness' probably depend on the brain's asymmetry. The human sensory-motor cortex gives great prominence to the hands and digits. This prominence far exceeds that found among other primates and is the result of evolutionary selection in favour of tool-use and complex social life. Progressively larger expanses of brain tissue have appeared with higher levels of dexterity, since the reciprocal relationship between increases in skill and biological success has ensured that modifications occurred in both the hands and in the governing cortical zones.

Incipient tools by their nature could never be recognised in an archaeological record, and the evidence for a tool-using and making way of life by the first hominids must remain circumstantial or be drawn from an analogy with other species. Stone tools appear to have been used, probably by the earliest species of *Homo*, between 3 and 2 million years ago. There are securely dated tools from the Omo valley in Ethiopia which are 2.4 million years old and others which may be 0.3 million years older at Kada Gona near Hadar (Lewin 1981). These tools occur over a million years after certain evidence for full bipedal walking and 4 million years after the probable appearance of the first hominids.

However a case can be made for tool-use, if not tool-making, in the rise of the hominids. New dating techniques have overthrown the old assumption that once the first tools appeared a rapid acceleration in technical progress ensued (Lancaster 1968). In fact from the emergence of the oldest tool tradition so far identified, the 'Oldowan' about 2.5 million years ago (Gowlett 1986), comes merely more Oldowan tools for a million years until this first industry begins to coexist with the second tool tradition, the 'Acheulean' about 1.5 million years ago, which in turn lasts for 1.3 million years. Small-brained hominids appear to have remained committed to technological forms and, one presumes, to cultural traditions, for inordinate lengths of time. In this primary phase of the Lower Palaeolithic era tools enabled an established way of life to continue on easier terms, but the immense conservatism seen here suggests that social change came much later. If such conservatism occurred after the probable event of stone tool-making, it is reasonable to suppose that a much longer tool-using phase preceded deliberate manufacture. Probably such a phase included occasional and haphazard tool-use and tool-making by a number of ape and hominid species. The early hominids may sometimes have reverted to ape-like characteristics, particularly those associated with tree life. However new opportunities were unfolding that were to give an omnivorous bipedal tool-maker with an interdependent mode of social organisation a huge selective advantage.

STONE TOOLS

Stone tools provided a sharp working edge which gave hominids access to a range of new materials and radically changed their relationship with the environment. Meat, bone, wood and vegetable matter could be modified with tools and become part of a growing material culture. Although stone is easily broken and lacks the flexibility of metal, it could be worked into a variety of tool types that, by the Upper Palaeolithic era 40 to 30 thousand years ago, were as versatile as a kit of bronze and iron tools used by a neolithic craftsman.

These cultural beginnings were associated with environmental change, manifested by a drier climate from about 2.5 million years ago, that led to the emergence of a new larger-brained hominid species whose behaviour comprised a set of adaptive shifts that included stone tool-making. The Kada Gona artifacts, which are classified as belonging to the Oldowan tradition, are dated to between 2.4 and 2.8 million years ago, and show that by these times hominids had mastered the basic principles of stone flaking, that a change in diet may have occurred that included a growing preference for meat and that a new social practice of congregating in groups at a 'home base' had been adopted. The first stone tools represent far more than a technological breakthrough. New forms of social behaviour that perhaps included social role differentiation linked with a cooperative subsistence strategy, that was based upon reciprocity and sharing, formed the basis of social cohesion. Certainly a shift in foraging behaviour is indicated by the presence of stone artifacts that were made to facilitate new activities that required cutting, scraping and piercing (Harris 1983). Many of these ideas, which are also found in the work of Glynn Isaac, have guided research in Lower Palaeolithic anthropology but they have also provoked controversy and here and in the following chapters we shall be concerned with the debates that revolve around these issues.

The uncertainty in fixing a firm date for the first stone tools serves as a reminder that these earliest known artifacts must be regarded only as a sample of an ongoing practice which was spreading in Africa between 3 and 2 million years ago. However this extent of time should not be allowed to detract from the ubiquity and suddenness of this innovation. Stone tools are found at sites across a moderately large region of Eastern and Southern Africa in this time span and this is after the fossil remains of at least two varieties of early hominid are found in earlier strata without the presence of stone artifacts. Isaac (1984) refutes the long-held assumption that the beginning of stone tool-making was gradual. A minimal acquaintance with the practicalities of stone-flaking by modern experimenters produces the understanding that the discovery of deliberate stone fracturing was a crucial threshold and, that once this had been achieved, a variety of artifacts became possible. These would have included both flakes struck from cores, which it is now known were not mere waste but were often directly used as tools, and a range of fully shaped core tools.

Momentous as these first steps in lithic technology were, their simple form invites scepticism. Can the loss of a few flakes from a water-rounded cobble constitute serious evidence of tool-making? It can definitely be shown that these seemingly crude modifications of pebbles, and even single flakes of stone, were in fact tools. Most commonly, stone artifacts are found to be out of their geological context. Only limited and relatively uncommon forms of stone such as flint, basalt, chert, obsidian and quartz or other fine grain rocks can be used for tool-making. The presence of such foreign material in a strata is in itself evidence of transport by hominids. The source of stone used in tool-making can often be accurately determined. Early hominids had carried stone at least ten kilometres to the sites at Olduvai Gorge (Gowlett 1984). But just as telling is the pattern of fracture in the stone and the presence of waste flakes which have been struck from the tool during manufacture. In tool-making, stone fractures in a characteristic pattern which, apart from single flakes, is not duplicated by nature. Deliberate conchoidal or shell-like fracture of stone produced by rapid percussion blows, leaves a set of characteristics including typical bulbar shapes and scar patterns on both detached flake and core. The reassembly of the tool and its chipped debris to reform the original nodule of weathered stone is an exercise which proves that only deliberate manufacture was responsible for the form of the artifact. Stone tools are often found in association with other cultural remains such as bone fragments and, by the Middle Palaeolithic, with hearths. Evidence is also available from microscopic analysis of both tool and bone surfaces (Keeley 1977). Particular edge-wear patterns can be seen on tools that originated in the working of meat, wood or hides. Close inspection of bones shows cut-marks made by stone knives.

Stone artifacts tend to have been made in accordance with a technologically standardised pattern and later with a cultural tradition both of which are easily recognised. J. G. D. Clark (1977) has proposed what has become a widely accepted typology of stone tools which is based on a technological progression. This five-fold classification, which includes cultural variations, represents the basic modes of tool manufacture presented in Table 6.1 and Figure 6.1. It should be added finally that stone artifacts, even from such ancient horizons as the Lower Palaeolithic, are not rarities. The persistent use and discarding of stone tools by hominids over millennia has resulted in a compacted accumulation which is revealed at many sites by scores of artifacts found in each few square metres of living-space. At the FLK site in Olduvai Gorge, for instance, Mary Leakey discovered 2470 artifacts associated with early hominids.

The appearance of stone tools seemingly provides clear evidence of important changes in psyche and in social behaviour. This opening of the cultural record represents far more than the attainment of new levels of skill by hominids. The new technology is a testament to the emergence of mental characteristics and patterns of group interaction which are almost unknown in the animal world. Foresight, calculation and the ability to evoke a conscious image of a result before it was to be realised in a material form are mental qualities which can be inferred from the first artifacts which surpass chimpanzee abilities. Just as significant are new learning-based modes of social solidarity and economic interdependence which are indicated by tools.

The earliest known tool types, the Oldowan and Acheulean industries which are classified as Modes 1 and 2 in Clark's typology, were made by direct percussion. A few rapid blows would detach debris from the core of an Oldowan pebble chopper to make available sharp cutting edges on both the tool and waste flake. The tool was held firmly in the palm of the hand and the ragged cutting edges used to process meat, wood, hides, bones and vegetable matter. For a million years no more sophisticated tools than these were used. About 1.5 million years ago the Acheulean hand-axe began to appear. These pear-shaped tools represent a leap in technological progress in the sense that both their manufacture and use suggest that a higher level of skill had been attained. Acheulean tools are themselves very large stone flakes which have been systematically shaped by secondary flaking on both sides to make a nearly symmetrical form. Direct continuity between the two traditions seems to have occurred. Bordes (1968) sees development from the Oldowan chopping tools by a process of extending the chipped face around both surfaces of the cobble and then to the circumference. The result was a thinner and lighter tool with a much longer and more regular working edge. Despite the clear advantages offered by Acheulean hand-axes, their coexistence with older Oldowan tools for at least a million years argues against notions of unilinear progress. In fact these two traditions – which are synonymous with the Lower Palaeolithic era – lasted until the first complex stone tools of Mode 3 began to emerge a mere hundred thousand years ago.

These Middle Palaeolithic tools involved the careful preparation of a tortoise-shell shaped stone core from which a ready made tool of predetermined type and size was struck. The convex flake tools produced by this technique used raw stone more efficiently. Oldowan and Acheulean technology has been found to coexist at this level with later tool modes, which serves to remind us that Clark's tool typology cannot be used as a simple evolutionary sequence. These two earliest tool traditions therefore fill 2.5 million years or over 95 per cent of humanity's cultural and technological past.

Toth has broadened and amended the prevailing view of early stone tools (Toth 1985, 1987 and Toth and Schick 1986). It had become accepted that Oldowan and Acheulean industries could be characterised by the different core tools which comprised a large range of implement types. Mary Leakey (1976) had accordingly designated a series of functional properties to some of the core tools found in the earliest cultural strata at Olduvai Gorge. She distinguished choppers, scrapers, burins, axes and awls from among the tool varieties found. In a fresh approach Toth argues that these core tools were probably not the principal forms taken by early technology. Experimentation with both the manufacture and use of Oldowan and Acheulean tools suggests that these artifacts have a different meaning. Toth found that core forms resembling Oldowan tools seemed to emerge spontaneously as flakes were produced during experiment at stone knapping. He suggests that too much emphasis has been laid on supposed core tools at the expense of stone flakes. The flake, once seen only as 'debitage' or waste, is in fact usually more useful than the core from which it has been struck. Early core tools have shapes which are likely to be incidental to their manufacture and do not indicate either the tool-maker's purpose or the tool's function.

Stone flakes used as tools were found to be well suited for cutting and far outperformed core tools in tasks such as butchery.

TABLE 6.1 *Palaeolithic Tool Manufacture*

Date	Mode and Stage	Industry Name(s)	Manufacturing Technology
	Mode 5	Capsian Azilian Natufian	Production of small 'microlithic' blades made by reworking blank flakes struck from a prepared core by the punch technique. These tiny triangular and trapezoid shaped blades provided very sharp cutting points in a wide range of composite tools. Microliths were used in gathering as blades fitted to handled implements such as knives, sickles, burins and scrapers. Microlithic technology provided hunters with arrow heads, harpoon barbs and spear points with better aerodynamic and penetrative abilities and are associated with a modern hunter-gatherer economy which utilises wider resources than earlier prehistoric peoples like fish and birds. Associated with *H. s.sapiens* exclusively. Figure 6.1 (e).
	Mesolithic		
10 k.y.	Mode 4	[Europe] [America]	Highly complex delicate blade tools represented by a great variety of diverse cultural traditions that changed with time and space. Thin blades were struck from a specially prepared cylindrical core using indirect percussion in the form of a stone punch and a wood or bone hammer and then reworked to become a range of tool types by pressure flaking techniques. This method produced more than ten times the amount of cutting edge from a given quantity of raw stone than the Mousterian technique. Associated with *H. s.sapiens* and late neanderthal populations. Figure 6.1 (d).
		Magdelenian Folsom Solutrean Clovis Gravettian Perigordian Aurignacian	
	Upper Palaeolithic		
40 k.y.	Mode 3	Mousterian	Complex multipurpose tool-kits made by the Levallois technique in which a carefully prepared stone core resembling a tortoise shell was made by removing chips from the top and sides. Whole flakes were then struck from the core to be reworked into a variety of specialist tool types. Some finished flakes were notched in order to be hafted to wooden handles to become the first composite tools like stone-tipped spears. Associated with *H. sapiens neanderthalensis* and archaic *H. sapiens*. Figure 6.1 (c).
	Middle Palaeolithic		
130 k.y.	Mode 2	Acheulean	Handaxe and flake tools with two cutting edges used as choppers and cleavers which developed as a symmetrical pear-shaped form. Originally made entirely with a hammer stone later handaxes were finished with a bone hammer to produce a more delicate tool with a longer and sharper working edge. Handaxes are associated with *H. erectus* and archaic *H. sapiens* and are found throughout Africa, Europe and in western Asia and India. Figure 6.1 (b).
	Lower Palaeolithic		
1.5 m.y.	Mode 1	Oldowan	Modified river cobbles and pebbles selected for their lithic properties and shaped by direct percussion with a hammer stone to chip flakes from a core. These chopper and flake tools are the oldest and most persistent form of stone tools that were used for over two million years. Associated with *H. habilis* and *H. erectus*. Figure 6.1 (a)
2.5 m.y.	Lower Palaeolithic		

SOURCE: Based on Clark's (1977) Fivefold Typology Of Technical Modes

k.y. = thousands of years ago
m.y. = millions of years ago

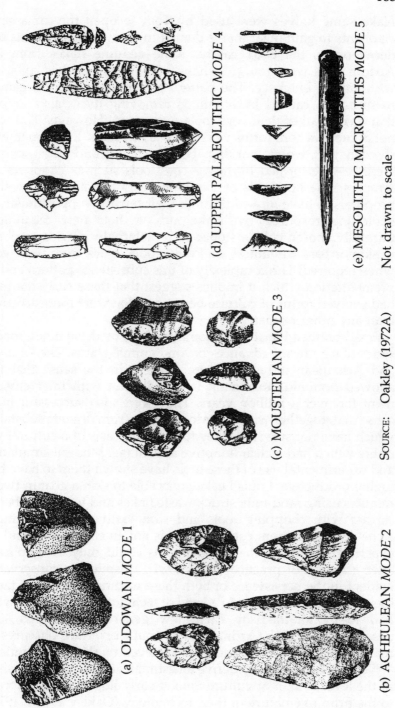

FIGURE 6.1 Palaeolithic Stone Tools (See Table 6.1)

(a) OLDOWAN MODE 1

(b) ACHEULEAN MODE 2

(c) MOUSTERIAN MODE 3

(d) UPPER PALAEOLITHIC MODE 4

(e) MESOLITHIC MICROLITHS MODE 5

Not drawn to scale

SOURCE: Oakley (1972A)

Flake stone knives were used by Toth to open the carcasses of elephants to gain meat that is denied to carnivorous animals until decay opens the body cavity. Some genuine flakes from East African sites were found to have microwear polishes consistent with animal butchery. The pattern of scratch marks on genuine fossil bones, caused by hominids removing meat, also suggests that flakes rather than core tools were used. However Toth does not argue that core forms were not used as tools. Experimentation showed that wood working, bone-breaking and hide scraping were best performed by heavy core tools such as choppers and scrapers. The hominids of 2 million years ago are seen in these proposals to have moved through their territory opportunistically exploiting resources with the aid of their new technology. Carefully chosen cobble stones were habitually carried, probably in skin or bark containers, and flake tools struck from these cores when required. The complexity of this subsistence pattern and the premeditation which it implies suggest that these early hominids had evolved forms of cultural behaviour that were more advanced than any other primate.

Small-brained Pleistocene hominids had, by these times, made a series of significant advances beyond animal status. The Oldowan and Acheulean industries were efficient in the sense that they allowed hominids to remain in equilibrium with their environment for over a million years. They were also successful in the sense that, together with other tools made from organic substances which have not survived, they formed the basis of a cultural way of life which had a clear adaptive advantage. Modern simulations and experimental use of these tools have shown them to have been highly productive. Louis Leakey was able to skin a goat in twenty minutes using randomly struck waste flakes and to dismember the carcass with chopping tools and stone knives. The making of Acheulean tools from cobble stones is not an easy matter and one experimenter found that proper tools could only be duplicated after a month of practice (Jones 1981). Despite the conservatism evident in the persistence of both these traditions, the very fact of their manufacture points to a level of technical achievement which had to be attained by each new generation. In this sense descriptions of these artifacts as uncouth or crude obscures the real importance of what had been achieved by the early hominids.

Nor is it reasonable to derive an estimate, in any precise fashion, of the level of skill or culture among early hominids by reference to the primitive nature of their technology (Oakley 1972A, 1972B).

Toth makes the same point with regard to cognitive ability. The organic elements of ancient material cultures are simply not available in the archaeological record. An ethnographic parallel makes this point eloquently. Modern native Australians frequently use naturally sharp pieces of broken stone to fell trees or to carve intricate wooden tools such as spear throwers. This use of unflaked stone as a tool – Mode 1 technology in fact – by a people with a highly complex lineage system and cosmology also serves to emphasise that it is unlikely that the first stone tools will ever be positively identified.

MATERIAL CULTURE AND SOCIAL LIFE

It seems reasonable to infer that culturally based behaviour, which included tool-use, probably existed well before the advent of stone tool-making. To suppose that such a novel practice as stone flaking emerged among non-cultural hominids, who had not had a long apprenticeship in using and perhaps making tools from organic matter, is implausible. After all chimpanzees use tools and have a culture of sorts. All hominid species are likely to have been culturally motivated to some extent even though only one branch of the lineage was to use culture as a vehicle for its advance. Simplistic ideas which suggest that culture provides a royal road to human status for any species who adopt it should be abandoned. Cultural behaviour in primate societies reminds us that at this level of human evolution culture was not self-generating. There were then no mechanisms for systematic social or technical development. Culture did not at once confer omnipotence for tool-using hominids or appease environmental pressures beyond a fraction.

Nevertheless a direction in hominid evolution can be seen. The first tool-makers began a technological tradition which increased in sophistication over a two million year period. After the Mode 3 industries already described, came the fine blade tools of the Upper Palaeolithic era 40 000 years ago. These Mode 4 tools were made to far higher standards of craftsmanship. Blades rather than flakes were struck from carefully prepared cylinder shaped stone cores by a variety of new techniques. The result was a slender blade with a far higher ratio of working edge to raw stone, which could be reworked into a variety of tool types. These tools and the small microlithic flakes of Mode 5, which began to be made 10 000 years ago at the end of the Palaeolithic era, were often hafted to organic materials. Wood and bone were more extensively used in these times and composite tools which combined materials, such as stone-tipped spears, sickles and perhaps arrows increased foraging efficiency.

Another inference is that social behaviour grew in complexity and flexibility during the first cultural eras as human ancestors gradually adapted to a mixed gathering and hunting economy. Stone tools made possible and probably initiated a new set of foraging strategies which were interconnected with behaviour. A shift from gathering vegetable food and protein from small animals, to a tool-based meat and vegetable subsistence, could only have occurred with parallel changes in social behaviour. In what is likely to have been a primary scavenging niche, new social skills which allowed mutual protection from rival animal predators, and rules to divide meat, would have arisen. The further development of cooperative hunting, in which animals were killed and eaten, occurred in a social context that was characterised by increased interdependence and division of labour. Social bonds at this time may have become synonymous with a rudimentary pattern of kinship organisation. By late prehistory specialised hunter-gatherers who exploited large animals had developed the full range of human behaviour patterns. The evolutionary product of this cultural complexity was a threefold enlargement of the hominid brain. Technical, social and cognitive factors were engaged in interactive evolution.

Irrespective of the initial advances over animal status presented by the first stone tools it is important to realise that neither the australopithecines, nor the earliest species of *Homo* would have been recognised as 'human' in the sense that they possessed a range of culturally based behaviour patterns equivalent to those of modern people. Whatever was distinctive about australopithecine culture is unknown, but it was marked by eons of stability which are a far cry from the rapid cultural innovations of modern *Homo sapiens*. It was not merely that early hominid culture was less developed, but that significant aspects of culture and essential 'human' abilities or characteristics which probably included language, were not present at all. The biological mechanisms which underpin 'human' status had at this stage of hominid evolution only partially appeared. From this it becomes possible to understand that human culture in its modern form did not arise as a unity but was the product of a series of accretions.

Intermediary stages between the primate condition and full human status are of course difficult to envisage. Modern people, living even at the most primitive level, are separated by an evolutionary gulf from animals. In human society food procuring activities can properly be described as an economic system since a division of labour and a related set of interdependent statuses coexist. The mode of isolated primate foraging, in which only infants are found food, contrasts strongly with human food-sharing practices. Communication, sexual behaviour and social control among humans are regulated by social rules that are only echoed by advanced primates. The habitual and dependent form of human tool-use, the occupation of camp-sites as a home base and the exploitation of large animals as a food source are again very distant from even the nomadic chimpanzee whose few tools and equipment are discarded after use and therefore never constitute possessions. The occasional meat-eating episodes of chimpanzees make no difference here since their game consists of small immature animals that are killed fortuitously. Chimpanzee behaviour is certainly flexible but life here is not based upon a set of variable social rules that may be justified by recourse to higher principles.

The early hominids then probably had behaviour patterns which were intermediate between ape and humans. The puzzle which confronts the social scientist here lies in discerning the pattern of natural selection which transformed these proto-human hominids into humans (Isaac 1978). The growing importance of meat in the hominid diet may have initiated a division of labour based upon sex. But why should meat have become a food source? It would be absurd to suggest that the carnivorous proclivities of hominids arose naturally. The early hominids were not biologically predisposed to hunting and could at best expect to gain meat from large animals only by scavenging.

THE TOOL MAKERS

Who was responsible for the first stone tools? The term 'early hominid' has been used here as a catch-all. In fact between 3 and 1.75 million years ago no less than a possible seven hominid species are currently recognised throughout Africa. Several of these species are known to have coexisted in East Africa at the time of the earliest known tools and conclusive attribution of authorship to any single species cannot be made with final certainty. It is *Homo habilis,* however, the first known representative of the *Homo* lineage, who is usually assumed to be the inventor of stone tools. Despite a huge growth in knowledge drawn from a new wealth of fossil evidence, the exact evolutionary interrelationship of the early hominids is unknown and remains a subject of considerable controversy punctuated by a shifting consensus. The theories and anatomical evidence used in physical anthropology to support arguments for particular branching patterns at these complicated junctions in the hominid evolutionary tree are often inaccessible to social scientists. The account of hominid evolution presented here will therefore be extremely simple and will avoid controversial areas.

We saw in Chapter 5 that *Australopithecus* is the earliest certain hominid genus. Since 1924, when Raymond Dart first identified fossil remains of this hominid, several hundred individuals of different species have been unearthed. Dart named his infant specimen, found at Taung in South Africa, *Australopithecus africanus* or 'Southern Ape'. However the australopithecine genus, although not *A. africanus,* was not as geographically restricted as Dart had supposed. Further discoveries have associated species of this hominid with a five thousand kilometre long north–south corridor running along the eastern side of Africa from Ethiopia almost to the Cape. Fossil finds from these sites, many of which are on ancient lake shores or near water courses in the Rift Valley areas, currently place the six diverse species within the australopithecine genus in an assumed time range of from about 4.5 to just over 1 million years ago.

The fossil hominid *A. afarensis* represented by 'Lucy', who was introduced in the last chapter, had an assumed time range of from four to three million years ago. Lucy's species had a face with greatly protruding jaws and long and narrow parallel lines of teeth, both of which are ape-like features. Holloway (1988) estimates that the *A. afarensis* brain was 310–400 ml in volume – only slightly larger than an ape brain – but this hominid was not as large and evidence of cerebral reorganisation is present. In most scenarios of hominid evolution, see Grine (1988) for instance, *A. afarensis* is seen as having given rise to *Australopithecus africanus* about 3 million years ago. This latter hominid was about four and a half feet tall and had a brain volume which averages 450 ml (Tobias 1981), which again is hardly larger than a chimpanzee or gorilla. When body size is taken into account, however, a significant increase in brain volume is evident in relation both to the apes and to *A. afarensis*.

The face of this fossil hominid is large and has a 'dished' appearance that is seen as a concave depression around the nasal region. The brow ridges are well developed, and the brain case is rounded without a proper forehead. The jaws are large but have no chin and project from the face rather like those of a chimpanzee. However the dental pattern of *A. africanus* is essentially human since a rounded V-shaped jaw resembling the parabolic arcade in *Homo* is present, rather than the characteristic U-shaped jaw form seen in apes. Although the surface area of the teeth is much greater than those of modern humans, the dental configuration is much nearer to that of humans than the apes. In fact the small canines and incisors – relative to those of apes – place this hominid on a trajectory that culminates in our species.

Between 3 and 2 million years ago an adaptive radiation began among the australopithecines. Essentially this involved the emergence of a robust lineage of *Australopithecus*, often referred to as *Paranthropus* or 'beside man', which separated with hominids bound for the *Homo* branch. The route taken here remains controversial but one possible pathway which represents this divergence is based on EHEP (1988) as Figure 6.2. In this proposal *Australopithecus aethiopicus*, a direct descendant of *A. afarensis*, gives rise to *A. boisei* and *A. robustus*. The latter species was restricted to Southern Africa while the former species and *A. aethiopicus* are East African variants. These robust hominids, which are not directly related to *Homo*, were the product of an adaptation to new conditions. At the margins of the Pliocene and Pleistocene epochs over two million years ago, warmer and more equable seasonal temperatures gave way to a climate with increasingly cold winters. *A. boisei* entered an environment of expanding grasslands as a herbivorous animal. Rather than developing the specialism of a tool-making, scavenging omnivore this hominid became committed to a diet of tough fibrous plant material that was consumed in great quantity. *A. boisei* probably ate roots, seeds, fruits and tubers all of which were foods that required processing for digestion by assiduous pulverisation and grinding.

The masticatory apparatus of this hominid is, therefore, more extensive than those of all other species belonging to this family. The demands of heavy chewing are met by massive molars and premolars which are set in very large jaws. In fact the amount of tooth material and surface area found in this hominid is twice that

170

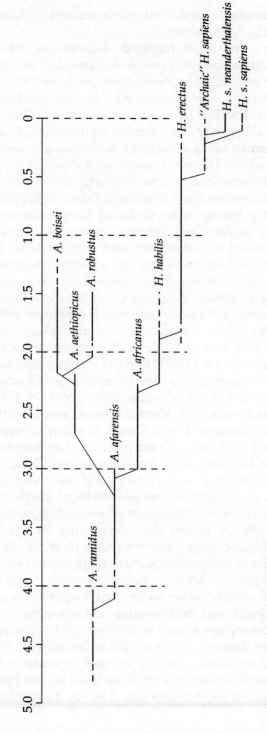

FIGURE 6.2   A Representation of Relationships in the Human Fossil Record

The phylogeny or family tree of the hominid lineage showing known time ranges in solid lines and possible range extensions in broken lines of the various recognised species. Light oblique lines indicate possible paths of descent

SOURCE: Based on EHEP (1988)

of modern humans and the cheek teeth are larger than those of any other primate. The musculature and bone structure required to produce such powerful bite-forces between *A. boisei*'s jaws are manifest in the huge crania. This low-domed head is dominated by the jaws rather than the brain case as in *H. s. sapiens*. The flattened skull has no forehead and in males is topped by a bony crest which runs along the mid-line of the head. The tension created in the large chewing muscles required strong attachments and these were provided by anchorage to the crest as well as to correspondingly well-developed cheek bones.

Sexual dimorphism, the characteristic differences between the sexes, reached extremes among the robust australopithecines. It is very likely that among all the australopithecine species males were considerably larger than females. Differences of this kind are typical of the primate order. Male orangutans, gorillas and baboons, for instance, weigh twice as much as the female, but as the *Homo sapiens* stage of primate evolution is approached this difference has become progressively reduced. Chimpanzee and human males are only ten to twenty per cent heavier than females. However among the compatriots of 'Lucy' 3.5 million years ago, this trend towards sexual equality had hardly begun. In modern apes and monkeys major differences in sexual dimorphism are correlates of extreme dominance, intense inter-male rivalry and competition for receptive females. This could indicate that the australopithecines lived rather like baboons in large troops ruled by dominant males who mated exclusively with a group of females. The robust australopithecines, a side branch in hominid evolution, probably remained closer to ape social behaviour rather than anticipating the life style of *Homo sapiens*. No indication of superior intelligence is evident even though the brain of *A. boisei* was about 500–550 ml (Grine 1988A), which is somewhat larger than that of *A. africanus*. However the difference can be attributed to a general increase in the body size of *A. boisei*, which is a characteristic of herbivores, rather than as selection for cultural behaviour.

The robust australopithecines disappear from the record leaving no issue at some time around one million years ago. Why this occurred after a period of considerable stability – fossil evidence indicates little change in over a million years – remains unknown. It is frequently assumed that competition from the larger-brained *Homo* lineage was the reason. However, in East Africa *A. boisei* coexisted with two species of *Homo*, in much the same ecological conditions, for about a million years. These extinctions should be seen as the result of competition from other herbivorous animals rather than from other hominid species who would have occupied different niches and been avoided (Foley 1987). Early *Homo* was in competition with a range of animals that included other omnivores and large carnivorous species – who were also evolving at this time – but adapted successfully by adopting a mixed foraging strategy. But as vegetarians, specialising in tough plant material, the robust australopithecines had entered an evolution-ary cul-de-sac. They were faced with severe competition from more efficient herbivorous animals who were rapidly evolving at the time of these hominids' demise.

The coexistence of three separate East African hominid species (*A. boisei*, *H. habilis* and *H. erectus*) can clearly only be explained by an adherence to different dietary and subsistence patterns. The diverse forms taken by ancient hominids represent separate adaptive patterns pursued in one general environment and a series of foraging niches are easily conceivable in savanna conditions. The hominids would have had a wide choice of foods available and distinctive specialisations in gathering would have been as characteristic as their physical forms. Direct evidence from the ancient environment in which the fossil record is located is clearly difficult to establish. However some connection is made between a reconstruction of East African lake shore areas and their associated fossil inventory (Behrensmeyer 1978). The fossils of both *Australopithecus* and *Homo* are found in the environs of such lakes. Behrensmeyer's sample was restricted to hominid fossils which lacked any evidence of post-mortem movement, making it reasonably certain that the position of death was close to the fossil site. Although her evidence is far from conclusive, some indication is given by this study that australopithecine remains are associated with gallery forest and bush along the stream and river beds which led to the lake shore. On the other hand, *Homo* appears to have lived around the lake margins and may have been experimenting with new open habitats.

At about the same time that *A. afarensis* gave rise to his robust cousins, that is about 2.5 million years ago, another branch of this radiation was characterised by hominids with enlarged brains. This branch was not a dead end, as was the former one, since it is by this route that *Homo* is reached. Most authorities agree that *A. africanus* was the ancestor of the *Homo* lineage which first appeared around 2 million years ago. The earliest species of *Homo* is in an intermediate position between their parent australopithecine population and *Homo erectus*, the hominid typical of the first three-quarters of the Ice Ages. Louis Leakey discovered this fossil hominid in the lowest strata of Olduvai Gorge in Tanzania and assigned it the name *Homo habilis* meaning able, handy and mentally skilful (Leakey et al. 1964).

This earliest and most primitive species of *Homo* possesses a cluster of features that allows it to be described as 'man'. The brain of the habilines is larger than that of a robust australopithecine, with a range of 600–800 ml (Rightmire 1988), yet *H. habilis*, who was only four foot three inches tall (Lewin 1984), was 25 per cent lighter (Fleagle 1988). The teeth are more human than those of *A. africanus* and suggest that a minor dietary shift had occurred although the wear patterns indicate that fruit was still the predominant item eaten. The hands and feet of *H. habilis* still resemble those of the apes, having curved phalanges for a strong power grip when climbing trees, but they are closer to those of modern humans. In particular there is evidence that a superior mode of dexterity had been attained and that therefore *Homo habilis* was able to manipulate objects in a precise fashion. In some ways the habilines anticipate later stages of human evolution since they seem to have begun trends in brain expansion, reduction in tooth size and restructuring of the face.

Homo habilis fossils have been found to be associated with stone tools at a wide variety of East African sites. But apart from the Kada Gona tools, which may possibly be 2.7 million years old, all the earliest stone artifacts are dated to a time span which corresponds with the appearance of both *H. habilis* and *A. boisei*. Who then made tools from stone? *Homo habilis* is by far the most popular candidate. However cultural remains cannot be unequivocally linked to a particular species of hominid until after the occurrence of ritual burial had become established in the Upper Palaeolithic era (Louis Leakey 1973). Many authorities see no means of conclusively distinguishing between the two hominids (Isaac 1984).

Evidence for *H. habilis*'s claim to have begun stone technology, though circumstantial, is strong. In addition to the habilines' superior brain size and, it has been assumed, more dextrous hands, no tools have been found to be associated directly with *A. boisei*. In fact fossil hands of *A. boisei* have not been recovered. However those of *A. africanus* and *A. robustus* indicate few if any modifications for tool manipulation according to Trinkaus (1988).

Further, stone tools continue to be made by the *Homo* lineage after the extinction of the robust australopithecines. However it would be unwise to dismiss *A. boisei* as a non-cultural tool-user and modifier (Tobias 1965) merely because *Homo* was to take culture to new lengths. Even with the status of only a walking ape, the australopithecines cannot be denied at least the same cultural level as chimpanzees.

Some evidence suggesting that the australopithecines may have been stone tool-makers, contradicting Trinkaus' findings, has been reported by Lewin (1988) and Susman (1988). They point to a study of the hand bones of the Southern African hominid *A. robustus* in support of this view. This hominid had hands with broad sensitive finger pads, and muscles that would have allowed a precision grip. Both of these characteristics, which were previously thought to be found only in the *homo* lineage, indicate an ability for sophisticated manipulation seen in stone knapping. Narrow finger pads are found in both the apes and in the first known australopithecine species, *A. afarensis*, but by 2.5 million years ago, at roughly the same time as the appearance of the first stone tools, broad pads and new muscles are present in the hands of both *Homo* and at least one australopithecine species. Susman concludes therefore that the robust australopithecines had much the same physical potential for tool-behaviour as modern humans and cannot be finally dismissed as non-tool users.

Perhaps the question should be kept open, but certainly the distinction between cultural and non-cultural tool-making is impossible to establish in primates. Regardless of the real advances in abilities which may have been present among the habilines it is not appropriate to assign culture to *Homo* alone. This would in fact represent an arbitrary attempt to differentiate between human and animal statuses. It is, of course, a matter of great importance that the first stone tool industries at times tend to conform with regular and evolving patterns but this does not alter the fact that culture is not an exclusive possession of humans and their immediate ancestors. One cannot conclude that culture has any singular direction. New technologies or social patterns that are adopted do not lead inexorably to cumulative progress. It is quite reasonable to hypothesise a succession of adaptations by hominids to cultural practices which were to be 'false starts', in that evolutionary pressures later induced a simplification of artifacts or gave less prominence to culture or even led to its abandonment.

A NEW WAY OF LIFE

The career of *H. habilis* was relatively short, compared with the australopithecines and later hominids, lasting about 0.5 million years until *c.* 1.6 million years ago when *Homo erectus* replaced the habiline population. This long-lived species of *Homo*, which had probably evolved nearly 2 million years ago (Rightmire 1988A) and was replaced by archaic forms of *H. sapiens* less than 300 000 years ago, had a geographical range which eventually reached beyond Africa into Asia and possibly Europe. European hominids during the later part of this era certainly resemble *H. erectus* but their classification as members of this species is not universally agreed. Besides leaving his natal continent, *H. erectus* has long been identified as the first hominid to use fire and engage in full-scale cooperative hunting. Until recently there was no doubt that *H. erectus* also came to develop an adaptive pattern based upon hunting and gathering. However it can be firmly established that *H. erectus* was a habitual tool-maker with an economy reliant upon tool-use and that this hominid lived in social groups which collected at camp-sites or home bases. The claim that *H. erectus* was the first hominid big-game hunter who initiated a way of life which resembles that of modern foraging peoples will be considered in Chapter 9.

Like *H. habilis*, *H. erectus* was an intermediate species which resembled both ancestors and descendants at either end of a sojourn lasting over a million years. The brain of *H. erectus* averaged just under 1000 ml and was 87 per cent of the size of a modern human brain, when adjusted for body size (Rightmire 1988A). The *erectus* population was about 10 per cent lighter than our species (Fleagle 1988). The *erectus* brain ranged from 800 to 900 ml in early specimens to more than 1100 ml in later individuals. Smaller teeth and reduced sexual dimorphism, two other long-term evolutionary tendencies, are also found here. The *erectus* skull has a backward sloping chin and a broad face over which prominent brow ridges give way to a low forehead and a brain vault, flatter than in humans, which tapers almost to a point at the back of the head. This last feature served as a means of attaching substantial neck muscles. The *erectus* skeleton, while basically human, shows significant robustness: thicker bones in both the skull and limbs indicate considerable strength required for a physically demanding life.

What selective forces initiated the creation of the *Homo* lineage and the habitual manufacture of stone tools? The appearance of tools over a wide area occurred with relative suddenness indicating that a new way of life had begun. The leading hypothesis attempting to explain this change concerns a shift toward meat-eating among hominids. Small game could easily be run down and killed but not easily eaten since, without the claws and teeth of carnivorous animals, a carcass could not be skinned and much of this food source would have been wasted, while the carcasses of large animals could not be opened at all. Hominid digestive and dental patterns certainly prohibit a meat oriented diet without the aid of tools. The broken pebble or stone flake therefore came to fulfil a new need.

Environmental changes are usually identified as the cause of this dietary shift. In 1960 J. D. Clark proposed that both the emergence of palaeolithic culture and subsequent cultural change can be understood in terms of climatic change. These changes were typified by dry periods which, because they appear to coincide with cultural phenomena, are seen in themselves as causes for rapid technological change. Clark's presentation argues that the beginnings of stone tool-making, the emergence of complex tools in the Middle Palaeolithic, the most advanced technology of the Upper Palaeolithic and the beginnings of farming in Neolithic times were all products of these drier climatic conditions.

A marked and widespread environmental change occurred in Africa between 2.5 and 2 million years ago as a result of the onset of glacial conditions in the northern hemisphere. Lower temperatures and rainfall produced an open grassland landscape in place of a woodland environment. In an 'evolutionary pulse', a whole range of animals began to form new species in this context. *Homo* was merely one of a number of species who adapted to become 'founder members' of Africa's new savannas. The major branching events among hominids, leading to the differentiation of robust australopithecines and the *Homo* lineage, occurred as ancestral populations such as *A. africanus* were subjected to fragmented habitats (Vrba 1985).

A new diversified environment containing many localised habitats, which can be seen as a fluctuating mosaic of forest and woodland dominated by savanna, was to be the location for hominid coexistence (Isaac 1976A). Evolutionary pressures within this environment may have caused the *Homo*–australopithecine division. Marked changes in food resources are found to occur during different seasons under savanna conditions. In wet seasons a huge range of food was available to omnivorous hominids, such as fruit, flowers, insects, eggs and the young of mammals and birds. However during dry seasons when only seeds and buried roots were available, hominids, like large herbivorous animals, would have experienced considerable dietary stress. These conditions were providing new opportunities for a growing group of carnivorous animals. Two possible strategies were open to hominids. Abundant food of low nutritional quality could be gathered using a digging stick with longer foraging and more time spent processing. Alternatively hominids could change to high quality food that was freely available in dry seasons in the form of herbivorous animals who had died or were in poor condition. Meat could be used as a high-yielding source of nutrition that would supplement failing plant food. The robust australopithe-cines adapted using the first strategy and developed huge teeth to eat large quantities of hard, dry, coarse vegetable matter. *Homo* represents the other evolutionary trend which went towards opportunistic omnivorous foraging of seasonally available food (Foley 1987).

The distribution of fossil hominids at the Lake Turkana site in Kenya confirms this dietary change (Lewin 1984). Before the coming of *H. erectus* the occupation density of hominids appears to have been similar to that of baboons in a modern woodland savanna. But with the innovation of scavenging and later hunting by hominids, territorial size increased by several orders of magnitude and fossil finds are correspondingly less common.

For many authorities the consequences of this climate-driven adaptive shift are to be seen in brain expansion, tool-making and new patterns of social organisation. For the social scientist, however, this form of explanation seems incomplete. Change is presented in this form of theory in terms of a simple challenge–response relationship. Less rain meant less forest and less fruit; hominids 'needed' to utilise the growing populations of herbi-vores as a food source and this necessity was met by stone tools as a means to process food efficiently. In this sense stone tools were a natural or inevitable product of a need. Environmental stress will certainly constitute circumstances which must be directly ad-dressed, but no necessary consequence or indeed innovation can be specified as a response. It is how the challenge of the environment is perceived and acted upon which is crucial. The

challenge of encroaching aridity, which has been a recurrent
feature of the earth's history, could have been responded to
fatalistically by retreat to the remaining forested habitats or
extinction could have occurred.

While it is quite reasonable to propose that climatic change had
a radical effect on hominid evolution, the emergence of complex
social behaviour and technology should not be seen purely as
the product of savanna ecology. The hominids of one and a half to
two million years ago were the most flexible and intelligent
animals of their time who lived in complex social groups. Before
the appearance of stone tools in the archaeological record
hominids had acquired the ability to transmit the accumulated
experience of previous generations. Socialisation would have
included learning complex social and technical knowledge. The
successful shift in ecological relations undertaken by what was to
become the *Homo* lineage, was possible only because of this mental
and social complexity. Human evolution at this time can only be
understood through the interaction of a changing environment
and hominid behaviour. Cultural practices, including a growing
dependence on a material culture, had become an established
means of adaptation for hominids by the beginning of the
Pleistocene epoch *c.* 1.6 million years ago. Taken together these
practices constituted an internal environment which reacted with
a changing ecology to produce both a cultural solution and a new
physical form.

Gowlett (1984A and 1986) urges us not to downgrade the
achievements of early hominids which, he argues, rested upon
considerable mental and cultural abilities for which there is often
only indirect evidence. It is reasonable to assume that hominids
capable of stone tool manufacture also worked in other materials,
that they had an extensive knowledge of the topography and
resources of their territories and were capable of social cooperation
and effective communication. Evident complexities in one area of
life are an indicator of complexity in other areas. Selection would
favour any form of social behaviour which improved fitness and
would advance the physiological apparatus which this behaviour
relied upon.

The complex mental machinery which underlies the total
cultural dependence of modern society rests upon a set of mental
abilities with an evolutionary history, Gowlett argues. Minds
make a given social system possible because they have developed
for this purpose. The mind is a practical organ which allows us to

make an internal reconstruction of the external world and provides the cognitive abilities to distinguish a variety of courses of action that are required to pursue a particular way of life. Stone tools are the mental products of early hominids and tool examination provides an understanding of a wider range of proto-human mental abilities.

Modern people possess the ability to internalise a detailed representation of space at the level of both forms that are to be created, or recognised, and in terms of geographical knowledge. Some evidence of this spatial sense of mind in early hominids is found in the distribution of tools in relation to their geological origin. It is often possible at Olduvai Gorge to trace the source of stone used in tools made 1.8 million years ago. These artifacts were in some cases transported eight to twelve miles to a living site. Tools from Kenyan Acheulean sites that are more than 0.7 million years old had been made from stone brought as much as 25 and even 40 miles. It is known that the Oldowan hominids carefully selected raw stone and discarded cobbles with defects that made them unsuitable for flaking (Toth 1987). This geological knowledge was part of a much larger internalised mental map of a territory that, in the case of Oldowan hominids, was probably more than 100 square miles in extent. A significant advance in planning and foresight is evident here. Chimpanzees will carefully select and shape twigs for termite fishing twenty metres from a nest. The early hominids, however, must have used a timescale of at least a day or more.

Another modern mental capacity that is shown to be present in early stone tools is the ability to pursue a chain of related activities over a prolonged period of time while maintaining an overview of the whole process. An image of a finished product or end can be projected forward, and a carefully monitored sequence of actions can be directed to achieve this. Analysis of Oldowan and Acheulean tools reveal that flakes were detached from a core in accordance with an operational procedure. Each step in the stone knapping process was performed in a given order and individual steps were subordinated to the ultimate goal: a tool with a desired form. Implicit in this act of design was the ability to project a mental image of the finished tool onto raw stone, and the concentration and manipulation necessary to achieve this. The Oldowan hominids of 1.8 million years ago had internalised both a mental template governing tool-form and a procedural template that directed manufacture.

Were hominids at this time so little different from animals that it is invalid to suggest that cultural behaviour constituted one of a number of causal factors influencing natural selection? The phenomenal persistence of standardised tool types might suggest a form of tool production similar to the stereotypical patterning found in animal behaviour such as nest building by birds. Lower Palaeolithic tools remained similar in form for over 1.5 million years. In an area which includes Europe, Africa and India little regional specialism is evident at all. While there is variation of tool-kit contents between different sites there are few, if any, differences in hand-axe design which differentiate one area or span of time from another. Is this perhaps evidence of proto-human conformity which occurred independently from social relations and is comparable only with instinctual drives?

Faced with such an obdurate rate of cultural change, this conclusion may seem tempting. However reproduction of material objects by hominids was a socially conditioned activity that depended upon both ideas and social organisation. The near identical technical systems which have survived do not imply identical forms of application to the environment which would be the case with an animal adaptation that featured construction as a specialisation. Indeed the differences in ecology across such vast regions of the world show that similar tools were put to different ends. At even a level of incipient human status this primitive technology allowed a variety of adaptations to quite dissimilar environments. Such standardisation indicates some form of instruction, and therefore enhanced modes of communication, among hominids. The earliest stone tools were the product of learning and experience within a social group. They were above all an outcome of a new form of group interaction in which different tasks combined to produce a social product.

There can be no precision in determining how a configuration of social mechanisms interacted with environmental pressures to produce a hominid behaviour pattern that came to influence future stages of evolution. But the principle that the early hominid social system induced, sustained and elaborated forms of cultural behaviour which in turn gave rise to tool-making, and subsequent development in foraging, is of central importance, and ecology alone cannot explain the rise of this new form of social life. Culture was a significant part of ancestral hominid societies by the time stone tool-making began and the biological basis of culture was established. The growing value of a cultural adaptation for hominids acted to ensure that culture was henceforth to gain a degree of independence from biology: culture had thus become self-generating.

7

Tools, Brains and Behaviour

In this chapter what is known of the mental abilities and social behaviour of early hominids are examined as they relate to tool-making. It is in this context that the emergence of intelligent behaviour will be considered. Brain expansion and a developing material culture have long been considered as two closely interrelated processes that are distinctive features of human evolution. Finally, the chapter looks at the idea that with the appearance of the first tool traditions a distinctive proto-human way of life had emerged among hominids that was focused on mixed foraging and the use of home base camps.

HOMINID INTELLIGENCE

How intelligent were the early hominids? Wynn (1979, 1981) claims that the Oldowan tool-makers had an intelligence that was not much different from modern apes. The crudity of Oldowan technology cannot be attributed to a lack of motor control since an impressive degree of dexterity is seen in small quartz flakes that had been worked into scrapers and awls. It was, rather, a low level of *conceptual ability* which limited the complexity of the first artifacts. Intelligence constrains behaviour, Wynn argues, and simple Oldowan artifacts are the products of a lower evolutionary stage of intelligence.

Wynn uses the developmental theories of the psychologist Piaget to analyse tools. In the mental development of modern humans Piaget distinguishes three sequential stages which culminate in 'operational intelligence'. Adult thinking processes are based upon 'operations' which are internalised schemes of mental action that are 'reversible' in that an operation and its opposite can be conceived of simultaneously, as in addition and subtraction, and are combined with other operations in an organised system. Operational intelligence is characterised by the ability to reason logically by drawing conclusions from a set of premises and by the further ability to manipulate propositions or ideas mentally. At this cognitive level it becomes possible to explore problems in a systematic way from a number of aspects and to mentally hold an overall view of all the variables involved.

Oldowan tools are seen by Wynn to embody no more than pre-operational intelligence. His analysis of the patterns present in these artifacts reveals that early hominids held very limited geometric concepts. The idea of proximity enabled adjacent flakes to be removed from a cobble; the concept of pairs allowed a cutting edge to be formed on a chopper, and in producing a straight edge on a scraper only the notion of direction had to be added to proximity. The cognitive structures available to an Oldowan stone knapper therefore provided only a simple internalised action-scheme which allowed trial and error tool-making. No more than the effects of one action need have been considered at a time while stone was worked, and all the tool varieties in an Oldowan assemblage could have been achieved by haphazard chipping. This level of intelligence would have permitted the behavioural basis underlying food-sharing and congregation at a home base – both novel practices that would have implied a radical break with ape behaviour – but the classificatory skills necessary for the delineation of a kinship system would have been beyond these hominids. While some progressive changes in intelligence had probably occurred by Oldowan times, the hominids of this era were not adapting by virtue of their cognitive skills since they had not moved significantly away from the mentality of the apes.

Real hominid operational intelligence did not start evolving until the first hand-axes appeared about 1.6 million years ago. These Acheulean artifacts were made to a predetermined shape by a process of deliberate chipping which at each stage would have anticipated the end result. The tool-maker was able to work toward the final hand-axe form by mentally reconstructing the unavailable shape yet to be made from the shape that was emerging, much as a traveller will be reoriented by mentally reconstructing the relationships between landmarks when a place is approached from a new direction. Wynn proposes that Acheulean tools made between 330 and 170 thousand years ago exemplify a series of complex geometric relationships such as symmetry and space–time substitution which are the products of early human intelligence. It can be inferred from these tools that the Acheulean hominids had the capacity to count, to classify and understand causality which would have permitted a more complex social life that could have included a true kinship system and creation myths. It was these advances in intelligence which allowed hominids to migrate beyond Africa into temperate latitudes, to pursue specialist hunting techniques over large territories as part of an annual migratory cycle rather than opportunistic foraging in a small area, and to exploit large animals by cooperative hunting.

Has Wynn established his case? The connection between Piaget's theories and stone knapping is by no means secure in spite of meticulous analysis. While Wynn is probably right to conclude that later Acheulean hominids, with larger brains, had a more advanced mentality than the Oldowan tool-makers, the obvious but important point that technology does not necessarily reflect cognitive ability must be repeated. Hunting and gathering peoples have been found to use only a Mode 1 stone technology, equivalent to Oldowan tools, in societies which draw upon modern intellectual qualities that are seen in their kinship and religious institutions. Practical work by Toth (1985) shows that since Oldowan artifacts were made from cobbles or chunks of raw material, it would have been much more difficult to shape these stones into a symmetrical straight edged form. Furthermore, whenever later Acheulean hominids worked with small cobbles the Oldowan form usually resulted. Wynn has argued that the earliest technology was made haphazardly yet the Oldowan tradition does contain standardisation, in the sense that a number of distinct tool types were produced, while some post-Oldowan industries seem to lack design content. The form of some early tools was dictated by the physical properties of the stone used rather than by a drive to impose a standard form. This is seen in the quartz tools found at Omo and Olduvai Gorge in East Africa where small, irregular, quartz flake tools are contemporaneous with standardised bifacial choppers. It is just as significant to note that the artifacts in the Choukoutien caves, that are associated with 'Peking Man', *Homo erectus*, and are dated as 460 to 230 hundred thousand years old – that is after over two million years of stone tool-making – lack the imprint of standardisation. Oakley (1972) notes that these tools would never have been recognised as such but for their context and association; they are 'among the most primitive known . . . Peking Man was a regular and systematic tool-maker [but] he made little attempt at standardisation; in fact many of his implements were evidently of an occasional type' (Oakley 1972: 37). And yet this is also the first hominid known to use fire and perhaps also the first to engage in ritual cannibalism (Johnston 1982).

Gowlett's ideas, presented in the last chapter, are contradicted by Wynn whose main conclusion, that *H. habilis* and early *H. erectus* had mental abilities which could be bracketed with apes, should be challenged. The first stone tools emerged from a new adaptive pattern and an important shift in social behaviour is indicated. It seems unlikely that tools were merely an appendage to an ape-like way of life among the early hominids. Stone technology marked a new set of needs and experiences of whole hominid communities. Nor should the achievements of Oldowan hominids be underestimated on the grounds that apes can be taught to make stone tools. In Chapter 4 the case of an orangutan (Wright 1972) was mentioned as an example of primate learning abilities. Wright's task, successful though it was, proved to be laborious as the orangutan struck only a single flake from a flint core after hours of exposure to the instructor's behaviour. If Toth's experimental work is correct, the Oldowan tool-makers frequently used single flakes but these hominids also made more complex core tools. Gowlett's work shows that trial and error knapping could not have produced Oldowan tools. Successful tools were the result of the correct choice of raw material and hammer stone and the right striking force applied at the proper angle. These conditions could not have been met by chance and they imply a far longer operational chain than found among apes. Gowlett's flow chart of the Oldowan and Acheulean production processes, (Figure 7.1) emphasises this point. He also shows that the hand-axes from an Acheulean site that is more than 700 thousand years old, perhaps as much as 0.5 million years older than Wynn's tool sample material, show a consistent geometrically accurate sense of proportion. To have produced such consistency of design – the same relationship between length and breadth is maintained on hand-axes of very different sizes – argues strongly that an image must have been present in the maker's mind. *H. erectus* one million years ago possessed mental abilities which allowed the pattern of a tool to be projected onto raw stone.

It should be added that Lower Palaeolithic tools, at least by late Acheulean times, suggest the emergence of an aesthetic sense.

FIGURE 7.1   *The 'Procedural Templates' involved in making Oldowan Tools*
             (above) and Acheulean Tools (below)

The flow charts indicate the actual technical and conceptual complexities
involved in producing the apparently crude Mode 1 and 2 stone tool industries of
the Lower Palaeolithic 2.5 and 1.5 million years ago

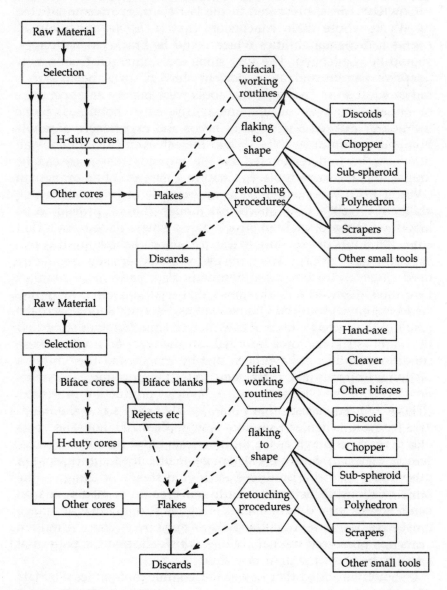

SOURCE: From Gowlett (1984A)

Although they still remained highly standardised objects, the design of later hand-axes sometimes includes a tendency to make the tool from special types of stone and to make its shape correspond with pleasing geometrical forms. Particular elliptical and oval shapes, implicit in the triangular pear-shape, seem to have been the aim of the tool-makers as much as were the functional properties. An exceptional hand-axe made 200 thousand years ago from Swanscombe, in Kent, contains a fossilised echinoid which the maker had kept in the centre of the tool by carefully removing surrounding flakes. The retention of this mollusc shell within the tool, and the existence of other hand-axes containing rare fossilised coral which could only have originated 120 miles away in Wiltshire, may be evidence of reflection and curiosity. One conjecture is that it represented the night sky (Oakley 1981). Another authority on stone tools confirms the idea that hominids '. . . of this time had a clear mental image of the object to be made before they set about making it' (Bordes 1968: 137).

Oldowan tools are undeniably simple and limited in variety but they nonetheless show the need for a mentality beyond that of apes. The very least that can be said is that the hominids of these times understood the need for careful selection of raw stone, that they had mastered the multi-step production routine their tools required, that they had internalised templates of tool form in three dimensions and that they were capable of conveying all of this to rising generations.

BRAIN EXPANSION AND SOCIAL BEHAVIOUR

These new mental abilities emerged concurrently with the other major episodes which comprise human evolution and they should be considered against significant changes in the size and organisation of the brain. This organ was evolving throughout the emergence of the hominids and its transformation is best understood within the context of a reciprocal feedback relationship in which social and biological factors acted as mutual stimuli for change. This meant that the growth of the brain was both the cause and result of new forms of behaviour and physical adaptation.

Variations in brain size within the human species have not been found to correlate with intelligence unless pathologically small brains are compared. Studies for races and social classes have failed to control for body size, and differences between the brain weights of the sexes become insignificant when they are scaled to body weight. But comparisons of brain mass between species, particularly between humans and mammals other than primates, show an important link between brain size and complex behaviour. Among other mammals the absolute brain weight of modern humans, which is 1444 gm for men and 1228 gm for women, is by no means impressive. The whale brain at 6800 gm weighs nearly five times that of humans, and the elephant brain at 4717 gm over three times as much (Steele Russell 1979). These animals, together with the dolphin which has 20 per cent more brain weight than humans, have large bodies and other organs which are correspondingly greater. When body and brain weight are placed in proportion, humans, with 1 part of brain weight to 44 parts of body weight, score over other large animals. For whales the brain to body weight ratio is 1 to 854 and for the elephant and dolphin 1 to 645 and 1 to 82 respectively. However this relationship is not constant since large animals have a much smaller brain to body ratio than small animals. This means that animals other than humans, all of whom are much smaller, have a more favourable ratio. We are in fact bettered by such creatures as the mouse, marmoset and squirrel monkey who have brain–body ratios of 1 to 38, 1 to 29 and 1 to 26 respectively, against our own 1 to 44 ratio. A test of true significance then is to examine the brain–body ratio of a given animal in relation to other species of similar weight. This approach is adopted in calculating an 'encephalization quotient' or EQ. Here the actual brain weight of different animal species is measured as a proportion of the size expected for an animal of equivalent body size. In EQ terms humans and other primates, with the exception of dolphins, predominate. The human brain is 7.4 times as large as that expected of a mammal of the same body size while the EQ for the chimpanzee and the dolphin are 2.4 and 5.3 (Steele Russell 1979).

It is important to note that while our brain is more than three times as large as that of a primate of the same build, primates as a whole, and the apes in particular, have a high EQ compared with other mammals of comparable bulk. This fact suggests that while a dramatic increase in brain size has occurred during the past 2 million years of hominid evolution, there is also a much longer-term, although less rapid, tendency toward brain expansion among larger primates as a whole. Environmental change during the past 5 million years enlarged primate brains for a greater learning capacity and flexibility of response. This tendency is considered by Lewin (1982) to strongly indicate that the common hominoid ancestor of both apes and hominids had a smaller brain than that of modern apes. The comparisons that have been made in the last chapter between the brain volumes of early australopithecines and chimpanzees is therefore not entirely fair. The trajectory which is presented by the tendency of hominid encephalization suggests that *A. africanus* probably had a considerably enhanced EQ compared with his hominoid ancestor. The early hominids were 'small brained' only in relation to later stages of human evolution but compared with the apes of the Miocene–Pliocene boundary they had advanced.

However it is the more recent dramatic changes in hominid brain size, a three to four-fold increase during 4 million years, which must be addressed here. Table 7.1 summarises these changes.

TABLE 7.1   *Brain Volumes of Hominid Species*

A. afarensis 310–400 ml (Holloway 1988)	4–3 million years ago
A. africanus 450 ml (Tobias 1981)	3–2.5 " " "
A. boisei 500–550 ml (Grine 1988A)	2 " " "
H. habilis 600–800 ml (Rightmire 1988)	2 " " "
H. erectus (early) 800–900 ml (Rightmire 1988A)	1.6 " " "
H. erectus (late) 1100 ml (Rightmire 1988A)	0.5 " " "
H. s.sapiens 1350–1400 ml	120–40 thousand years ago

Some increase in brain volume can be attributed to a larger body size but most of this expansion is real. Our brain is about 3.1 times the size expected for a primate of similar body size (Falk 1980).

This magnitude of expansion demands interpretation. In terms of human metabolism the brain is a very expensive organ which represents only 3 per cent of body weight but requires the continual use of over 20 per cent of blood supply (Holloway 1981, 1983). The degree of rapid encephalization seen in the hominid lineage could only occur therefore as a response to strong evolutionary pressures which made large brains adaptive. Clearly a large commitment has been made by natural selection to brain development which can be seen to have been under way in the first known hominids. *A. afarensis* had already advanced beyond the chimpanzee level in terms of size and in terms of cerebral reorganisation. Holloway's work on the brains of fossil hominids is accomplished through his technique of taking impressions or 'endocasts' of the outer surface of the brain encased by its lining tissue which enables volume to be determined and allows some understanding of the configuration taken by neural anatomy. It is of course not possible to infer types of behaviour from superficial features – Holloway is not engaging in phrenology – but a guide to brain evolution in general is provided. His examination of the endocast of *A. afarensis* shows that the pattern of brain organisation typical of chimpanzees was not present and that by 3.5 million years ago the hominid cerebral cortex had significantly changed in relation to the apes. From the *Australopithecus* stage a human pattern of brain organisation is evident which is most clearly manifested as cerebral asymmetry. This feature is strongly correlated with right-handedness in modern humans and it is of great importance that it should appear at the same time as stone tools and a new way of life. If these observations are correct it can be established that brain reorganisation occurred before the stage of rapid expansion. These findings have been challenged with regard to the australopithecines but there is a consensus favouring human brain form after the appearance of *H. habilis*.

What could have initiated this process? Lying behind brain expansion were a cluster of new social practices and behavioural adaptations that were linked together with biological mechanisms which would accommodate phenomenal growth. Ideas of a 'prime mover' or an 'initial kick' which could have begun to reorganise and enlarge the brain are found in profusion since the nineteenth century. Falk (1980) has surveyed hypotheses which revolve around the themes of warfare, language acquisition, tools and labour, hunting and heat stress. However it is Holloway's ideas which comprise a comprehensive evolutionary theory based upon social behaviour which will be given scope here.

It is in the area of social behaviour, he argues, that the most significant evolutionary change has occurred, for social life has acted both as the driving force of physical transformation and has itself been changed by new attributes exclusive to humans and their ancestors. Irrespective of the continuity which joins us to other primates we are set apart from animals by virtue of our overwhelming predisposition to use culture symbols in all facets of life. It is through our use of arbitrary symbols that we impose structure on the environment. Our symbol systems, seen at work in language and in all our social institutions, designate a meaning and purpose which act to integrate what would otherwise be atomised individuals into a social whole. Cultural life implies the transformation of environments in accord with our symbol systems, and the creation of an environment that is dominated by cultural practices to which, in turn, we are obliged to make further accommodation.

Cultural complexity, seen in more sophisticated tool-making, and brain enlargement can be seen to correlate positively but this is not a simple causal relationship. Stone tool-making is but one manifestation of an emerging set of perceptual, cognitive and motor abilities which all stem from a new adaptive matrix of social behaviour. Tools should therefore be seen only as clues to this behaviour. He continues:

> In other words, it was not the tools themselves that were the key factors in successful evolutionary coping. Rather, the associated social, behavioural and cultural processes, directing such activities as tool-making, hunting and gathering were basic.
>
> (Holloway 1981: 288)

In this theory the growth of the brain is tied to whole socio-cultural processes which have given rise to new behavioural forms that are witnessed, at the outset of this evolutionary phase, as tools made to a preconceived or standard pattern. This behavioural matrix, which was present before the dramatic stage of encephalisation, was a complex and growing set of elements which relate to social interaction. Natural selection came to favour this collection of social traits, and working through the endocrine system has come to produce larger bodies and brains.

For Holloway, behaviour is the link between physical structure and natural selection. With the success of the new pattern of social behaviour a dialectical process of positive feedback operated in which all elements involved in the evolutionary system interacted to produce directional change. Brain enlargement and increased socio-cultural complexity interacted to confirm the validity of the evolving complex of elements. Each variable was itself altered by the process of positive feedback, and the direction of change continued until constraints, such as the size of the female pelvis, limited foetal brain growth. The early hominids began a process of environmental transformation with stone tools in social groups, with the 'supporting apparatus of social rules and symbolic codes' (ibid. p. 294) which had the effect of inducing further technical proficiency and social change. The environment thereby expanded for hominids in terms of their awareness of its greater potential and its increased complexity.

It was the whole collectivity of variables linked together in a dialectical relationship, rather than a single variable in a cause–effect relationship, which was the source of the increase in cranial capacity.

> The tools were used in different environments, and cooperative social behaviour was also basic in formulating and facing the rigours of a hunting and gathering existence. Hunting is a complex organisation and sequence of acts, requiring not only perceptual and motor skills, but intelligent learning of the surrounding plants and animals, terrains, spoors, tracks, habits of prey, seasonal effects on these, anatomy, butchering techniques and perhaps storage.
>
> *It is the total range of cultural activities that must relate to brain increase, and the complexity of stone tools relates only partly to that whole.* To the extent that the hunting of large animals involved cooperative enterprise, selection would certainly have favoured structures which facilitated the increasing symbolisation of language, and this would have meant an increasing complexity of social interaction, involving cues from social and material environments along with control and inhibition of immediate responses . . . Another way of putting this is that tools do not 'makyth' man, but symbolic communication does, and men 'makyth' tools only in the context of their ever increasing awareness of their environment, a matter dependent upon symbol systems, social organisation and brains.
>
> (Holloway 1981: 296/301, original emphasis)

A 1000 ml increase in brain size within 2 million years – a spectacular evolutionary leap that rivals any episode in natural history – occurred with a major expansion of the cerebral cortex. This involved a change in the size and density of neurons, a more complex branching pattern between cells and an increase in the basic number of cells. Again Holloway takes social behaviour patterns as the 'initial kick', which he comments:

> . . . can be defined as a transition made to a type of social ordering based on different components of aggressive control within small groups. This type of social ordering would have

included the following phenomena: sexual division of labour in the food quest, cooperative sharing between and among sexes and social nurturing of offspring; decreased dimorphism in so-called epigamic structure (related to sexual signalling) such as permanently enlarged breast, fat and hair distribution; raised threshold to aggression within primary groups; permanent sexual receptivity of the female and male; and a new way of transferring information about the environment, through language and gesture ... the 'initial kick' was a complex reticulation of anatomical, physiological and behavioural changes ...   (Holloway 1981: 298/300)

One might wish to challenge Holloway on a number of points. Is he really correct in claiming that cultural behaviour is restricted to hominids? Surely no final distinction between cultural and non-cultural tool-making may be drawn. At least some capacity for culture, in the sense of arbitrary symbolling, exists among our primate relatives. It has already been argued that standardisation of tools is no real guide to cognitive ability and one would like more than circumstantial evidence for the use of language in this stage of human evolution, even though the endocasts do not rule out language in these times. Were the Oldowan hominids true hunter-gatherers? This problem will be dealt with in a moment. However Holloway has provided a comprehensive and holistic theory in which he properly tries to combine all elements of the evolutionary process within a dialectical schema. His theory is therefore of great value and must be preferred to more partial reductionist explanations.

ARCHAEOLOGY AND HOME BASES

A comprehensive reconstruction of hominid social life, drawn from the evidence found at East African sites, has been provided by a number of archaeologists among whom the late Glynn Isaac figures largest. In what remains of this chapter Isaac's model of early hominid social behaviour will be reviewed.

At a number of sites, including Olduvai Gorge, substantial cultural material has been found which is associated with the remains of *A. boisei*, *H. habilis* and *H. erectus*. A full sequence of cultural phases has been established in the Oldowan strata which span a 1.5 million year period. Deposited in the lowest strata at this site, designated as Bed I, dated as 1.83 million years old, are the remnants of a living-floor used by hominids. The material here, as in other sites of similar age, documents a new form of adaptation comprising a collection of unique behavioural patterns that were directed by culture. The juxtaposition of animal bones and stone tools within a localised area can be taken as direct and inferred evidence of the use of camp-sites or a 'home base', food-processing with tools, a growing reliance upon meat, and perhaps food-sharing and a division of labour. These characteristics suggest that a rudimentary mode of proto-human social organisation based upon gathering, including scavenging, and some hunting was being practised by hominids and that a major behavioural shift away from the adaptive pattern of other primates had already occurred two million years ago.

Reconstruction of the ancient environment indicates that hominids had a propensity to establish camp-sites close to water. Such sites were located near lake shores or in river channels among trees which provided shade and may possibly also have served as the dormitory area of the hominid band as well as its means of retreat. The validity of claiming that these locations were in fact camp-sites rests upon the pattern and concentration of artifacts and bones. The accumulation of food litter and stone tools in an area with a definite boundary occurred as a result of deliberate transportation and prolonged occupation of sites. Although sites were probably frequently revisited, and the evidence of occupation deposited intermittently, each sojourn must have been for longer than those of non-human primates since even the most elementary cultural activity practised by hominids precludes the perpetual nomadism seen in ground-living apes and monkeys.

The scatter pattern of material across sites shows a configuration that can only be associated with a hominid living area since there are clear outer limits and areas of specialised activity devoted to stone knapping and bone breaking. Analysis of the density and distribution of bone refuse and stone artifacts led Isaac (1971) to suggest that some sites may be differentiated according to their primary functions. Varying ratios and quantities of archaeological material indicate that a site may have acted as a quarry or workshop, as a transitory camp, a butchery site or as a camp-site. Mary Leakey (1976, 1978) has shown that the Oldowan living-floor in Bed I contained a low pile of stones which demarcated a semi-circle of ground which was construed as a wind break. The existence of this stone construction also makes the use of thorn or brush fences around living areas seem to be a likely feature of the first camp-sites. Evidence for this is inferred by Leakey from the discovery at another Oldowan site of a narrow barren zone devoid of artifacts that enclosed a living-floor beyond which were only sporadic clusters of objects that had perhaps been ejected.

The discovery of hominid camp-sites of this age is of considerable importance. What is implied here is a radical change from the behaviour of all contemporary non-human primates whose social groups feed as they range. In establishing a home base a primary social innovation was initiated, since this spatial focus allowed members of a foraging band to move about independently, pursue different tasks, and rejoin their compatriots. The home base in this sense attests to what may be the earliest known form of division of labour and it is indeed this behavioural transformation that was responsible for accumulating what is now archaeological evidence.

Even at such a primitive technological level several momentous strides away from the primate condition had been achieved. The hominid subsistence pattern was one which exploited several ecological niches and as such was a behavioural adaptation which conferred distinct advantages. Tooth wear patterns from fossil hominids unequivocally point to omnivorous foraging which was an activity pursued in dispersed and distinctively different localities. By gathering both vegetable and animal foods with the aid of tools and by postponing consumption until this food was transported to a home base, several novel possibilities that are denied to other primates became available to hominids. Separate sub-groups of the hominid band, perhaps differentiated by sex or age, could engage in specialised forms of foraging. The products of gathering, rudimentary hunting of small animals and scavenging from the carcasses of larger animals could be combined. This departure from the singular mode of food collection practised by a primate hoard would have resulted in a considerable enhancement of food supply. For instance, deep-growing tubers and meat of large animals – a unique combination – became available only to hominids. A fixed point of reference given by a home base allowed unproductive members, particularly the young and infirm, to remain as part of the band. The contrast here with baboon societies where injured members are abandoned during the incessant movements of the troop is marked.

For fully bipedal hominids 2 million years ago, confinement and childbirth would have posed a greater obstetrical risk than that experienced by other primates even though the major advances in brain enlargement lay in the future. Lovejoy (1988) has shown that for 'Lucy' (*A. afarensis*) birth would have been more difficult than for a chimpanzee. Lucy's pelvis had narrowed, to accommodate adaptation for bipedalism, and her infant's head would pass through the birth canal only if it first turned sideways and then tilted. To be sure, Lucy would not have faced the same anatomical difficulties encountered by large brained humans, but the need for assistance while giving birth cannot be ruled out. In this sense a home base was as much a part of the essential conditions for the existence of hominids as it was a source of efficiency and a route to reproductive success.

Stone tools, the physical manifestation of new forms of social behaviour, which caused the archaeological record to start forming, are in themselves also evidence of a growth in cultural sophistication. In surveying progressive technical change during the whole Palaeolithic era, one criterion employed has been economy of raw material. Leroi-Gourhan (quoted in Holloway 1981) has estimated that from a single kilogram of unworked rock, 40 centimetres of cutting edge could have been achieved in making an Oldowan chopper. However, 60 to 120 centimetres of cutting edge could have been gained by producing an Acheulean hand-axe and as much as 400 centimetres created by the makers of the Mode 3 flake tools typical of Middle Palaeolithic times. In the Upper Palaeolithic the same kilogram of rock might yield over ten metres of cutting edge. But evidence of change is apparent even in the opening phases of the stone age. Mary Leakey has contrasted the artifacts of Bed I at Olduvai Gorge with those of Bed II, that lies immediately above, and which began to form 1.6 million years ago. In Bed I, bone remains are predominantly of small mammals, birds and fish. However the repeated discovery of the remains of large animals who seem to have died under identical conditions in Bed II, in this case antelopes with depressed skull fractures – by stones that had been thrown perhaps – suggests that hominids were at this stage exploiting a food source that was rarely available to their forebears. This substantial increase in the presence of large animals in Bed II is complemented by an advance in stone tool technology which, it is reasonable to conjecture, was itself a product of more complex social organisation. The Oldowan chopper accounts for more than three-quarters of all tools found in Bed I but only for about one-third of the contents of Bed II. A more differentiated and versatile tool-kit is present by Bed II times. In fact an increase by more than one-third in the number of basic tool types is found in contrasting the two cultural levels. In place of the ubiquitous chopper comes the configuration of tools called 'Developed Oldowan' which gives equal prominence to scrapers and stone spheres. This latter tool-type, which is an accurately and carefully shaped stone ball often weighing less than four ounces, is suggested by Leakey to have been used as bolas weights or as missiles for hunting.

More recently Barbara Isaac (1987) has argued that accurate and forceful throwing is a well developed human skill that has been put to good use by modern hunting peoples but is poorly developed in the apes. The hands of *A. afarensis* show that these abilities were present in pre-Oldowan times. Some early hominid sites contain large quantities of unmodified stones that are not suitable for tool-making but could have been used as missiles. The energy investment implicit in these accumulations could perhaps be due to the value of throwing stones in both foraging and defence. Certainly the coordinated stone throwing of a band of hominids would have presented a formidable defence against carnivores in the pre-tool era. It is even possible that this activity improved competition for carcasses and that it enabled hominids to cross a threshold between scavenging and hunting.

Glynn Isaac has suggested that although the Oldowan tradition appears to 'represent opportunistic, least-effort solutions to the problem of obtaining sharp edges from stone' (Isaac 1981: 184), experimental use of such tools shows that they are quite adequate to perform all of the basic functions of a human way of life that are recounted by ethnographic studies of modern hunting and gathering peoples. Oldowan tools are sufficient to cut and sharpen wood for spears or digging sticks, or to form containers and carrying devices such as bark trays and even to dismember very large animals such as elephants and hippos. Ignorance remains concerning the precise mode in which different Oldowan tool types were employed. However micro-wear analysis of tools provides evidence that a whole variety of applications was fully operating among the Oldowan hominids which represented a diversified subsistence pattern based upon cultural use of a range of materials.

The presence of a rising level of technical competence in Oldowan tools indicates that hominids were becoming dependent upon a material culture and that tool-making had become a habitual activity. It is as yet unclear if this progress was the result of gradual cumulative development or whether it comprised a series of innovatory great leaps which occurred after technical thresholds had been crossed. However Isaac (1976B) considers the former hypothesis highly improbable. It seems more likely that a number of crucial technical problems were solved and that thereafter tool-making came to explore new possibilities that had been created.

Another tendency revealed by the East African sites is for the density of artifacts to increase during the first stages of the Palaeolithic era. This evidence suggests that an intensification of the new adaptive strategy was being evolved, which implies that camp-sites were occupied for longer periods, and that better stone technology was complemented by improved carrying devices. The greater quantities of tools and bone refuse on sites argues for the use of baskets and bags to transport food and stone.

What significant changes in the proto-human hominid's social behaviour and adaptive strategies are indicated by the presence of fossil bone material? The obvious response to this question should be resisted. There is a body of opinion that reaches back to Darwin which holds that hunting constituted the formative element in human evolution and that essential social organisation was constructed to accord with the demands of this activity. But this position overlooks the importance of an interdependent cluster of factors that were responsible for the formation of human society within which may be placed only the most rudimentary form of hunting. The hunting hypothesis is misleading because it ignores the fact that human hunting patterns were the outcome, and not the primal cause, of physical evolution and cultural development. We should ask about the role of meat as a new element of diet and as one factor within a model of pre-human physical and behavioural evolution.

There is considerable evidence for carnivorous behaviour by hominids. The 2 million year old bone refuse can only represent the residue of meat-eating. Even though the amount of animal protein consumed by hominids remains unknown it is inconceivable that it was anything but a lesser part of a diet that was dominated by gathered vegetable foods. In this sense meat taken from the carcasses of dead animals was an addition, albeit an important and significant one, to an older gathering subsistence pattern. But the regular consumption of meat by hominids, regardless of the quantity involved, was an elaboration of this pattern since it marked the beginnings of an adaptive shift – an accommodation to an available niche – which was to remove hominids yet further from the condition of other primates.

It has already been mentioned that an increase in the amount of bone material and the size of the animals represented occurs between Beds I and II at Olduvai Gorge. By Bed II times, hominids were accumulating bone refuse from animals far larger than themselves: antelope, horse, hippo and even elephant remains are found along with a far greater quantity of deliberately broken bones. These findings point to a collection of practices without precedent among primates. Foraging expeditions that sought out dead or infirm animals, butchery of the carcass and transport of the meat to a camp-site had clearly become a part of a novel way of life. Hominids were becoming accustomed to a highly diverse diet that comprised an ever broader range of vegetable and animal foods. Isaac (1980) has likened this strategy to a widely cast subsistence net in which social cooperation and tools allowed access to a rich omnivorous diet, the variety of which was itself a source of security.

There is now important archaeological evidence that supports these contentions. Potts and Shipman (1981) have established that bones found at a number of sites, that are at least 1.8 million years old, bear cut-marks produced by stone knives. They have shown as well that such bones also include the teeth marks of carnivores with whom hominids were in direct competition for animal carcasses. Bunn (1981) has found specific evidence of butchery practices including the removal of skin from bone and dismemberment of limbs. In many instances it has been possible to reassemble bone fragments that had been shattered by hominids in order to retrieve marrow. Hammer stone marks, which could only have been caused by hominids, are found on these bones. A butchery site in Kenya excavated by Isaac (1978) provides powerful evidence for accepting the thesis of hominid meat-eating. The site, which dates from Bed II times and is at least 1.6 million years old, contained the bones of a hippo together with over 100 stone artifacts including hammer stones used to strike flakes to make core tools. The animal originally lay in a stream bed. The conclusion that hominids had transported stone to this site, had manufactured tools and butchered the hippo would seem to be inescapable. Evidence concerning the importance of vegetable food in the hominid diet is not given support by archaeology. However since the technology of hominids in early Palaeolithic times was extremely crude compared to that of modern hunting peoples – whose diet in tropical areas never includes more than 40 per cent meat – it is inconceivable that the ancestors of modern humans could have gained similar levels of animal protein.

CRITICS OF THE HOME BASE MODEL

How valid is this model? Were the Oldowan hominids beginning to sketch out a human-like adaptive pattern? Considerable controversy has arisen here. It has been objected that Isaac's view of early hominid behaviour is anthropomorphic, that the data does not support the home base thesis and that key elements of social behaviour, such as food-sharing or a division of labour, cannot be inferred from the archaeological evidence.

Isaac's strongest critic is Lewis Binford who, in his book, *Bones, Ancient Men and Modern Myths*, complains that an unhealthy, fable-generating tendency has long existed within archaeology that, in this instance, casts the early hominids as modern hunter-gatherer peoples. He asserts that hominid social life seen as one of organised bands engaged in hunting and food-sharing is merely fashionable speculation which assumes that Palaeolithic societies had behavioural patterns that were midway between humans and great apes. Isaac is guilty of postulating a list of behavioural contrasts here. The listing of universal human traits such as tool-use, meat-eating, division of labour, camp-sites, female–male pair bonding and social rules which then give rise to questions about when, and under what circumstances, this behaviour arose is unacceptable. 'This model is a speculation as to what the past was like. It is conjectural history and involves no theoretical arguments that I can recognise' (Binford 1981: 250). Isaac is merely interpreting the past in terms of the present and there is no validity in postulating what cannot be substantiated by archae-ological evidence. No real proof of 'base camps' established by hominids is provided by the East African evidence; the bone and stone accumulations could have resulted from other causes which Isaac has failed to investigate and it is only 'wishful thinking' which allows them to be interpreted as living-floors. Nor does Binford see any necessary association between the bone remains and stone tools: they are in fact a series of palimpsests or multiple overlays that were formed because of separate activities. Hominids were not hunters, they merely scavenged marrow bones from kill sites, and other scavenging animals also contributed to the bone accumulations.

A more serious critique of the home base thesis, which draws directly on the archaeological evidence and offers a modified alternative explanation, is provided by Potts (1984). An ideological bias is detected in Isaac's work which has prevented him from distinguishing between the antecedents of hunter-gatherer behaviour, which are present, Potts argues, and a fully human economy. The discovery of apparent home bases formed by early *Homo* implies a long-term continuity with modern humans which suggests that our evolutionary past has programmed us for a natural condition. It confirms ideas that culture, technology and meat-eating were the principle reasons for hominid ascendancy and it suggests that the modern human pattern of adaptation, which stresses social institutions for learning, exchange and reciprocity had commenced in the early Palaeolithic era.

Unlike Binford, Potts accepts that the Oldowan sites were formed by the activities of hominids, but it is also evident from tooth marks that carnivores were competing with hominids. Bones exhibit marks which would have been caused as hominids rapidly defleshed selected parts of animals having abandoned considerable amounts of meat and marrow. Animals seem not to have been completely processed in the modern sense of skinning and butchery for full utilisation of a carcass, and therefore inferences that the sites were a focus of social activity are invalid.

The accumulation of tools were not formed at home base areas; rather they were carried and left at prearranged localities for later use. Potts proposes that the East African sites are in fact a series of stone tool caches that are strategically situated across the landscape in a pattern which represents a least-effort, optimum expenditure of energy for carrying stone and meat. Hominids would have carried parts of animals, gained by hunting or scavenging, to the nearest tool cache, thereby reducing confrontation with carnivores. A unique and novel mode of foraging was in use here but this is a pattern of behaviour that has similarities with the nut-cracking habits of West African chimpanzees who leave hammer stones in strategic positions. These sites then are thought to have preceded true home base camps and their formation represents behaviour which is antecedent to contemporary hunting peoples.

Finally, Shipman (1986) has examined the cut-marks on the fossilised bone of large herbivorous animals in Bed I at Olduvai. She has attempted to distinguish between hunting activities like those of modern hunting peoples, which would show evidence of disjointing, skinning and carnivore tooth marks made after butchery when the bones had been discarded, and scavenging which would not show this form of carcass treatment. Her results pointed to a pattern of cut-marks that are most likely to have originated in scavenging rather than hunting. Oldowan hominids were well adapted for scavenging. Efficient bipedal walking, a high line of sight for locating carcasses, an arboreal adaptation which provided a retreat and diminished competition from other carnivores, plus fruit as an alternative food source, provided a good basis for scavenging. Shipman then argues that scavenging, as a mode of foraging, must cast doubt on the idea that archaeological sites represent home bases where food was brought and shared. Without fire, domesticated dogs or good projectile weapons hominids would have been at great risk living and sleeping close to carcasses. Like Potts she proposes that the sites were in fact either safe areas such as trees around which stone tools were kept for repeated use and where hominids retreated separately to eat scavenged meat, or that they were tool base areas to which bones and meat were brought for processing.

Isaac's response to these critics was to elaborate, revise and rigorously test his model. However the basic outline of features which comprised the new hominid behaviour patterns, that revolve around what he came to refer to as 'central place foraging', remain intact and are reproduced here as Figure 7.2 from Isaac (1983A). Hominids were more likely to have been scavengers than hunters, but with stone tools they held an advantage over all other carnivores in having access to the carcasses of tough skinned pachyderms, such as elephants and hippos, which would have supplied prodigious quantities of meat (Isaac 1984).

Binford's charge of anthropomorphism has had effect in stimulating a re-examination of the causes of site formation. An earlier climate of opinion had unequivocally accepted a close link between hominids and modern hunters as unproblematic. In Gowlett (1984: 50), for instance, an illustration reconstructing a hominid hut appeared. This was the famous Oldowan 'wind break' of basalt blocks – the earliest known shelter – that was thought to resemble the foundation stones of temporary structures

FIGURE 7.2   *A Possible Configuration of the Social Behaviour of Tool-Making Hominids Two Million Years Ago*

The diagram indicates the potential relationship between technical performance and conceivable forms of social conduct which gave support to tool-making. Behavioural elements that are firmly backed by archaeological evidence are shown in italicised capitals, and are linked by solid lines. Other hypothesised but likely interrelated forms of behaviour are in lower case and joined by broken lines

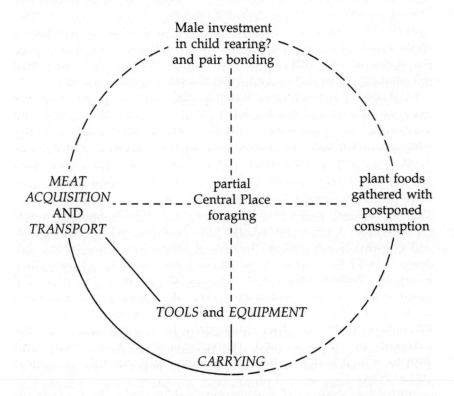

This system is envisaged as set in a matrix of *BIPEDALISM, a partially ENLARGED BRAIN,* some prolongation of infant dependency and some enhanced communicative abilities, but lacking full language abilities.

SOURCE: Adapted from Isaac (1983A)

erected by contemporary nomads. Isaac has now questioned this feature as an artifact. Potts' (1984) explanation of the stone circle is that it was produced by the radial distribution of tree roots which are known to lift and break rock. The matter has not been resolved and remains as an area of dispute (Toth and Schick 1986). Isaac looked into the possibility that the stone/bone accumulations occurred by natural causes. Site formation by water, a 'hydraulic jumble' of bones and artifacts can be ruled out. Sediment size shows no evidence of fast-moving streams; tools and marrow bone pieces can at times be reassembled and a full diversity of skeletal parts are present showing no instance of the pattern of differential sorting by water. It is also possible to eliminate carnivores as the agents responsible. Cut-marks and hammer fractures on bones show that hominids carried parts of carcasses to central locations for butchery. Whatever the contribution made by carnivores, hominids were mainly responsible for these accumulations.

Evidence of cut-marks on fossil bones, seen in studies with an electron microscope, has forced Binford to ameliorate the force of his attack; however, the controversy remains. Isaac and his colleagues still hold to a view of small mobile hominid groups making tools that were used to dismember carcasses which they carried to 'central places'. These may also have been camps and were probably social foci situated in groves of trees close to water. The social and cultural implications for this behaviour were considerable. A premium would have been placed upon learning and cultural transmission. Prolonged infant and juvenile dependence would have been duly selected for, and the larger brains which this implied had the consequence of helpless young and females being more vulnerable during pregnancy (Isaac 1986).

The question of the source of hominid meat, whether from hunting or scavenging, remains open. But the fact of scavenging does not unduly damage the argument that this activity had powerful social consequences. Hominids had adopted a new adaptation which entailed the use of the first material culture and they had also encroached on the niches of carnivores. It would be absurd to imagine that either this new technology or behaviour had no social support or context in which learning, rules, agreements and reciprocal understandings underlay a tool-based way of life or that the first tool-makers had not generated cultural practices which regulated their activities. Movement between caches of tools with scavenged meat was dangerous for small, slow-moving hominids and meant competing with carnivores. This alone would have been sufficient reason to engender social cooperation and rules for stockpiling, use and reward. Blumenschine and Cavallo (1992) argue that scavenging fits Isaac's model of cooperative foraging. Hominids could well have exploited scavenging opportunities by taking meat and soft tissue within bone cavities that was left after lions, leopards and sabre-toothed cats had eaten. Hominids would have effectively increased their food supply during the critical dry season when tool-aided scavenging for marrow and brains would have provided far more calories than were currently available from plants. Many human social skills such as planning, cooperation, a division of labour and accurate mental mapping of a whole foraging range would have been selected for as hominids came to rely upon a combination of technology and organization to gain food from a broad range of habitats.

Whether sites were formed in the manner Potts suggests, by the tree-feeding proposal of Shipman, or as in Isaac's original home base, food-sharing hypothesis, new forms of social behaviour are as much in evidence as the technology they sprang from. But we should not see the early hominids as diminutive humans. *H. habilis* and *H. erectus* were perhaps not on a linear trajectory leading from the apes to modern humans. They may have produced telling patterns of archaeological evidence through forms of behaviour unknown among any primate including modern peoples. But as Isaac concluded in his last publication, unhuman as they were these hominids had begun to engage in some forms of behaviour which are characteristic of humans alone. He continues:

> What is important about the archaeological evidence for tool-making, for meat-eating and perhaps for central-place foraging is not that these traits made the early enactors human, but that the traits helped to establish a situation in which individuals were exposed to natural selection patterns which transformed their descendants, over two million years, into the complex human beings that we now are.   (Isaac 1986: 238)

A NEW ADAPTIVE THRESHOLD

How can the evidence presented so far be used to provide a model of the origins of human society? Let us summarise what has been argued here and in the last chapter. At least 3 to 2 million years ago bipedal hominids were making stone tools, and by 2 million years ago they were living partly on meat and organising their foraging patterns around a home base or a series of central places. In many ways these hominids can be considered to have taken a path that led to a human way of life. The value of a broadly based, flexible subsistence pattern directed by culture had been confirmed by an enriched food supply. This economic success was due as much to a change in behaviour as it was to the habitual use of tools. And in turn such changes induced formative selection pressure for further change, in particular 'for social mechanisms such as emotional control, cooperation, reciprocal relations and communication' (Isaac 1976B). Much of the morphology of these hominids, including the post-cranial skeleton and dental pattern, was already distinctly sapient and the hominid brain, which at this stage ranged in size between 600 and 800 ml, had a human pattern of organisation.

A significant adaptive shift had occurred but in no sense can these hominids be described as human. A proto-human adaptive pattern which was to serve as a precondition for the form of behaviour of modern peoples had been established. But as we shall find in the next chapter the early hominids were not gatherers and hunters in the modern sense since they lacked complex tools and the large brain necessary for such pursuits. The absence of other human practices, particularly art and ritual, must also be assumed. There is no evidence of family organisation beyond the likelihood that a division of labour and food-sharing may have provided a matrix for such relations. Perhaps the rudimentary pair-bonds between the sexes are part of the inheritance of the hominids: for a bipedal female and her offspring, birth and the post-natal period posed acute hazards making support and protection a necessity. There are, of course, a variety of possible solutions to this dilemma. However kinship, together with the other essential elements of human status, is a human invention which arose during a prolonged interaction between biological and cultural systems. The rise of kinship was not a process in which novel behaviour patterns were merely superimposed upon a life style now exemplified by the apes. There was certainly an expansion of traits which are at least echoed by other primates, but the emergence of the defining characteristics which are basic to the human condition represents the development of dimensions unknown in other species. In the following chapters the evolutionary conditions for this transition will be examined.

8

A Foraging Economy

Any reconstruction of the economy and society of the early hominids is fraught with problems, not least of which is the absence of surviving representatives of these species. Evidence from both animal studies and anthropology can be helpful here but cannot be used as more than an analogy for the social behaviour of human ancestors. While chimpanzee social life may apparently suggest significant parallels with proto-human society, it must not be forgotten that, despite their genetic closeness, chimpanzees have a separate evolutionary history of at least 5 million years and have successfully adapted to a forest rather than savanna ecology. Modern hunter-gatherers are anatomically modern people whose economies have been developed in the face of cultural, environmental and historical pressures that are quite different from those encountered by the first species of *Homo* (Leacock 1978). Such societies, and those of hunters in the recent past, cannot be assumed to represent an ancestral state. At best they may approximate only to the economies of the final phase of prehistory. Neither model then provides more than an oblique access to the world of early *Homo*.

Yet to reject these models because of their limitations is to simultaneously reject evidence of evolutionary and social continuities. Because little is known directly of early hominid social life it is legitimate to use closely related species and the foraging behaviour of modern hunters as a basis from which the range of options available to *Homo* 2 million years ago may be examined. Hominids at this time faced new adaptive opportunities that could be exploited by developing an appropriate social life and it is reasonable to suppose that they drew upon at least some of the behavioural repertoire observed in apes. In a recent study Zihlman (1989) suggests that the pygmy chimpanzee is a useful role model for understanding the emergence of new forms of social behaviour among hominids, including the pattern of foraging, food-sharing, social relationships and communication. Some modern hunter-gatherers have occupied ecological zones of Africa similar to those of early *Homo* and, if hunting tools are excluded, have used very simple forms of technology in their gathering practices. Lee's description (1979) of gathering and carrying tools – a digging stick, carrying skin and leather bag – would surely have been possible for hominids with Acheulean tools. Perhaps, too, a certain social logic that included sharing and band formation was engendered by the new pattern of subsistence.

But what was the nature of this subsistence? In this chapter and the one that follows we will be concerned with the evolution of hunting among hominids. Previous assumptions that hunting was of the same strategic importance in early prehistory as it was to become among sapient peoples living in extant societies have been effectively challenged. In place of the earlier view, in which the pursuit of large game was seen as a dominant activity, has come a less dramatic picture of more generalised foraging from a range of food resources. Contemporary hunter-gatherers have forms of social and economic organisation that vary starkly from those of pre-human hominids. This distinction must be respected yet it is also worth inquiring whether the occupation of foraging carries with it certain ecological imperatives that tend to induce similar forms of economic and social behaviour.

HUNTING MAKETH MAN?

Until the 1970s hunting was assumed to have been the basis of the hominid economy at least by *Homo erectus* times. It was also assumed that hunting, and its associated activities, constituted the single most formative selection pressure encountered by hominids. Proto-human hominids came to the threshold of human status because hunting was seen as the fundamental matrix for both human nature and society. The Man the Hunter image, projected by an eponymous conference in 1966, and subsequent publication (Lee and DeVore 1968) typified hunting as the touchstone which initiated the human evolutionary journey.

Washburn and C. S. Lancaster, for instance, argued that 'In a very real sense our intellect, interests, emotions and basic social life – all are evolutionary products of the success of the hunting adaptation' (Washburn and C. S. Lancaster 1968: 293). The persistence of hunting as a way of life during 99 per cent of the human and pre-human past is taken in itself as the determining criterion of characteristics fundamental to mankind. Killing large animals with spear and axe required a high degree of coordination and cooperation among males. Hunting was the stimulus to increased technical skills and to a wider knowledge of botany and zoology. Similarly sharing, planning and a division of labour are seen as products of hunting.

Even the human physique, for males at least, has been conditioned by selection for hunting efficiency. Washburn and C. S. Lancaster follow Brues (1959) here, who argues that the body build of hunters will become optimised for the use of certain weapons. Hominid evolution, like that of other carnivorous animals, will be guided by success in gaining meat. Body build, or rather the length and size of limbs and the configuration of crucial muscles, will constitute the critical factor in the use of hunting technology. Bludgeoning prey with a club would favour a large body size, but with the invention of the first projectile weapon, the spear, a premium would be placed upon leverage to achieve maximum velocity and penetration. At this technical stage of prehistory, therefore, a long build or ectomorphic physique would have been favoured. This, incidentally, is a body build denied to some hominids in late prehistory such as Neanderthals and Eskimos whose bodies display a rotund endomorphic tendency because of a fundamental necessity for adaptation to extreme cold. The body mechanics which are most compatible with the use of the bow, however, are quite distinct from those of linear spear-using hunters. An optimum body build for the archer is one with short, thick muscles, the mesomorphic body type, and it is no coincidence, argues Brues, that this physique is synonymous with the western aesthetic view of the male. This degree of specialisation, which is still manifested as a diversity of body builds among modern racial groups, is apparent in the choice of hunting weapons used in the recent past. Most African peoples still reflect selection for spears while East Asian peoples have adapted for use of the bow. In this thesis, then, the ability of hominids to attack prey species constitutes a major driving force of natural selection.

Washburn and C. S. Lancaster's elaboration of the 'hunting maketh man' theme and its creation of psychological characteristics requires close attention. Besides arguing that the range of a hunting band implied continual territorial exploration which led, at length, to the conquest of the Arctic, Australia and the Americas, and that hunting was the necessity behind both social and technical invention, there is also the insistence that modern appetites for aggression may stem from our past as hunters. 'The extent to which the biological bases for killing have been incorporated into psychology may be measured by the ease with which boys can be interested in hunting, fishing, fighting and games of war. It is not that these behaviours are inevitable, but they are easily learned, satisfying, and have been socially rewarded in most cultures' (Washburn and C.S. Lancaster 1968: 300).

Washburn and C.S. Lancaster's ideas are not laden with the fatalistic form of sociobiological determinism that is characteristic of Ardrey (1976) which will be considered in Chapter 9. Their presentation of prehistory as a feedback process in which cultural developments and hominid evolution are mutually interactive should be endorsed. But their focus upon hunting alone leads Washburn and C.S. Lancaster to a narrow conception of the human adaptive pattern that is focused on a single activity and to misconceive its results. Their work does not differentiate between the stages which led to modern hunting techniques nor does it provide a discussion of chronological development. Nor is the particular variety of hunting that was supposedly most formative given attention. The subsistence pattern of prehuman hominids during the past 2 million years cannot be taken as an undifferentiated whole. What requires attention is the entire changing configuration of social and economic behaviour pursued by both sexes which served to progressively increase the importance of culture and initiate the beginnings of human society, rather than consideration of one specialised activity which, in the form recorded by ethnography, almost certainly emerged in late prehistory when early modern hominids had established a variety of adaptive forms that included the hunting of large animals. The legacy of this process in which the social and material forms of behaviour engaged in by hominids were beginning to change the terms of natural selection, is in evolutionary terms a large and complex brain which has allowed an increasingly refined and profound accommodation for the human organism with culture. A deeper capacity for cultural development and greater versatility of behaviour is the heritage of this era. There are no elemental blood lusts or innate delights in the chase encoded in our species as race memories. Explanations of aggressive behaviour are to be found in the life experience of individuals as they connect with the social structures of societies. It is not legitimate to propose that modern humans possess a 'biological basis for killing' which stands as an unreformed mental residue of a savage past. Finally it is interesting to note that in a review of Brues' hypothesis by Frayer (1981), who examined the skeletal remains from Upper Palaeolithic and Mesolithic populations in Europe, body size and weapon usage were not consistently correlated. Since the reduction in body size that occurred during those times was not accompanied by a change in limb proportions, it is more likely that overall body forms were the results of selection for a lower metabolic rate found among smaller individuals with reduced food needs.

MODERN HUNTERS AND EARLY HOMINIDS

No purpose is served by assuming that an undifferentiated foraging way of life has existed since the inception of *Homo*, and it is crucial that a clear distinction is made between modern and archaic hunting peoples. In fact a new picture of early hominid subsistence patterns, which will be presented in Chapter 9, has emerged during the 1980s and this has cast considerable doubt on hunting as the primary means of adaptation. Even the most simple generalisations taken from observations of modern hunter-gatherers immediately demonstrate the huge gulf between these societies and those of early prehistory.

One striking feature which recurs in ethnographic writing on hunters is the social distinctiveness of these societies. Foraging peoples live in societies which comprise a series of autonomous bands that share a unique culture, language and belief system which includes rites, rituals, transcendental interpretations of the world, laws and taboos peculiar to each society. They also have well established artistic traditions which contain a series of distinctive images and concepts. A separate identity, derived from a consciousness of the society as a distinct corporate whole, is experienced by foraging peoples even though there are no coercive state institutions acting to maintain social boundaries. This sense of ethnicity is often linked to a flexible notion of territory which implies some form of demarcation with neighbouring peoples and consequently a variety of relations with outsiders. Contact between the Kalahari !Kung Bushmen, for instance, and surround-

ing Herero and Tswana pastoral peoples has been mostly benign
for over a century. Considerable technical diffusion has occurred
with the adoption of iron by the !Kung leading to abandonment of
their stone-tool technology. Intermarriage, trade and even employ-
ment have also marked this hunting society's contact with
outsiders (Lee 1984). Among native Australian peoples, however,
contact could swiftly lead to conflict. The formalities of crossing
territorial boundaries could involve delicate negotiations and
specially appointed heralds complete with appropriate body paint
and carved message sticks who acted as officials in these
proceedings (Coon 1976). Between the Ihalmiut Eskimos and
Dene Forest Indians in the Canadian Arctic persistent hostility
fuelled by violent feuding and numerous fatalities has kept the
two communities rigorously apart (Mowat 1989).

In each instance a strong collective identity differentiated and
maintained these societies through the social symbols embodied in
language, art or mental concepts. Contact between societies was
interpreted in terms of these social meanings. Neighbouring
societies could not simply merge or disperse like a herd or flock
because of substantial social barriers. In this sense modern hunter-
gatherer societies are like all other modern societies since social
coherence is a product of collective symbolic meaning. But the
social world of early *Homo* was certainly not as structured and
determined by cultural symbolism. There is no evidence for either
art or language at this time which implies that the development of
distinctive societies would have been severely inhibited and social
cohesion would have depended far less on culture.

Hunting and gathering in recent times has not precluded some
reliance on domestication in what are basically foraging societies.
Dogs are used in hunting by the !Kung, for traction and for
sniffing-out seal breathing holes by polar living Inuit peoples
(Steward 1968), while the G/wi Bushmen cultivate melons as a
means of water storage (Marshall 1955). These innovations almost
certainly arose recently in historical times or at least since the
inception of farming during the early Neolithic cultural era 10 000
years ago (Coon 1976). These are perhaps yet more examples of
diffusion from pastoral societies and can be ruled out as being of
any significance to the hominids of 2 million years ago.

Contemporary foragers have been observed to possess kinship
systems that have complex formal rules that can be far more
intellectually demanding than those found in the advanced
societies. Each individual among the !Kung for instance success-

fully internalises a unique social position in relation to all other members of society. Because kinship terms are applied to both kin and non-kin this society possesses a remarkable degree of cohesion. The special place of a person in this society is complicated by the fact that, in addition to the usual kin relations between parents, children and siblings, the !Kung use both joking and avoidance categories and a naming system to regulate all social relations. Joking and avoidance categories, which prescribe different modes of behaviour typified alternately by informality and respect and by appropriate forms of address, are determined among kinsmen by genealogical position. A man, for example, may joke with his brothers, grandparents and grandchildren (of both sexes) but not with his children, parents or sisters. But among non-kinsmen these categories are fixed according to given names. The limited number of names used by the !Kung – 36 for men and 32 for women – means that a high number of namesakes will exist in both the band of an individual and in those beyond it. The !Kung apply the same family relationship to all people with the same name as that of a family member. Thus a man will refer to all men who possess his father's or brother's name as 'father' or 'brother' and will accordingly place these non-related people in their respective joking and avoidance categories. He will also direct his behaviour in a manner appropriate to each category. This system will affect marriage because a person may not marry someone in the avoidance category even though they are unrelated. A woman therefore cannot marry a man with the same name as her father or brother since this would be considered incestuous. As a whole this system serves as a powerful means of social integration because all members of !Kung society, including distant strangers, are united by an intricate series of personal ties with the same emotional force as bonds within a family.

However these complex rules can lead to contradictions between members of different generations whose name systems may clash with the kin-based rules for joking and avoidance. But these problems are resolved by reference to a further series of logical rules used by the !Kung to construct their social world. The difficulties reported by the anthropologist in comprehending these rules (Lee 1984) are not faced by members of this society. The !Kung appear to understand their evolving universe of kin and social relations, which change for an individual as different stages of the life-cycle are encountered, as if they were the internalised rules of a formal grammar.

This mode of social integration must surely have been beyond the intellectual abilities of the first species of *Homo*. It is possible that *H. erectus* may have begun to regulate and channel sexuality by social rules – movement away from the condition of the apes is apparent in other spheres – but there is no direct evidence for this. The social recognition of marriage occurred therefore in times that are unknown. However, it must be presumed that complex forms of kinship reckoning and their social meaning and consequences can be used to mark off modern and archaic foraging societies. Another example here is one pattern of marriage prevalent in most hunting cultures, although not among the !Kung, that revolves around cross-cousin marriage. In this practice a person is encouraged to marry either their father's sister's child or their mother's brother's child, that is the children of two siblings of the opposite sex. Cross-cousin marriage serves to unify society because cross-cousins are more likely to be the children of aunts and uncles who have left the band of their birth on marriage because of the rules of residence. This form of marriage therefore ties the separate bands of a hunting society by creating renewed kin bonds in each generation (Service 1979). Marriage to parallel-cousins, who are the children of two siblings of the same sex, that is the mother's sister's children or the father's brother's children, is forbidden even though the genetic relationship is identical to that between cross-cousins. This distinction, which is almost universal among foraging peoples (Coon 1976), suggests a level of conceptual finesse unlikely in the early stone age.

Modern hunter-gatherer utilisation of the environment is anything but haphazard, in fact subsistence activity often features a bias in the exploitation of certain animals or plants. These foraging economies, which depend upon advanced hunting technology, exacting ecological knowledge and social organisation, have attained a high level of resource extraction. They have also allowed an enhanced carrying capacity for the environment to support a larger population than any conceivable in the early stone age. Arctic, desert or coastal environments which include islands, that are now the homes of modern hunters, were barred to palaeolithic peoples (Johnstone 1980). Modern hunters then have evolved an economic efficiency which far outpaces the early hominids. Specialist adaptations have featured a staple plant food source such as the !Kung's mongongo nuts (Lee 1968) or game animals like seal and caribou hunted by the Netsilik Eskimos (Balikci 1970). Bows and arrows, spear-throwers and Mode 5 stone tools which include the production of micro-blades (J. G. D. Clark 1977) attached to projectile weapons, together represent a technological breakthrough made during Upper Palaeolithic times. Long-distance seasonal migration, planning and social cooperation for what can be prolonged periods of hunting or fishing, sometimes followed by food processing, are activities requiring the intelligence, skill and patience of biologically modern peoples (Jenness 1975). It could also be claimed that while the early hominids probably relied upon a broad spectrum of vegetable food sources, their conscious understanding and use of their environment was more general and less exacting than that of foragers today. Authorities have made detailed inventories of plant resources used by Bushmen societies in the Kalahari. Lee (1979) and Silberbauer (1981) show for the !Kung and G/wi respectively that survival depends upon a knowledge of the season, location and conditions for growth of a range of fruit, bulbs, seeds, tubers and roots. The !Kung know and have named over 200 edible plant species and the extensive botanical knowledge of the G/wi includes a precise taxonomic system which, in some instances, is said to be more discriminating than that of a modern botanist.

Irrespective of these significant contrasts it is still legitimate to ask whether modern foraging societies can in some instances serve as models for understanding the economies of stone age peoples. The biological differences between archaic hominids and modern humans alone suggest that extreme caution should be exercised here. Isaac (1981A) argues that as time recedes into the distant past it becomes increasingly difficult to assume that the basic organisation of our ancestors' socio-economic behaviour was structured according to principles which obtain universally among foraging people today. Because of displacement by agrarian cultures, modern hunter-gatherer societies represent by definition adaptations to extreme environments that cannot be tolerated by farming peoples. Desert, rain-forest, arctic or tundra-living hunters do not now occupy the same ecological niche as early *Homo*. The occupation of the Kalahari by the !Kung along with the conquest of the New World and Australia occurred relatively recently. Yellen (1977) points to the fact that archaeological sites dating from the Pleistocene are highly diverse with no single pattern being evident. Nor is there a typical form of modern hunter society from which to draw comparison with early *Homo*. It is of course impossible to know the full extent of the material culture used even by recent foraging cultures purely from the archaeological record. Orme (1981) makes this point graphically when she compares illustrations of artifacts used by modern hunter-gatherers with those showing the meagre non-organic contents which might become represented on an archaeological site. Even where ancient and modern foragers occupied similar or equivalent ecological zones in Africa, the fact of invasion by food producing peoples has meant radical change to the size and nature of animal and plant populations. But even if these ecologies had remained intact we must ask whether similar environmental pressures will produce similar cultural responses because different hominid species are involved in the comparison (Freeman 1968). However simple the technology involved, a uniform pattern of adaptation cannot be assumed. To be sure modern prehistorians have been divided on the utility of interpreting archaeological evidence on the basis of simple ethnographic parallels (Orme 1974). The comparison of living communities with the remains of ancient hunters has been seen to be of little help on matters of cultural content, and similarities found in subsistence practices may serve to obscure significant behavioural differences. On the other hand there is a growing tendency among authorities to treat archaeology and ethnography as integrated parts of a single mode of investigation which together serve to stimulate interpretation of the variety and diversity of possible cultural responses.

HUNTING AND GATHERING AS A TYPE OF SOCIETY AND ECONOMY

These distinctions between ancient and living hunting societies are important and demand caution, but it must be acknowledged that both early hominids and modern pre-agricultural peoples practised foraging for a living, albeit of different varieties. In this sense it is reasonable to consider that at both ends of a huge time scale broadly similar forms of social organisation may have operated. Societies which are diverse in time and space can contain similar social and economic practices (Dalton 1981). Irrespective of their specific historical location some societies can be placed in similar sociological categories because they possess a general coherence in the means by which they regulate their material provisioning. Many contemporary hunting societies, for instance, use an exchange of gifts as a means of preventing the accumulation of wealth among single individuals or groups (Lee 1984). A set of discernable rules, principles and institutional procedures will emerge to produce an overall socio-economic structure, which conforms to the logic of similar types of production even in unrelated societies. Even though archaic and modern peoples are involved here it is likely that cooperation in different forms of production will lead to a predictable set of structured social relationships.

This idea has a familiar ring for sociologists. Marx's concept of a 'mode of production' – a conjoined set of social relations which embrace the totality of economy and society – is used as a means of categorising diverse societies in a schematic outline of world history. A mode of production, seen as a complex unity of the forces and relations of production, specifies how labour appropriates from nature, how the means of production and social product are distributed in society and how the labour process is organised (Hindess and Hirst 1975). The idea of hunting and gathering as the first mode of production, 'primitive communism', also provides an understanding of how particular constraints and parameters faced by production in this context will shape a distinctive type of society that will give rise to similar forms of social organisation and behaviour. It may therefore be appropriate to envisage foraging as a type of society united by a series of general principles of adaptation. Even though no extant society can begin to be representative of early prehistory, some observed behaviour among existing foraging peoples might be seen as the product of current needs which were also experienced by human ancestors.

The mode of production among hunters has been seen as one with 'instantaneous production' where nature's bounty is there to be taken independently of any previous investment (Meillassoux 1973). This basic fact of existence comes to structure the relations of production. In practice the immediate realisation of the social product means that social cooperation during hunting expeditions need only be short-term. Once the spoils have been divided no lasting forms of dependence are binding for band members who do not need to remain continually in the same group. The simple artifacts which constitute the forces of production, which may be owned by individuals, and the absence of any form of fixed investment in the economy absolves hunters from the need to have a managerial hierarchy dominated by the old or by males only. Sharing the output within the cooperative band means that the social product does not come to constitute private property nor does it undergo any form of circulation typical of a commodity, since production is always for need rather than exchange. Sharing in effect means that the whole community appropriates its own surplus labour. This means that the social product is rapidly dispersed or redistributed among band members without regard for individual economic contribution or pre-existing rights to property. This form of production and redistribution does not therefore create any form of permanent material dependence between age groups or permit social inequalities to develop. In this type of society there is no basis for power or wealth to become centralised or associated with a specific group such as the intelligent, expert hunters or a gerontocracy.

One consequence is that women can enjoy considerable freedom in hunting societies. It is usual to find far greater levels of sexual equality among hunters than among sedentary farming communities. In foraging societies, marital ties are weaker and women are not used as property to be bartered in marriage deals. Full expression of a woman's sexuality is found among the !Kung where bawdiness and sexual joking is a normal part of female behaviour (Shostak 1990).

Can this type of society be called 'primitive communism', a universal stage in Marx's scheme of historical materialism? Anthropologists have frequently played down or dismissed the idea. Service (1979), for instance, finds the concept meaningless since property is confined to the family in foraging societies. However Lee (1988) shows that the facts of communal life-styles among hunters have been well known since the nineteenth century. It was Morgan, a non-Marxist, who was to describe in detail a communistic state of society among foraging peoples which Marx came to theorise. It is indisputable that this type of society has always been found to be classless and that all members have collective rights to basic resources which are seen as communal property. The absence of any form of hereditary status, authority or form of systematic dominance is also typical. Power is diffused among a community whose politics are based upon consensus and this relatively democratic system is founded upon a communistic form of economy. This linkage between economy and society is also reflected in everyday behaviour which contains numerous levelling devices that assert norms of equality and restrain greed and arrogance. In short, as Leacock and Lee (1982) show, the similarities of social organisation and subsistence found among foraging peoples throughout the world suggest that this type of society is grounded upon a mode of production.

In recent work it has become common to reject the use of analogies drawn from modern hunter-gatherer societies in the study of early hominid behaviour (Foley 1988). However authorities remain divided on this issue. While no simple lesson may be drawn here some insight may be gained. Even within the African regions that were natal to early hominids, contemporary hunting societies present us with specialist adaptations in marginal environments (Musonda 1991). Pleistocene hominids lived in grassland, swamp, river valley and woodland habitats close to permanent water sources, not in the arid areas frequented by Bushmen societies. The abandoned camp-sites of foragers today contain evidence of activity patterns that were not present 2 million years ago. Early hominid sites are without hearths, pits or tool-making locations separate from other areas. On the other hand the division of labour by sex and age, in which women gather and men hunt, could also have been present among the early hominids and may have given rise to semi-permanent relationships between the sexes. Food-sharing, which is a universal practice for modern foragers, cannot be assumed for hominids. However both early hominids and their descendants would have been subject to seasonal food shortages and this nutritional stress would have been a powerful incentive to evolve sharing behaviour as a survival strategy.

Perhaps the problem needs to be posed more precisely. To present the early hominids as full practitioners of hunting and gathering as it is known today is simultaneously to reject the overwhelming evidence that our ancestors had not reached human status and has the effect of promoting the idea that existing foraging societies are living fossils. Both claims are absurd and are easily falsified. But to suggest that some cross-cultural evidence can be valuable in providing an understanding of an archaic life-style is valid. Even used only as a negative case-study, modern foragers provide us with a wealth of knowledge. If all hunting societies had vanished unrecorded several centuries ago we would have little inkling of the meaning and significance of the first cultural remains. The idea that modern foraging societies retain some aspects of the prehistoric economy and life-style is reasonable given the level of technology and environmental constraints. But this claim does not imply any form of identity of social organisation between modern and ancient societies.

Observations of hunting societies and experiments with their techniques can provide an understanding of the range of options that were open to hominids – cutting into the carcasses of dead pachyderms was an example discussed in Chapter 7. Modern hunters can be used as a data base for reviewing archaeological material (J. D. Clark 1968). The productivity of gathering as opposed to hunting can be measured in terms of calories gained per hour, the pattern of daily ranging from camp and movement between camps with the seasons observed, the components of diet and the ratio of meat to vegetable foods recorded, and the relative ease or difficulty in gaining a living through foraging together with the likelihood of encountering periods of dearth understood. Considerable access to the life-styles of ancient hunters can also be seen through studies of the configuration of activity areas within a camp, selection and use of stone and other raw materials, tool-making together with modes of tool-use, wear patterns and rate of discard.

Glynn Isaac (1989) points to the hominid exploitation of savanna ecology as the key factor shaping the new adaptive pattern. Hominids had evolved an 'optimum foraging strategy' in which nutrient-rich animal and vegetable foods became available in abundance because of the combination of technology and social cooperation. Both meat and underground bulbs and tubers were available with new equipment designed for cutting, carrying and digging. But hominid tools were only effective in a social context in which mutuality and reciprocity were predominant forms of behaviour. In effect this meant that hominids would share the labour of scouting a wide area for animal carcasses and that they would share this information and the food gained thereby. Isaac envisages a pattern of central place foraging in which the rudiments of camp life may have developed as hominids began to practice the congregation and dispersal movements typical of modern hunters. The suggestion that food-sharing may have been part of the complex process of cultural development was also discussed in Chapter 7. There is good evidence that the hominid diet was omnivorous. In addition to the stone cut-marks on animal bones, the wear polishes on stone tools show that they were used to cut meat and vegetable matter while wear patterns on bone tools indicate their use for digging. Yet other tools were used for cutting wood, suggesting that the stone tool was the means by which other tools were made. The hominids of 2 million years ago were clearly engaged in a complex series of cultural activities, and the idea of some rudimentary form of division of labour, which was unified by sharing practices, seems to be worth considering.

Isaac's idea of an optimum foraging strategy is important. This concept, which was developed to explain hunter-gatherer socio-ecology, interprets a given mode of foraging in terms of the balance between energy expenditure and food returns. In an overview of these strategies Durham (1981) shows that the economic behaviour of hunters is not random but accords with the need to optimise resources. Hunting and gathering represents an adaptation to a specific set of environmental constraints in which survival and reproduction demand that foragers modify their behaviour to conform to a pattern which will maximise efficiency and reliability and minimise waste. Strategies for optimum resource procurement will favour certain aspects of social behaviour as much as the methods used to exploit potential resources.

Foraging societies will normally allow their environments to regenerate and will therefore deliberately allow hunting grounds to lie fallow rather than deplete resources. A modal group size will be maintained and the balance of human population with the environment regulated by a range of social practices (Hayden 1972) which may include dispersal or infanticide. The fact that female children are systematically killed in a range of societies, including those of early modern Europe (Langer 1972), is well known. Shostak (1990) and Marshall (1960) have shown this practice among the !Kung, and Freeman (1971) suggests that it is adaptive among Inuit peoples while Douglas (1966) surveys the practice of infanticide among primitive societies in general. To be sure, this method of population control has given rise to debate. Divale (1972) sees infanticide as the principal means of regulation in both modern and Palaeolithic times. Denham (1974) considers the practice to have been overemphasised by anthropologists. High infant mortality and frequent catastrophic incidents required only ordinary social practices such as abstention, post-birth taboos on sex or restrictions on the age of marriage to maintain a stable population. On the other hand sociobiologists, like Bates and Lees (1979), argue that population regulation is a myth. There are no regulatory mechanisms in human populations that have been convincingly shown to operate at the level of the social group, and thus female infanticide should be interpreted as a means by which parents attempt to maximise their genetic contribution to the next generation. Nevertheless, whatever the mechanism or the motivation of hunter-gatherers, there is a consensus that these societies maintain optimum group sizes that accord with the carrying capacity of their foraging territories.

Is it reasonable to suggest that sharing may have been part of an optimum foraging strategy among early hominids and that this was a form of behaviour that served to minimise risk, like other aspects of resource exploitation? The economic interdependence proposed by Isaac might have arisen along with a change from the individual mode of primate feeding and become the foundation of a new pattern of social relations. Early *Homo*, still a 'walking ape', was far from human and there is no means of knowing when such a behaviour pattern first emerged in prehistory. But during the career of *H. erectus* a series of relatively complex activities were performed which strongly suggests that social cooperation was a vital part of the adaptive strategy and it is therefore reasonable to hypothesise some form of accompanying behavioural shift which included sharing.

Sharing behaviour is a primary and universal element of foraging societies today. The !Kung are fully aware of the enormous social value of their food-sharing practices and regard the idea of a family eating in secret with horror (Marshall 1960, 1961). The meat of large game is shared among a whole band in accordance with customs designating the division of a carcass, while meat from small animals is distributed as the hunter wishes. This is not to say that concepts of property and ownership are absent from this type of society. Dowling (1968) notes that while meat is shared, a single individual is still designated as the carcass owner. This may be a member of the hunt or, as among the !Kung, the owner of the first arrow to hit the animal who, because arrows are habitually traded, may be infirm or a non-hunting elder or woman. Vegetable food, which is the main part of the diet, is normally shared only within the family circle. However women who are pregnant will be given vegetable food and sharing will also occur in any case of need (Woodburn 1982). The Netsilik eskimos have evolved a series of formalised sharing partnerships in which each hunter had particular cuts of seal-meat reserved. Partnerships were deliberately made between hunters who were not closely related, and throughout their lives the partners' reciprocal obligation was reinforced by using the body part of the seal guaranteed for each man as the term of personal address (Balikci 1970).

The equity which sharing behaviour promotes in such societies is vigorously asserted in everyday life (Woodburn 1982). However the social order of the hunting band is frequently disrupted by individual challenges to equality. But these attempts to acquire power, prestige or wealth are themselves challenged as band members reassert collective values. Most hunting and gathering societies are characterised by an 'immediate-return system' in which food is consumed on the same day it is obtained rather than stored or processed. This type of economy usually coexists with a form of social organisation which has flexible and changing band membership where individual choice guides decisions about residence and partnerships made for foraging. Nor is there any dependence on dominant individuals or groups for access to basic resources, since all social relationships stress sharing and mutual obligations. Woodburn found that these aspects of social organisation were common to all the six foraging societies with 'immediate-return systems' that he had studied. He also found that no society with a 'delayed-return system', that is a society in which property and investment in buildings, land or herds, that automatically led to inequalities of political power, had egalitarian social relations comparable to those of hunter-gatherers. All of the 'immediate return system' societies acted to systematically eliminate distinctions of wealth, power or status which might arise. In fact mechanisms to restore equality came into play for all signs of social distinction except those between the sexes (Woodburn 1979).

Hunters and gatherers lack stores or investments in fixed apparatus and own all resources collectively. Their societies contain fluid nomadic groups who associate voluntarily, often on a short-term basis without incurring obligation, with all individuals having full rights to use common territory and its resources. This pattern of organisation has the effect of undermining authority whenever it might spontaneously develop because individuals are not bound by specific groups, areas or resources. Disagreements can be solved by dispersal of the parties involved but conformity to collective norms will be maintained by the common knowledge that no effective protection from ambush while away from camp can be given to a person whose behaviour is persistently anti-social. The potentially dominant, acquisitive or arrogant band member will encounter minor sanctions on a daily basis: however since no system of long-term dependence exists such a person may simply be withdrawn from. In effect this means that neither property nor powerful personalities have the ability to create relations of dependency which would threaten equality.

Generalised reciprocal behaviour within an open society in which none are excluded is fundamental to modern foraging peoples and is deeply embedded in their moral ideas. It is this form of behaviour which serves to integrate and coordinate a small-scale economic system (Price 1975). The social structure as a whole is animated and made to operate in harmony through multiple acts in which goods and services are distributed without any immediate calculation of returns. From an ecological point of view it is important to reflect upon the fact that, as we saw in Chapter 4, primate sharing is highly restricted, being at best only a form of tolerated scrounging (Blurton Jones 1987). It is in fact among the order of animals described as social carnivores that sharing and cooperation is most prevalent. Gathering dispersed vegetable foods requires far less cooperation and reciprocity than gaining meat from animals. The initiation of such group-based activity by hominids must have led to considerable behavioural change, but the question of hunting or scavenging as the principal activity in the era of early stone tools requires attention.

By way of a conclusion it is worth considering that even though we are dealing with different but related species, the societies of both the early hominids and modern foraging peoples may have had significant shared features that derived from similar ecological and social relations. The necessity of combining to appropriate from nature implies small groups living as bands of twenty to eighty individuals at the most. Maintaining a primate troop structure would have been difficult in the context of mixed foraging. More likely would have been a fluid mobile community which dispersed and aggregated according to resources. We cannot know whether reciprocity and sharing form the bedrock of social relations in prehistory, but it is clear that social groups engaged in this way of life rarely survive without these key features. Pre-human hominids would have had a greater need to follow a risk minimisation strategy than their modern counterparts (Yellen 1986). Smaller brains, less sophisticated communication and environmental knowledge meant an inferior ability to predict the availability of future resources. Nor would the less extended social networks of the early hominids have been as effective in evening out variations in the local environment by facilitating the movement of people and food. The necessity for social cooperation in all subsistence activity strongly suggests a communal economy in which free exchange was a condition of survival. Communal hunting has been found to be more efficient for both searching and killing animals and in providing greater quantities of meat (Driver 1990). Perhaps some residue of an ape-like dominance hierarchy persisted among early *Homo*, but strong pressures to evolve an egalitarian social system based upon sharing and equity were also present. In both epochs existence itself may not have been as arduous and insecure as was once thought. The standard picture of hunter-gatherers in British archaeology, which remained uninformed by anthropological insights until the 1960s, was one of an ignoble savage whose depravity was directly proportional to a ceaseless need to engage in food-finding (Wheeler 1956). This conception has now been found to be contradicted by reality. Foraging implies only a short working week and provides abundant leisure (Sahlins 1972). Perhaps observations that modern hunter diets usually consist of about one-third meat and two-thirds vegetable matter also apply to our first omnivorous ancestors. There are also intriguing parallels and questions concerning relations between the sexes, reproduction and the family, but these are better left for Chapter 12.

9

Man the Hunter?

The idea that hunting has been an inseparable part of the human condition since at least the appearance of the earliest stone tools has become deeply imbedded in anthropological thinking. This conception has nurtured popular ideas that hunting is an expression of an intrinsic animal nature that lies behind our cultural clothing. These assumptions have recently been effectively challenged and the debate that has ensued is reviewed in this chapter. But it is equally important to consider what the demands for a successful economy based upon hunting and gathering comprise. Was fire a part of the technology possessed by the early hominids? Did they in fact live by hunting or were they reliant upon scavenged meat for animal protein? Other questions which relate to the connection between hunting and aggression and to the environmental consequences of hunting are also issues that are raised here.

FORAGING IN ECOLOGICAL AND EVOLUTIONARY PERSPECTIVE

There are definite prerequisites for a life based on the communally organised hunting of large animals. One might specify the criteria of ecological adaptation and knowledge of the environment, technical skill and social cohesion as minimal demands here which distance the early tool-using hominids from later species. The early hominids remained confined to their African homelands until *H. erectus* migrated to adjoining continents a million years ago. The conditions which prevailed in glaciated Europe and Asia imposed new constraints that had not been experienced in tropical regions. Hominids were obliged to seek cultural adaptations to new ecological zones that would save energy by retaining body

226

heat and food conservation. Clothes, shelter, carrying devices and food preservation methods are essential for these purposes and, in general, presuppose the use of fire. Perhaps the controlled use of fire was the precondition for the hominid exodus from Africa. But despite recent suggestions that hominids may have used fire at Kenyan sites 1.5 million years ago (Gowlett et al. 1981) there is doubt that it was used first in Africa (Isaac 1982A). There certainly could have been no use of caves or exposed open sites until night-prowling carnivores had been evicted or frightened, and the conclusion that before fire the early hominids used trees as sleeping places is tempting.

Did the early hominids use fire over a million years ago? Until the 1980s the consensus was overwhelmingly negative. But with new tentative evidence from African sites the possibility that the controlled use of fire rivalled and perhaps accompanied stone tool making has to be considered. There can be no doubt that fire would have provided enormous adaptive advantages (Toth and Schick 1988). In addition to protection, warmth and light, which extended the active day, cooking served to preserve and detoxify meat and vegetable matter, remove parasites, reduce bacteria, yeasts and mould and provide a wider range of digestible food sources. The technological benefits of fire-hardened tools and fire used in foraging would also have been of inestimable value. Advantages are of course not sufficient as explanations and we are forced to confront ambiguous and controversial evidence.

A fundamental problem lies in distinguishing unequivocal evidence for fire in the archaeological record. If there is no exactitude in dating the rise of hominid tool-making there can be even less certainty in estimating the emergence of fire as an artifact. Until recent doubts were raised it had been long accepted that *H. erectus* made fire in China almost half a million years ago. But the remains of ancient fire are rarely well preserved and do not provide conclusive proof that it was used and controlled by hominids much before 150 thousand years ago (Toth and Schick 1988). For those who propose late models of humanisation there is no sound evidence of fire until after the appearance of the Neanderthals some 20 000 years later (James 1989). There are certainly no deliberate hearth structures that can be identified with the early hominids in Africa and any fire use in these times has to be inferred from baked clay, charcoal or burned remains.

In 1981 Gowlett and his colleagues reported the discovery of nearly 1000 Oldowan artifacts that were found in situ with faunal remains and hominid fossil fragments at Chesowanja in Kenya. Burned clay deposits, which had been heated by fire to 400°C, which is the expected temperature of a camp-fire, were also found. The authors claim that since the clay deposits could not have been introduced into the site after its formation, and since they are not the results of bush fires which are rarely hot enough to bake clay, there is now evidence which associates hominids with the control and use of fire almost 1.5 million years ago. This point in time is contemporary with early *H. erectus* specimens found elsewhere in Africa and it is this hominid which the authors favour as the main occupant of this site.

Clark and Harris (1985) espouse the most systematically presented case for fire as a part of the early hominid life style. Natural conflagrations caused by lightning, volcanism or spontaneous combustion were common features of the savanna ecology natal to hominids. With the coming of drier conditions 1.7 million years ago a series of significant evolutionary and behavioural changes were initiated that included the emergence of *H. erectus*, the occupation of more open treeless areas and places above 2000 m, a shift toward hunting and an increase in the complexity of stone tools. Accompanying these changes is evidence that may also indicate the use of fire by hominids at a number of sites. At Chesowanja a series of discoloured baked clay rings that are associated with stone tools have been dated as 1.42 million years old. These clay clasts have been found to be the results of fire on the basis that local disturbances in the magnetic field have been detected within the discoloured area and that the clay had been heated to temperatures consistent with those of camp-fires (Barbetti 1986). Suggestive as this evidence may appear, it may be no more than the remains of burned tree stumps caused by natural forest fires. However Clark and Harris pursue the idea that fire may have been a catalyst in the formation of hominid society and that it ranked equally with food-sharing, meat-eating and new modes of sexual behaviour. It is not the case that all animals have a total aversion to fire, they argue. Chimpanzees for instance do not fear fire and have been observed to jump through gaps in bush fires and to sit around the remnants of human camp-fires while many carnivores will take advantage of an advancing bush fire to catch prey. Hominids then can be considered to have been fire tolerant and may conceivably have exploited naturally occurring fire sources. This habituation to fire in the environment and its opportunistic use in foraging served as the basis for predetermined use seen in the intentional maintenance of fire for a variety of purposes.

Many of these ideas are seen as a misinterpretation of the evidence and are dismissed as speculative by James (1989). The burned clay clasts are probably fortuitous since they can be found independently of artifacts and no strong association exists between bone remains and ancient fire. As a whole James sees no convincing evidence of purposely controlled fire in Lower Palaeolithic times, and the most parsimonious explanation of the data presented by Gowlett, Barbetti, Clark and Harris is natural burning. However the nature of early hominid sites is unlikely ever to provide the level of proof demanded by James. Natural processes such as erosion and the movement of sediments over a million years old can be expected to have obliterated traces of ancient fire. Early hominid fire use may have left distinctly different remains from the hearths and cooking pits of modern foragers, and the fact that evidence of burning exists at so many African sites suggests that the question should remain open.

Beyond Africa, subsistence in general and hunting in particular was conditioned by the stark seasonal variations of a glacial climate. A savanna ecology is, to be sure, affected by annual climatic variations, particularly of rainfall, but the form of stress imposed on hominids by long northern winters during the Pleistocene had no parallel in Africa. In such circumstances the constant gathering of vegetable foods would not have been practical and, at certain times, only the meat of large animals would have provided the bulk of food needs. Binford (1985) has claimed that hunting arose as a major activity in places with a shorter growing season, basing this argument on the fact that hunting peoples eat more meat as a proportion of diet as one moves from the equator to the poles. This presumes some knowledge of the habits and anatomy of prey species. An understanding of the migratory habits of large herbivores and the configurations of their vital organs would seem to be essential. Rich in game as the temperate habitats of these hunting peoples were, the killing of animals much larger and more ferocious than their modern domesticated descendants demanded a high level of planning and social cooperation. Adaptation here was directed to ways of overcoming sharp discontinuities in food supply: there were few chances to engage in the haphazard foraging or opportunistic hunting practices available under savanna conditions.

The technical prerequisites of hunting large animals constitute at least the existence and skilled use of the spear. Spear making, however, is not a simple matter and was probably beyond the proficiency of the Oldowan hominids. The manufacture of this weapon requires a spoke-shave tool to fashion a balanced shaft. It seems unlikely that the early hominids had any mental access to the concept of balance and could never have made a truly aerodynamic projectile. While these hominids possessed tools that would sharpen sticks that were sufficient for digging up roots and tubers, the points they achieved would have been quite inferior to the fire-hardened tips that were possible by late *H. erectus* times. Hafted tools, which consist of a wooden lever and a stone cutting edge, do not appear as artifacts until about 100 thousand years ago and can be discounted as part of the habiline or even *H. erectus* tool-kits. It is important to be aware that the existence of many advanced specialist hunting peoples of the Upper Palaeolithic era 35 000 years ago, and in the recent past, has been based on such a technology. Apart from poison, only accurately aimed stone-bladed spears will produce sufficient penetration and haemorrhage in a large animal to allow successful hunting. In arctic conditions the killing of Kodiak bear, the largest modern terrestrial carnivore with a weight twelve times that of an Eskimo hunter, is possible only with the deep and precise penetration of specially pointed spears (Laughlin 1962). The hunting of some herd animals with a group defence, such as Musk oxen, became possible only after the late appearance of the spear-thrower in Europe some 17 000 years ago. This tool extended the range of a spear by up to 75 per cent and allowed hunters to penetrate the centre of herds. *Homo erectus'* inferior technology would have achieved a lower level of hunting efficiency while among the early hominids tools were just adequate for butchery rather than killing big game.

A cohesive social organisation capable of planning, allocating responsibilities and distributing meat is another condition of success in this way of life. Suggestions that hunting represents an extension of the pre-human state in humans or that it is an amplification of an essentially animal nature or even that a continuity with the animal world is manifested by this activity, should all be rigorously rejected. Hunting by hominids is very much a cultural contrivance which rests upon social solidarity and has no parallel among carnivorous animals. For most animals finding food is not usually a cooperative pursuit. Among primates life is usually contained within a troop but feeding remains a solitary activity and social cohesion has no direct economic effect beyond protection. The problem to be faced is how and when these preconditions were met by hominids.

SCAVENGING OR HUNTING?

It is important to understand that the advent of hominid hunting represented a radical break with a primate tradition and cannot be seen as a simple extension of animal behaviour. There are, to be sure, evolutionary continuities with primates which are significant, but hominid hunting is a bio-cultural adaptation without precedent among other species.

Most social carnivores, such as big cats or wolves, hunt at dawn or dusk and lack the persistence of modern humans whose hunt may last for several days. These animals engage in cooperation and sharing largely because such practices are products of an innate behaviour pattern. But this 'cooperation' does not imply a group strategy or a division of labour. Coordinated hunting by these species consists of a pack of animals each replicating the behaviour of another animal, differing only in the direction from which they approach their prey. Among small carnivores, such as hyenas, group hunting occurs only because the animals preyed upon are larger than their assailants and are capable of surviving the assault of a single attacker. The hominid specialisation of long-distance walking, which had perhaps developed from the practice of extensive foraging, was later allied in human hunting with strategic planning that might involve preparations of a ritual, technical or migratory nature as much as several months in advance of a hunt. Similarly there is no animal who, having made a kill, returns to a camp and divides food on the basis of social rules or engages in exchanges derived from a division of labour. Carnivorous animals do not have butchery practices nor do they differentiate between taboo and permissible prey species or even reject portions of a kill as taboo or make ritual propitiation to the spirit of the animal that has been killed.

Evolutionary adaptations for hunting by hominids are at best poor. The digestive tract is long and ill-suited to an all meat diet, especially if cooking is not used to kill parasites. The teeth suggest no specialisation for meat; in fact an omnivorous diet is indicated. Speed and strength are unimpressive and, without technology, are ineffective. To match the speed of other social carnivores bipedal hominids would be obliged to sustain a running pace of over thirty miles per hour – equivalent to a two minute mile – while carrying equipment and perhaps an infant. The significant absence of women hunting among all known hunting societies in the recent past is also at variance with the behaviour of all carnivorous mammals whose females feed both themselves and their young from kills. 'The only form of division of labour by sex found universally is that women do not have established roles or equipment for hunting' (Hayden 1981: 419). The relevance of this fact is that no carnivorous animal has evolved with a sexual bias in food finding. The preconditions for hominid hunting and the importance of meat-eating then rest upon a culturally conditioned social group not upon natural propensities.

What form of foraging can reasonably be associated with the earliest known hominids using an Oldowan technology? During Bed I times, *c*. 2 million years ago, there is no evidence that either *Australopithecus boisei* or *Homo habilis* practised cooperative hunting of big game. There are, however, a variety of animal remains at Olduvai and at least three antelope skulls that show clear signs of accurately placed fractures which were perhaps caused by clubbing (Mary Leakey 1980). But it is unlikely that the hominids of this lake-side camp were capable of successful big-game hunting and much of their meat was probably gained by scavenging or came from small game. The opportunistic hunting of sick, injured or trapped animals is also indicated by the presence of stone tools among the butchered remains of a hippopotamus and even an elephant. These small-brained hominids still retained adaptations for tree climbing and were only four to five feet tall with an estimated body weight of 40– 100 lb (Shipman 1984, Blumenschine and Cavallo 1992), and therefore it is unlikely that they possessed the speed, strength and social organisation necessary for the hunting of big game. It had long been thought that these requirements together with appropriate weapons were not present until at least 0.5 million years ago, but even in these times full cooperative hunting is now seen as questionable.

Big-game hunting, therefore, arose after considerable brain enlargement had occurred, and was a result and not a cause of the mixed foraging strategy developed by the early hominids. The probable causes of the shift away from predominantly vegetarian fare in favour of meat as an important dietary element were discussed in Chapter 6. But until recently an evolutionary continuity between hominids and other primates has been used to explain this change. A number of field studies of baboons and chimpanzees had established the facts of meat-eating among primates.

A challenge to the previous exclusively vegetarian stereotype that had been imposed on primates until the 1970s was provided by evidence that several species engage in opportunistic hunting. Meat-eating is known to occur among more than one-third of all the species which comprise the primate order (Harding 1981: 207). One field study revealed that in the diet of seven baboon troops observed in Kenya, animal protein, including invertebrates, amounted to only two per cent. This food was gained by males pursuing young grazing animals (Harding 1975). But chimpanzee hunting and division of the carcass was suggested by Suzuki (1975) to be a precursor of the behaviour of proto-human hominids. Goodall (1976, 1979) has seen this continuity in terms of male groups acting in cooperation to run down small animals and share the meat. There is certainly a real liking for meat among chimpanzees which under some conditions has included cannibalism (Goodall 1990). Instances of cannibalism in which infants were taken from their mothers and eaten by other females have been recorded by Goodall. Whatever the reasons for this behaviour it is now clear that meat gained by active hunting does form a small but much desired part of the chimpanzee diet.

Ancient bone remains suggest that the early hominids also included animal protein, from a variety of sources, in a fruit and root dominated diet. As we saw in Chapter 6, the increased importance of meat in the hominid diet, which is evident from Bed I at Olduvai, could have begun as a means of overcoming seasonal food shortages. Dunbar (1976) has suggested that like modern baboons, who resort to meat-eating when arid conditions occur, the australopithecines would have supplemented a seasonally deficient vegetarian diet with increasing quantities of meat. Their existing gathering tools could also have been used in meat processing. Foley (1987) has suggested dietary stress in the dry season may have been a reason for the emergence of *Homo*.

Irrespective of these hypotheses the savanna environment contained a varied stock of animals and offered the hominid social group an important opportunity to broaden the basis of its niche. The early African hominids lived in what Butzer (1977) has termed a mosaic environment that included lake shore and marsh, plain and forested country, plateau and hills all of which could be marked by a range of vegetation, fauna and local climate. The importance of this is that hominid evolution, 'took place mainly in relation to multiple inter-fingering ecological opportunities' (Butzer 1977: 577). It is from this broad range of habitats, which provided a wide choice of foodstuffs, that the subsistence pattern typical of the Oldowan hominids emerged.

Food gathered from small animals in these different zones must have formed the major protein component of the early hominids' diet. But in addition to this, medium-sized animals could be pursued to the point of exhaustion or the meat of large dead animals taken ahead of other predators. Several objections have been raised against scavenging as a significant part of the early subsistence pattern. Goodall (1976) and Washburn and Moore (1974) point to the very real dangers of sharing a kill with carnivorous animals before tools and hunting techniques had been developed. Chimpanzees and baboons usually ignore the carcasses of dead animals and meat eaten is restricted to animals which they have killed.

There are, however, good reasons for considering the Oldowan hominids to have been opportunistic scavenger-hunters whose use of scavenging increased with the size of animal that was to be eaten. Studies of the patterns presented by the bone refuse at Southern African hominid cave sites suggest a chronological sequence in which a primary phase of bone accumulation occurred, because of the activities of carnivorous animals, which was then followed by a second and third phase which represented hominid scavenging and then active hunting (Isaac and Crader 1981). These assumptions, which are based on the proportion of juvenile animals in bone remains and the presence of stone tools in strata, are tentative. In practice, however, the distinction between scavenging and hunting is often unreal in the sense that a savanna environment frequently contains animals who are too weak or disorientated to resist predators and may be easily vanquished. By exploiting these animals, as well as those already dead, an emphasis would have been laid on cooperation and sharing. The difficulties of killing large animals, even in a crippled condition, maintaining vigilance for carnivores attracted by the same prey and rapidly butchering and dismembering carcasses ahead of competitors could only be overcome by developing a coherent group strategy. Given sufficient opportunity to scavenge and hunt in this way, hominids would cultivate a working style in which tasks were sub-divided for a shared reward and the successful pattern would become reinforced and adopted as a perpetual element of cultural behaviour (Blumenschine and Cavallo 1992). None of this, however, means that the hominids of *c.* 2 million years ago were hunter-gatherers in the contemporary sense.

A scavenging niche is not common among mammals and there is no large mammal with this adaptation (Lewin 1984A). In most circumstances such a food source is likely to be uncertain. It follows, therefore, that no mammal relies exclusively on scavenging. Schaller and Lowther (1969) argue that scavenging by hominids could only be successful if it was allied to a vegetarian diet. A series of brief participatory experiments by these authors in East African savanna game reserves, which approximate the conditions once encountered by the Oldowan hominids, have suggested that considerable scope for opportunistic scavenging and hunting existed. During the birth season, new-born gazelles were both plentiful and easy to catch. At other times there appeared to be a rich source of animals who had died from disease, drowning or old age. Similarly, sick and injured animals and the remains of carnivore kills were also plentiful. In total a considerable quantity of meat was found available for an active hominid band. But there is the question of constancy in food supply throughout the year and there must be doubt as to the probability of a hominid's survival in an occupation that would frequently involve competition with carnivorous animals. Nevertheless tool-using hominids had available a food source, in the form of bone marrow and brain matter, that was normally denied to carnivorous animals. All areas of soft tissue would have been specially attractive to hominids whose teeth were poorly adapted for uncooked meat but, unlike the visceral organs, there would be few competitors for the highly nutritious contents of bone cavities. It is, in fact, the long bones of large herbivores which comprise a major part of the cultural remains at hominid living sites. These bones may have been collected from kills on which carnivores had already fed since the tooth marks of these animals are clearly distinguished. Few carnivorous animals are able to feed on the tissue within bones and hence the ability to split such bones with a stone hammer conferred a selective advantage on the early hominids.

Objective evidence from the fossil bone remains at Olduvai Gorge has been used to test these ideas. In Chapter 7 the work of Shipman and Potts was introduced in the context of hominid home bases. In a number of papers (Shipman 1983, 1984, 1986 and

Potts 1983, 1984A) it has been argued that this evidence strongly suggests a scavenging niche allied to foraging rather than hunting and gathering as the basic form of hominid subsistence.

Bones accumulated at hominid sites indicate meat-eating but not active hunting. The pattern of cut-marks made by stone tools on bone surfaces does not show a form of carcass treatment typical of modern hunting peoples. Ancient cut-marks tend to occur predominantly on the mid-shaft of bones rather than close to the joints which is the reverse of modern practices. This suggests that hominids rarely disarticulated a whole animal carcass or had butchery practices which involved removing large amounts of meat from a long bone. There is some indication, however, that hominids may have been utilising carcasses as a source of skins or tendons which could have been used for a range of purposes such as carrying devices. In some cases the marks also indicate that hominids used their stone tools after animal gnawing had left tooth marks. As a whole this evidence points to a more opportunistic and less systematic use of carcasses in which early hominids, still living on a mainly vegetarian diet, operated in competition with carnivores. However these conclusions represent only one facet of an ongoing debate and there are telling arguments against this thesis, which reinstate at least some capacity for hunting among these hominids, to be considered later. But we can all agree that the hominids of *c.* 2 million years ago were not hunter-gatherers in the contemporary sense. We must next consider whether *Homo erectus* initiated this way of life.

NEW ASSESSMENTS OF HUNTING IN PREHISTORY

Homo erectus living sites are found as far apart as China, Java, Africa and, if one interpretation is accepted, in southern and eastern Europe (Howells 1980). These sites contain characteristic Mode 2 tools such as the Acheulean hand-axe and those of the East Asian chopping tool industries. An extremely slow but distinct developmental progression can be seen in cultural deposits during the *H. erectus* time span from *c.* 1.6 to about 0.4 million years ago and until 250 thousand years ago in China. Better selection and working of raw materials occurred. Tool types grew in number suggesting that the activities and behaviour of hominids were becoming more complex while the form taken by hand-axes tended to become finer. Stone for tool-making was carried greater distances and there is every indication of greater skill and labour being invested in tools. Camp-sites are larger and appear to have been occupied for much longer periods suggesting than an increase in the size of the hominid band had occurred. The social group of *H. erectus* could possibly have been as large as fifty individuals (J. D. Clark 1976: 39) which would have represented more than twice the habiline band size. There are also greater densities of stone artifacts at Acheulean sites indicating further a growing reliance on material culture and dissemination of tools within a band.

Taken as a whole this evidence strongly points to a previously unparalleled economic achievement. *Erectus* populations had led hominids out of Africa at least a million years ago. *Homo erectus* had developed a range of highly successful mixed foraging strategies in a whole series of different ecological zones of the world. Preservation patterns almost exclusively favour bones among the food remains of Acheulean people and there is no doubt that the meat acquiring prowess of *H. erectus* was growing. But for *H. erectus* and previous hominids the utilisation of insects, invertebrates, deep-growing tubers and starchy roots, as important dietary elements if not as staple food stuffs, should be taken seriously even though these are poorly preserved or non-existent in cultural remains. Is it possible that a division of labour based upon acquiring vegetable food and game may have developed in this era? These sites are the products of social cooperation that was a necessary part of an economy based on diversified foraging in which sharing and group cohesion would have been rewarded. Some direct evidence of the gathering of vegetable foods is, in fact, preserved. Various fruits, seeds and nuts are found at Kalambo Falls in Tanzania and at *H. erectus* sites in China, while in France the pollen of the shrub *genista* has been found in fossilised coprolites. In addition to this evidence, the inference drawn from the study of contemporary hunting societies points to an omnivorous diet for early man. But until the 1980s it was *H. erectus'* achievements as a big game hunter that were thought to be most significant.

Much less emphasis has been placed on hunting as a central agent of human evolution in recent work. As doubt has grown about the predatory competence of *H. erectus*, hunting has increasingly been seen as a product of later prehistory. One exception here is Hill (1982) who proposes that hunting was the key behavioural variable in a comprehensive evolutionary scenario which included the achievement of hominid status itself together with an elementary sexual division of labour, the loss of estrus and continual sexual receptivity by females who were provisioned by hunting males. The development of weapons and tools, brain expansion and the evolution of the modern hand and dental pattern followed. Finally, improved parenting came with a more reliable food supply, a longer period of juvenile development and dependency, increased longevity and the menopause allowing a non-fertile grandmother to care for her daughter's children.

Hill's thesis is now overshadowed by a new consensus which relegates hunting to much later times. The early hominids had tooth wear patterns that indicate a fruit and vegetable diet in which scavenged meat was a small additional extra. In fact the configuration of australopithecine jaws and teeth implies continual and prolonged chewing of fibrous and tough low-quality vegetation expected in a savanna environment. There is also new evidence that these hominids still had a short period of growth and dependency, like the apes, and that therefore no developmental threshold had been crossed. To be sure, Hill's claim that small game – birds, snakes, small mammals and molluscs – would have been easily gathered items of diet that could be acquired without real weapons is to be taken seriously, but this fare is the product of mixed foraging rather than hunting.

The extreme reverse of Hill's thesis is presented by Binford (1985) who argues that true hunting began only about 50 000 years ago. The killing of large animals, providing more food than a single hunter could consume, has only a short history and was not properly developed until just before the appearance of anatomically modern humans. Many basic human traits such as a division of labour, separate economic roles for each sex, food-sharing and foraging from a central location have their origins only in these times when evidence for modern forms of butchery and transport of meat appear. In this scenario, hunting arose simultaneously with modern humanity to become a part of the human condition rather than acting as its cause.

Binford was convinced that the European Mousterian sites, which mark the opening of the Middle Stone Age about 125 thousand years ago, contained evidence that pointed to a radically different form of subsistence from that found among modern hunting peoples. If these makers of Mode 3 stone tools were to be denied hunting there could be no question that earlier hominids could have followed this occupation. Mousterian technology is far more sophisticated than that of the earlier Acheulean hand-axe users. Before *H. s.sapiens* evolved there can be no assumption that stone tools and animal bones are the products of hunting and still less that a major behavioural change is indicated. Binford's work at the Middle Stone Age site of Klasies River Mouth in Southern Africa led him to conclude that scavenging was still a regular and significant part of a subsistence strategy pursued by archaic *Homo sapiens* less than 50 000 years ago. His study of bones from much earlier sites which date from as much as 400 thousand years ago, shows far less evidence of tool-use. Hominids in these times were not dismembering carcasses or removing meat in any quantity to another location: they were however gathering bones for their marrow content but took little part in gaining this from live animals.

Apart from fire and better tools little seems to have changed for Binford even after the demise of *H. erectus*. Some of these conclusions are accepted by Foley (1988) who argues that hunting and gathering as the type of society that we know of today, emerged even later during the past 10 000 years. Before this, hunting certainly occurred among pre-sapiens but was not a part of the same social matrix observed among contemporary foragers. The early hominids, however, were not hunters. Their growth patterns show a markedly precocious rate of development similar to that of the apes. Work on the incremental growth of tooth enamel (Beynon and Dean 1988), which is known to be formed at a constant rate, has shown that both the australopithecines and *H. habilis* grew to maturity much more rapidly than modern humans. At the late *H. erectus* and archaic *H. sapiens* stages body size increased considerably producing a robust thick-boned muscular hominid quite unlike gracile modern humans. Foley detects no particular dietary specialisation and suspects that these hominids were generalised omnivorous animals with a static and unvarying technology. The Oldowan and Acheulean traditions indicate hominids with inflexible behaviour patterns who were committed to a uniform cultural practice for over a million years across three continents. In short, these hominids were substantially different from modern humans and lacked the social organisation of true hunter-gatherers.

With the emergence of anatomically modern humans and their spread around the world by 30 000 years ago, new evolutionary pressures arose which finally promoted the modern foraging adaptation Foley argues. A significant reduction of sexual dimorphism and robusticity occurred suggesting that with more equality in body size there was less difference in male and female foraging patterns and less competition between males. This in turn may have meant that in the latter part of the Ice Ages males were provisioning dependent females who were dominated by male alliances. The heyday of hunting was to end, however, in the post-Pleistocene period when animal resources diminished in *both* size and quantity and a new strategy of exploiting plant food in increased quantities arose and promoted greater sexual equality. In this era two new adaptations emerged almost simultaneously which consisted of agriculture and modern hunting and gathering. An evolutionary basis existed for these new strategies and both food production and modern hunting and gathering are adaptive responses to environmental change. In Foley's thesis Man the Hunter is not an archaic or ancestral way of life but merely a development that occurred in parallel with the rise of agriculture.

In a survey of predation among human ancestors, Trinkaus (1988) argues that we can no longer assume that human evolution was directed by hunting. He agrees with Foley and Binford that hunting as it is known in extant societies is a recent phenomenon associated with anatomically modern humans. Few of our distinguishing characteristics are the products of selection pressures for hunting. In fact bipedalism, manual dexterity, our large brains and extensive material culture should be seen as products of a generalised foraging system.

The partially tree-adapted australopithecines remained vegetarians. Apart from the consumption of small animals, *H. habilis* was a scavenger, but with *H. erectus* more widespread opportunistic foraging is indicated. The more pronounced nasal equipment of this hominid functioned as a more efficient means of conserving moisture from exhaled air and reducing body heat than was possible in earlier hominids with flatter noses. *Erectus* populations were able to engage in extensive foraging over larger territories in arid areas, which enabled their geographical spread and stimulated their technology. The increase in *erectus* brain size and slow technical progress suggests an increase in meat consumption but the muscular build of this hominid points to prolonged foraging that probably included small animal prey. Among the archaic *Homo sapiens*, who evolved from *H. erectus* between 500 and 300 thousand years ago, this trend continued. The bodies of our archaic ancestors were built for prolonged ranging and had a strength and endurance not seen in modern humans. This, Trinkaus supposes, was to allow them to spend a high proportion of the day engaged in unplanned opportunistic foraging rather than hunting. These hominids had a shorter life-cycle than modern humans. They also used more strength to perform similar tasks – they regularly used their teeth for manipulation for instance – they had crude tools and, judging from the absence of art objects, had a poorly developed ability to use information about the environment for their subsistence activities. Only the more gracile *Homo sapiens sapiens* who arose between 50 and 30 thousand years ago along with superior technology can be awarded the accolade of true hunter. The pronounced change in body form seen here is synonymous with a planned foraging pattern based upon extensive environmental knowledge and marks the end of the aimless wanderings of earlier hominids. The arrival of representational art, the first habitation in high latitudes, and sophisticated stone tools all suggest that less physical demands were made on humans in the Upper Palaeolithic era. Only these anatomically modern humans, with higher cognitive abilities, were able to prey upon large and dangerous animals.

HUNTING AND ARCHAEOLOGICAL EVIDENCE

The effect of these new theories has led to a reassessment of the archaeological evidence associated with *H. erectus* and archaic *H. sapiens*. Until the mid-1980s unequivocal evidence for hunting large animals appeared to be present at a number of sites containing the remains of these hominids, and the idea of hunting at this stage of prehistory was widely accepted and disseminated: for instance Wenke (1980), Johnston (1982) and Richard Leakey (1981).

At the Zhoukoudian (Choukoutien) cave site in north China, associated with 'Peking Man', a sequence of cultural developments across a 230 thousand year period which began 460 thousand years ago has been excavated (Wu Rukang and Lin Shenglong 1983). During this timespan the brain of this East Asian sub-species of *H. erectus* grew from 915 to 1140 ml, which is a size close to the upper limit for this species. Technology and hunting methods were thought to have advanced during this period and both of these practices occurred in conjunction with fire. A series of ash layers and hearths within the cave constitute what was widely accepted as the earliest instance in the cultural record of the controlled use of fire. At one time it was also widely thought that ritual cannibalism was practised at Zhoukoudian (Bergounioux 1962). A number of hominid long bones had been split open for marrow and the base of several skulls had been broken around the *foramen magnum* to extract the brain in a fashion similar to that of tribal peoples in New Guinea who eat small portions of their dead in funeral rites.

The bone refuse indicates that Peking Man's prey species included large and fast moving animals, in particular deer which form about 70 per cent of the bone remains, but over 40 other species are also represented including sheep, zebra, buffalo and even rhinoceros. There are also remains of vegetable foods in the cave. Hackberry seeds that had been roasted, and other nutritious products of a temperate forest ecology, such as elm, hazelnut, pine and rose are present in the form of preserved pollen. In a standard Chinese interpretation of the 1970s, Chia (1975) suggests that the daily fare of these hominids came from plant food and small game. Fossils of frogs, bats, hedgehogs and hares occurred frequently in the ash layers. Despite the difficulties of killing animals with crude tools, hunting is seen as a rare but important element of subsistence that was performed with clubs, rocks and fire. The Zhoukoudian evidence was therefore taken as proof of *H. erectus'* ability to utilise meat from large animals. This hominid was thought to have broken new ground by competing successfully with other carnivores and had given human ancestors access to a new energy-enriched food source.

A radical reappraisal of the Zhoukoudian material is presented by Binford and Ho (1985) in which almost all of the claims made by archaeologists who have worked at this site since the 1920s are discounted. No conclusions that Peking Man was a cannibal, a hunter, that he made fire or even that he inhabited the cave as a home can be substantiated. They argue that this hominid was not culture-based but was governed by 'a tool-aided, somatically transmitted and conditioned behavioural system about which little is known' (Binford and Ho 1985: 429). This idea is left unelaborated but presumably they have in mind a form of object-use similar to that of birds or beavers. Most of the earlier interpretations of *H. erectus* behaviour represent only a diluted version of culturally-adapted modern humans projected backwards onto a primitive hunter-gatherer state. A principal source of error on the part of earlier workers was the assumption that the presence in the cave of animal and hominid bones, stone tools and fire remains were all the products of a hominid behaviour system.

The case for ritual cannibalism and the assumption that this shows spirituality in *H. erectus* is attacked in the light of the natural agencies which modify bone. Hominid long bones were, they argue, split apart by weathering or stained by minerals to give the misleading impression of having been burned in a fire. Breakage of the cranial base and skull face is to be expected since these are the weaker parts that are frequently destroyed by animal gnawing or during deposition. The remaining evidence does not point to Peking Man as a cave-dwelling fire-maker who ate hunted animals. Since the coprolites of hyenas are associated with hominid bones it can be assumed that the cave was a carnivore den area. Bones from a variety of animals, including *Homo*, were probably brought to the cave by scavengers and were not the household refuse of hominids. Nor were the ash layers produced by hominids roasting meat, since they are not true hearths. They are rather the results of natural conflagrations of huge guano deposits that burned by spontaneous combustion. Finally they claim that the mass of deer bones were found in strata that lack hominid remains and could not have been accumulated by hunting because no evidence of hominid use of the cave can be found at this time.

Binford and Ho's attack has changed the existing consensus, particularly with regard to Peking Man's alleged cannibalism, but it has not won over all scholars. Their work was not done at Zhoukoudian but, as they say, 'at a distance', and involved a re-examination of the stratigraphic records and photographs of various excavations. A number of inadequacies in their analysis are apparent. Stone tools were found in the cave together with raw stone material and the debris from tool manufacture. Since these items *are* associated with the deer bones the hypothesis that animal remains were introduced only by carnivores and that hominids did not use the cave in a habitual fashion is difficult to sustain. Apparently there has been no examination of the bone remains of some 3000 deer found in the cave for cut marks and, without the fire evidence, the possibility that *H. erectus* was only a small-time forager-scavenger cannot be ruled out. However evidence presented against hominid hunting rests only on the inadequacy of the known *erectus* technology and a priori arguments about animal scavengers. Similarly the argument that the ash layers were once guano deposits and not hearths is not backed by more than circumstantial evidence. The presence of charcoal granules in the Zhoukoudian ash layers, which would not result from burning bird droppings, and the presence also of burned hackberry seeds and carbonized redbud wood, found only in the ash layers, has convinced authorities that Peking Man was a genuine fire user. Animal bones showed clear evidence of having been in a fire since they were found to be partially carbonized, curled and cracked in heat which could not be reproduced by mineral staining. An abundance of quartz tools were in fact found in the ash layers indicating the presence of hominids. Along with this material were many fossilised deer antlers, that would have provided little meat for animal scavengers, and these had a breakage pattern not found among wolves or hyenas. The idea of *H. erectus* following a non-cultural, tool-aided adaptation is unexamined since the advent of 'humanisation' and culture are seen as synonymous. Yet this is a hominid whose brain grew by about twenty per cent, a process which is known to be physiologically costly, and whose technology shows a degree of change and refinement during a sojourn of nearly a quarter of a million years at Zhoukoudian.

At European sites occupied by Acheulean peoples in the same time span, similar doubts have been raised about hunting abilities and debate has been joined. These sites, which may have been formed by *H. erectus* but more probably by archaic *H. sapiens* (Brooks 1988), seemed to offer firm evidence that large and dangerous animals were preyed upon by hominids. Freeman (1973, 1981) has made a survey of Spanish sites from the end of the Lower to the Upper Palaeolithic eras and concludes that a growing sophistication of hunting techniques can be detected. One hypothesis holds that hand-axe using hunters and their prey were attracted northwards into Europe during an amelioration of the climate (Wymer 1984). The evidence of hominid occupation has been taken to suggest that mobile hunter-gatherers had worked out an adaptive pattern that took advantage of the annual migrations of herd animals (Butzer 1982).

About one-third of a million years ago bands of hunters in Spain and France were thought to be exploiting an abundant supply of plains-living herbivores. A wide variety of species were killed including horses, wild cattle and rhinoceros but extensive excavations of elephant-hunting middens provide the most dramatic evidence of an emerging subsistence pattern. At Torralba and Ambrona in central Spain, elephants were driven into marsh land, killed and butchered on site. These drives were likely to have been the work of large groups of hunters, representing several bands, who worked in coordination for large returns of meat. One estimate (Freeman 1981) is that one such hunt would have yielded over 13 500 kilograms of meat which would have fed 100 adults for 60 days. It is reasonable to suppose that these hominids must have used food-preserving techniques. The elephant carcasses show that extensive removal of flesh had occurred and nearby butchery sites have been found where meat was cut into small pieces. Drying and smoking methods of preservation, together with all the social organisation that this would entail in terms of management and ultimate division of meat within and between bands, can be surmised. These bands certainly possessed fire, which had been used as the means of driving the elephant herd, and had probably also had clothing and rudimentary shelters.

Many of these suppositions are substantiated by a contemporary French site at Terra Amata on an ancient beach location in Nice. The findings here (de Lumley 1969) are of a hunter-gatherer band's seasonal camp which, the pollen record reveals, was occupied during eleven separate spring–summer seasons. Traces of 20 oval huts, containing hearths and tool-making areas, have enabled accurate reconstructions of what are the oldest known buildings. These crude, rapidly erected, draughty structures, which consisted of a series of branches braced by stones and stone piles which served as windscreens for hearths were surprisingly large; they ranged from 26 to 49 feet in length and 13 to 20 feet in width. The huts' occupants were mobile foragers with a mixed subsistence pattern. There are signs of plant gathering and that small game and seafood was eaten. Stag, wild boar, elephant, rhinoceros and ox are represented among the bone remains and the hunting of these animals was probably the reason for the camp's existence. The hunters had a wide knowledge of the habits and location of a large number of animals and plants. It is clear that they had a large hunting territory within which they moved throughout the year since Terra Amata contains tools made of volcanic stone that originated thirty miles away. Compared with the Oldowan industries of the Lower Palaeolithic there are a considerable variety of tool types present and the cultural use of raw materials had been extended. Tools made from bone are present and remains of what was probably a wooden bowl is indicated. The use of animal skins as clothing can be assumed from impressions made on the hut floors. About 300 thousand years ago Terra Amata served for a few days each year as the spring quarters of a hunter-gatherer band whose annual movements across large tracts of land during different seasons reveals an ability to utilise the environment far more successfully than early *homo*.

In this view, which prevailed until the mid 1980s, progressive developments that anticipated modern humanity occurred in these times. Butzer's model (1982) of Acheulean peoples is one of mobile hunter-gatherers who adjusted their subsistence patterns to accord with the seasonal movements of animals. It is no accident that the Torralba-Ambrona sites are situated in a mountain pass which gives access to plains on either side since this was a natural migration route between northern summer steppe grassland and southern winter pastures. Such a location would have provided a perfect place from which combined bands could engage in organised game drives in which they ambushed gregarious animals in spring and autumn. In winter and summer the lack of game would have forced the bands to disperse to temporary camps close to water where raw stone was extracted and tools made. Archaeological evidence supporting this model includes high concentrations of bones and stone artifacts within the Torralba-Ambrona pass area, the presence of flint from locations some tens of kilometres distant and the remains of migratory birds in food middens.

The hominids of these times had explored a wide variety of habitats in Africa and Eurasia and there were few ecological zones which had not been penetrated. These societies were firmly based upon hunting big game since it was this activity that had stimulated hominid migrations and had enabled effective adaptation (Wymer 1984). Little is known of the actual hunting weapons or techniques used beyond a few wooden spear fragments recovered from a number of sites, and conjectures that game was driven into marshlands or rocky cul-de-sacs can be drawn from the position of some sites. It is possible that the bolas was used to stun animals since spherical pebbles have been recovered from Acheulean sites, but if traps, pits or poison were used no sign of these has been found. Evidence for the use of fire, though rare, is certainly present at Terra Amata and at African sites. At Hoxne, in Suffolk, there are signs of cooking meat, and the pollen record and presence of charcoal in lake sediment suggest that it is possible that hominids here were using fire to clear forested land. Hoxne has also produced a series of stone tools with micro-wear polishes that indicate a meat diet. Hand-axes and single flakes were certainly being used to cut meat, for butchery, skinning and hide scraping besides being applied to plants and wood. A whole range of activities are indicated here that include production of various artifacts, clothing and shelters. At the French cave site of Lazaret, which held Acheulean occupants 130 thousand years ago, some modification of living space is found in the form of a piled stone barrier which was perhaps used to demarcate different areas. Refuse was excluded from the living and sleeping zones, a drainage channel dug and dried grass and seaweed found in the cave were probably used as bedding. The fact of fire placed at the centre of a dwelling, as in this cave and at the much earlier site of Terra Amata, implies a considerable social advance over Oldowan peoples and suggests that some elements of the human condition had arisen.

However, the image of hunting as the foundation of Acheulean societies in Europe and the idea that hominids were developing foraging patterns similar to those of anatomically modern people now faces criticism and revision. As at Zhoudoukian the assumption that stone tools and bones signify active hunting is attacked. Klein (1988) dismisses the idea that firm conclusions can be drawn from the Torralba-Ambrona material and argues that natural causes, such as the work of water in creating bone concentrations and animal scavenging, must be examined. These doubts are given substance by Klein's claim that abrasion patterns on the bones were produced by movement in streams. In fact the bone assemblage represents a predictable collection of skeletal parts caused by fluvial sorting. Hyena coprolites were intermingled with the bones and there was a lack of clear cut-marks on bone surfaces produced by artifacts. Compared with other Acheulean sites there were few stone tools present – only 785 artifacts were actually recovered – and the age-at-death pattern of the animals, which could be estimated from the teeth, indicates that natural causes rather than hunting caused their demise. Perhaps scavenging by hominids occurred at Torralba and the possibility that hominids in middle Pleistocene times engaged in limited forms of predation is held open, but it was only in the Upper Palaeolithic era that a quantum advance in hunting occurred and therefore the proficiency of earlier peoples is rendered questionable. Many of these conclusions are shared by Binford (1988) who, in performing another 'distanced' re-examination of the archaeological record, joins with Klein in attacking the 'earlier romantic view' of hunting by pre-modern hominids. He accuses Freeman, who has led excavations at Torralba, of a series of unsupported interpretations which derive from poor methodology. Claims that a 'living-floor' can be identified are unfounded, and organised game drives were not the reason for the bone concentrations; indeed hominids played only a minor part in their formation during opportunistic scavenging.

This critical onslaught has led to a more intensive examination of archaeological material from all phases of the Palaeolithic era but, while it has led to a modification of earlier theories, it has also stimulated a counter attack in which Binford's ideas have been the principle target. It has been accepted that bone material has a complex history involving many interacting natural agencies besides those of hominids. However, the reality of meat-eating throughout Palaeolithic times is held to. Bunn and Kroll (1986), for instance, argue forcefully that early hominids at Olduvai Gorge about 1.75 million years ago, in the opening phase of the Stone Age, were using tools to systematically butcher the carcasses of a range of small and large animals, that they ate substantial amounts of meat and bone marrow and that these aspects of diet came from a subsistence strategy that involved both scavenging and hunting. They assert further, in support of Isaac, that the large quantities of meat involved here strongly imply social cooperation and food-sharing on a scale unknown among apes.

Toth's experimental work with stone tools, which was discussed in Chapter 6, pointed to single unmodified stone flakes, struck from Oldowan cores, as important artifacts in themselves. Flakes have been found to be highly efficient cutting tools and their abundance at early African sites, complete with meat-working polishes, seen by micro-wear analysis, is strong evidence for hominid meat-eating. But Bunn and Kroll also found that gnawed bones from Olduvai formed a considerably lower part of the total bone assemblage than would be expected had they been accumulated by carnivorous animals, and the presence of cut-marks suggest that animal scavengers were active *after* hominids had finished feeding. They dismiss Binford's notion of hominids as marginal scavengers by showing that hominids had full access to the carcasses of small and large animals. The most meaty portions were brought for consumption to the site and Binford's claim that only less nutritious cuts were available 'should be regarded as a myth that is contradicted by the archaeological facts'

(Bunn and Kroll 1986: 439). They also argue against Shipman and Potts' findings, which were introduced in the last section of this chapter, that hominids did not engage in full-scale disarticulation and butchery. The clustering of cut-marks on skeletal parts could only have resulted from systematic butchery in which disarticulation and defleshing occurred after skinning. The question of how the carcasses were obtained should not be reduced to a simple dichotomy between hunting and scavenging they argue. The experimental work of Schaller and Lowther (1969) showed that even the carcasses of large animals who were sick or injured were available for hominids provided they gained the meat ahead of other carnivores. However the high proportion of adult animals represented in the Oldowan bones is inconsistent with death from natural causes which would be more likely among the youngest and oldest animals in a population. Passive scavenging was therefore less likely to have been the reason that these bones were transported by hominids. It is also unlikely that our ancestors collected the residue of carnivore kills because little meat would have survived. The fact that highly nutritious limb bones were found in abundance at Olduvai complete with evidence of defleshing and marrow extraction points to the possibility that active scavenging and hunting occurred. In this case hominids may have confronted and driven off carnivores at kill-sites or have dispatched large animals in a vulnerable condition while they may also have actively hunted smaller animals. There is as yet no archaeological means of distinguishing between these possibilities but in each instance there was a paramount need for social coordination and group solidarity to enable participation in such a dangerous mode of subsistence.

If the possibility of hunting and active scavenging by early hominids 1.75 million years ago is to be seriously considered there is reason to suspect that by mid-Pleistocene times this was an established part of behaviour. Evidence for hunting large animals at European sites in these times is presented by Svoboda (1989) who attacks the scepticism of Binford and his colleagues as excessive and overcautious. At the Spanish site of Aridos, which is contemporary with Torralba, hominids were clearly exploiting elephants. Villa (1990) claims that while the Torralba evidence is too poor to support hunting, the Aridos material allows us to infer at least active scavenging in which a degree of planning and anticipation is evident. This is seen from the fact that hominids had gained access to the elephants ahead of scavenging animals which were represented by only a few rodent tooth marks on bones. Ready-made tools and several kilos of raw flint had been carried to the site from a location at least three kilometres away and tools had been made in an unhurried manner close to the carcass. Several visits to the site occurred and tools were resharpened as butchery proceeded. Even though hunting cannot be proven here, this evidence does point to a more methodical form of subsistence than the ad hoc type of marginal scavenging proposed by Binford.

In the same time period the first firm evidence for the use of wooden implements is found in the form of a possible spear point from Clacton, England (Oakley et al. 1977). This 15 inch piece of preserved yew wood was found in association with stone tools and the remains of deer, horse, bison, rhinoceros and elephant that are dated to 400 to 300 thousand years ago. The wood had been worked to a point and stone tool marks are seen on the shaft. No indication of either fire-hardening of the point or scraping, to produce an aerodynamic spear, is evident. However the long tapered point would have taken considerable time and effort to achieve and would have been more appropriate as a thrusting spear, that was able to penetrate thick hides, than as a stake or digging stick. Yew wood is capable of being worked to a very sharp point and while it is not clear what type of tool was used here the micro-wear polishes found on Clactonian implements show that a whole range of wood-working activities occurred.

The Clacton point has been compared with a fire-hardened spear from Lehringen, Germany, found between the ribs of an elephant. This spear, which was over two metres long, is 120 thousand years old and had its point of balance at the lower end of the shaft, which would have made it difficult to throw with any accuracy, and is therefore more likely to have been used for thrusting. Similar thrusting spears are used by modern African hunting peoples for large and dangerous game such as hippos and elephants. These ethnographic comparisons suggest that the Clacton point was also part of a heavy large diameter thrusting spear. The Clacton point does not have the polish found on digging sticks and suggests that active scavenging and hunting was a part of hominid subsistence in these times.

The case for hunting as a recent adaptation confined to the Upper Palaeolithic is also opposed by Chase (1987, 1989) whose work on later Middle Palaeolithic sites, c. 110 to 40 thousand years ago, provides evidence for the regular killing and butchering of large animals. Hominids in these times, he suggests, were likely to have been occasional scavengers but could not have relied on such a haphazard food source. At the French site of Combe-Grenal the absence of the largest meat bearing bones from deer carcasses and the concentration of cut-marks on the meatiest parts of horse and bison remains, suggest that hominids here were hunters rather than scavengers. By taking the best parts of the carcass hominids had shown that they were ahead of other carnivores and that they were not forced to rely, like scavengers, only on marginal body parts such as the outer limb joints. All the tool marks on the horse bones were produced by cutting through fresh soft flesh with no indication of hacking or bashing at a carcass already stiffened by rigor mortis or frost. The age profile of the horse remains also points to hunting because of the absence of the old and young who could be expected to have died from natural attrition. In fact none of the horse samples from Combe-Grenal are consistent with an expected scavenged population. Evidence from other Middle Palaeolithic sites shows that hominids were hunting dangerous animals such as leopards and wild boar for their skins, and numerous middens of bison remains found throughout Europe point to a high level of predatory competence.

Hominids of the European Middle Palaeolithic were no less specialised hunters than their descendants in early Upper Palaeolithic times in the sense that a single animal species was exploited as a major food source throughout the year. The evidence that large numbers of herd animals were consumed by Middle Palaeolithic peoples can reasonably be taken as an indication of cooperative hunting. Planning, foresight and food-sharing are also reasonable inferences since these sites were not just kill and consumption locations. Killing and butchery sites as well as hunting camps can all be distinguished archaeologically which implies the regular transportation of food and cooperation for activities like driving large game over cliffs. Bone remains often tend to favour a single migratory herd species suggesting that seasonal hunting trips were made to key locations. The bone evidence could not have been produced by random or opportunistic killing, in fact it demonstrates a pattern of social organisation which '. . . involved a degree of foresight and probably [of] cooperation which, archaeologically, is indistinguishable from those in modern hunting systems' (Chase 1989: 334). In short, Chase argues that there is no evidence to sustain Binford's claim (1985) that hominids were incapable of hunting large game or that they lacked the ability to engage in long-range planning and food-sharing before Upper Palaeolithic times.

The issues here remain unresolved. The economic behaviour of prehistoric peoples, and hunting in particular, have always produced controversy perhaps because of the variety of inter-disciplinary evidence and popular myths which are involved. The problems at stake here – hunting competence and sharing behaviour before the emergence of anatomically modern humans – cannot, it appears, be resolved by archaeological or fossil evidence alone. It is reasonable therefore to propose that ideas from the social sciences can make a useful contribution. In Chapter 8 it was proposed that both modern and archaic foragers were likely to have responded to ecological pressures with a broadly similar pattern of social organisation. A distinctive type of hominid society with some behaviour patterns that had coherence between species is also likely. A technologically dependent economy, in which a mixture of hunting, active scavenging and gathering formed the subsistence base, could only be sustained through cooperation and reciprocity. There is nonetheless a body of folklore which embraces a perception of humans as an inevitably selfish and destructive species with aggressive propensities that exceed those of other animals. These traits, it is thought, can be traced to our experience of hunting in prehistory. In one such myth our desire for meat marks the beginning of our fratricidal career, and in another hunting is seen as part of a compulsive process that irrevocably leads to environmental devastation and the extinction of species. We shall examine two varieties of these myths, using in turn the perspectives of archaeology and anthropology.

HUNTING AND VIOLENCE

In the last chapter the idea that hunting alone formed the basis of the foraging economy was reviewed, and in this chapter the possibility that the cooperative hunting of large animals was not practised until late prehistory has been discussed. The foraging economy was shown to be a complex adaptation involving a varied series of inputs among which the pursuit and killing of large game was usually a significant but relatively small part. But, as we saw in Chapter 8, there is a considerable body of ideas that attribute destructive and violent behaviour in humans directly to predatory experiences during prehistory. Popular belief in this area has often been fuelled by anthropology and here we shall look at one such influence which is drawn from the work of Raymond Dart whose conception of early hominid behaviour arose from his excavations at australopithecine sites in South Africa.

Dart discovered and named the australopithecine genus in 1924 having made an important find at the Taung cave site. In his later work, he developed the idea of hominids as 'killer apes' whose evolutionary success was based upon their violent carnivorous proclivities. The australopithecines were seen as tool-making hunters with a developed pre-stone material culture (Dart 1953, 1957, 1959, 1967). Dart extracted a mass of fossil bone evidence from the South African cave sites to support his claims for an 'Osteodontokeratic' – literally a 'bone-tooth-horn' – culture that included daggers, clubs, blades, knives and saws that had been fashioned by early hominids as hunting tools. The transition from ape to hominid had occurred because of the ability to kill large fast animals, and accompanying this predatory drive was a strong urge to behave violently that Dart described vividly. His australopithecines battered their quarry to death, drank its blood and tore it limb from limb. There were no inhibitions about devouring the writhing flesh of animals or in hunting other primates. Small parallel puncture holes in baboon and hominid skulls had, he claimed, been caused by the thrusting of bone daggers. The presence of so many depressed skull fractures was evidence of systematic clubbing. In this era lie the origins of our contemporary desire for hunting and the gratuitous killing of our own species. Human cruelty and murderous urges derived from these times, and are a necessary by-product of a common blood lust that is based upon an implacable addiction to red meat. This most basic of human traits underlines all cultural manifestations of vegetarianism and is a basic part of our essential nature. 'The blood-splattered, slaughter-gutted archives of human history . . . [include] . . . universal cannibalism, systematized animal and human sacrificial practices' which together with head-hunting, mutilation and necrophilia, place the 'marks of Cain' on our species (Dart 1959: 198/9).

Dart's ideas feed straight into crude forms of biological determinism and became widely disseminated in the hands of his populariser, Robert Ardrey. Ardrey (1961, 1976) presented the australopithecines as battling apes whose evolutionary progress came from competition with established carnivores. Many features that are central to modern society such as drives to achieve male dominance, acquire property and status together with territoriality and hostility to neighbours and strangers were all seen to have derived from ancient animal traits which became fixed in our species because of their utility as part of hunting behaviour.

Ideas concerning the role of conflict in human evolution are frequently encountered. Bigelow (1975) for instance presents a hypothesis that brain enlargement occurred because of competition between hominid groups, which led to the greater need for cooperation within groups, making both unity and warfare essential aspects of the human condition. However the difficulty of connecting such theories with substantiating evidence is considerable. In a survey of the evidence for killing among hominids throughout the Pleistocene, Roper (1969) is unable to produce convincing evidence of violence in early prehistory. Only the now questionable instances of cannibalism at Zhoukoudian are cited as examples of murder in Lower Palaeolithic times.

Even before the emergence of taphonomy, the specialist branch of palaeontology devoted to the fossilisation process, many anthropologists were sceptical of Dart's main arguments. Washburn (1957) pointed to the unfounded assumptions of tool use and carnivorous diet that Dart had made. Later studies have confirmed that A. africanus was a herbivore whose teeth markings indicate a diet of fruit and leaves. The food middens that Dart had assumed to be the results of hominid hunting were much more likely to have been produced by carnivores, such as the hyena, and the australopithecine bones were merely the remains of their meals.

In the years following Dart's claims, taphonomy has provided an understanding of the principles governing the natural processes by which bone becomes fossilised and accumulates. There have also been studies which have reconstructed the stages of cave formation at the australopithecine sites. On this basis it has become possible to make a reliable reinterpretation of the Osteodontokeratic artifacts that form the basis of Dart's case. In a series of papers Brain has established that there is a predictable pattern of survival and modification of bones which comprise a carcass. Dart had commented on the fact that some animal parts were rarely present at the australopithecine sites. He assumed that the missing bones were not brought into the caves because the occupants had put them to specific uses. Tails served as whips or hunting signals, segments of backbone as projectiles and the long bones as clubs. However, Brain's experimental work has shown a close match in terms of the composition and type of damage sustained in both the ancient fossil bones Dart had excavated and those resulting from a modern simulation study in which experimenters monitored goat bones as they were consumed and discarded by scavenging carnivores. Natural processes ensured that some bones with tough physical characteristics, such as those from the jaw or skull together with teeth were much more likely to survive and become fossils.

Dart had also made assumptions about australopithecine tool-use on the basis of the suggestive form taken by some bones which he took to be implements. In a study by Shipman and Phillips-Conroy (1977), however, it was found that natural breakage patterns reproduced similar forms in bone. It was possible to find fragments of bone in a savanna location that had recently been chewed, crushed or regurgitated by hyenas that closely resembled Dart's Osteodontokeratic tools. There is no need then to invoke hominid activity to explain either the disproportion of skeletal parts that had become preserved or the forms taken by these remains.

Finally, Brain has provided a plausible explanation for the presence of the bone material discovered by Dart within the caves. By reconstructing the probable land forms in the vicinity of one site at Swartkrans it was possible to determine that Dart's finds had come from what had been the base of a 50 foot vertical shaft that originally rose above the modern cave to an entrance at a small depression or sinkhole. It was down this shaft that animal and hominid bones had been washed over the millennia. The bones came from baboons and australopithecines but they were overwhelmingly from antelope species: a composition which gave the appearance that they were carnivore food remains. The concentration of bones around the sinkhole is best explained by the fact that it was often only in such locations that trees grew in a fire-prone and waterless dolomitic landscape. In this type of environment sabre-toothed cats, particularly leopards, are known to place carcasses in tree-dens to protect them from non-climbing hyenas. The combination of leopard and hyena feeding produced both the bone-forms and skeletal composition that were typical of Dart's Osteodontokeratic artifacts. The leopard hypothesis is considerably strengthened by Brain's findings that the teeth marks of these animals are found on baboon and australopithecine skulls. The parallel puncture holes above the eye sockets that Dart took to be a sign of murderous attack by hominids on their own kind have the same spacing as a leopard's canine teeth. Leopards are known to drag their prey to a feeding place in this manner and this more prosaic cause for the injuries and accumulation of bones is more convincing than Dart's sensational scenario.

HUNTING AND THE ENVIRONMENT

What were the long-term effects of hominid hunting on animal populations? Little, if any, evidence is available to determine the ecological impact of the early hominids on African fauna and we must turn to later prehistory when the undoubted predatory abilities of our own species were well established to discuss this issue. It will be productive to follow a controversy on the scale and effects of hunting in these times if only to appreciate the relative differences in resource utilisation by foraging peoples living in opposite extremes of the Stone Age. This excursion into the Upper Palaeolithic era monopolised by *H.s.sapiens* is distant from the world of the Oldowan and Acheulean tool makers but it can also be justified on the grounds that the relatively enhanced efficiency of our recent ancestors provides us with a yardstick to assess the ecological effects of the early hominids. If the impact of modern peoples following a hunting and gathering way of life can be shown to have been largely benign, the early hominids, with their smaller populations and more primitive technology, must have posed little environmental challenge. The social sciences provide us with models of foraging as a type of society that are useful in determining the likelihood of food-sharing among hominids. Similar insights from the social sciences can be usefully applied to a debate about the effects of hunting in recent prehistory. In this instance it was proposed by Martin (1973) that predatory efficiency was so successful that when innovative hunting technology was introduced into North America by the first human occupants about 11 000 years ago significant numbers of animal species became extinct.

The 'Overkill hypothesis', as Martin's claim became known, held that at this time an unprecedented number of large-bodied animals vanished as new hunting cultures overexploited resources. These 'Megafauna' included several varieties of elephant and mammoth, camels, horses, the giant bison, llamas, a bear species, giant rodents including the ground sloth and carnivores such as sabre-toothed tigers. The extinctions seemed to be concentrated within a few centuries 11 000 years ago and the only coincidental event in these times, argues Martin, was the appearance of big-game hunters in North America. A hypothesised first crossing of the Bering land bridge between Siberia and Alaska 1000 years before this gave Asian hunting peoples access to a previously untapped food source in America. A lethal swath was cut into animal populations as hunters moved south through a temporary ice-free corridor in an advance across a broad front. Animals in a previously uninhabited continent lacked effective defensive behaviour and were unprepared for humans. The invaders had developed what has come to be known archaeologically as the Clovis and Folsom hafted spear points that seemed to have been specifically designed for big-game hunting. Devastation of large herd animals was followed by the demise of their dependent carnivores and scavengers. The new stone projectile points allowed an expanding population of hunters to move through the continent engaging in an orgy of slaughter for a few brief centuries. The animals had survived previous climatic changes in earlier Ice Age times and once extinct were not replaced by new species or by migrants from territory untouched by hunters. Because large-bodied animals reproduce at slower rates than smaller animals the sudden toll taken by human predators could not be absorbed by a population, and whole species were overwhelmed. Martin estimates that an annual removal of thirty per cent of animals occurred by hunting. The abrupt and final disappearance of over half America's megafauna can only be explained by the coming of an explosive wave of human hunting cultures. Computer simulations and mathematical models (Smith 1975, Mosiman and Martin 1975) have shown the plausibility of the hypothesis, but the proposal has initiated a lively debate and the idea of overkill has been attacked on all fronts by a range of disciplines.

Can a connection between hominid hunting and animal extinction be substantiated? An earlier attempt to argue for overkill in Africa by Martin (1966) with the emergence of hunters using developed Acheulean technology 60 to 40 thousand years ago has been rejected because of insufficient evidence. Webster (1981) has seriously undermined the Overkill thesis in Africa. He points out that there is no indication that greater hunting efficiency would have resulted from enhanced Acheulean tools in this era and anyway hominids were exploiting large game animals well before these times. Furthermore climatic change leading to deterioration of the once fertile Saharan environment is now known to be synchronous with the loss of some species, and the lack of subsequent extinctions in the presence of much more efficient hunting cultures remains to be explained by Martin.

The Overkill hypothesis in North America has become vulnerable to new lines of evidence that question firstly the chronological relationship between human entry into North America and the timing of the extinctions. Dates for human occupation of the Americas well before the last 12 000 years have damaged the overkill case as they accumulated, despite some scepticism reported by Patrusky (1987). These new dates seriously weaken Martin's claim that the Clovis peoples were the first Americans who swept into the continent just before the extinctions. Any earlier human presence would nullify his contention that animals were overwhelmed by the sudden appearance of a new super-predator against whom they had no innate defences. Extremely crude stone tools from an Andean cave site in Peru dated to 22 000 years ago (MacNeish 1971), which show none of the finesse seen in Clovis technology, compromise Martin's case but have not gained universal acceptance. However recent detailed evidence of a well-established community of hunter-gatherers, living in a dozen huts at a long-term camp-site at Monte Verde in a forest environment in southern Chile 13 000 years ago, is provided by Dillehay (1984). Since no serious alternative has ever been proposed for entry to the Americas except by the Bering Strait the presence of hunting peoples so far south must indicate a much earlier date for the first colonisation of North America. Jennings (1978) argues that hunters were well to the south of the ice sheets considerably before 11 000 years ago. In fact the first well-dated human presence in America can be placed 27 000 years ago. He shows that the Bering land bridge was open on four occasions during the past 60 000 years for periods which lasted from 5 to 10 thousand years. Hunting bands using a Eurasian tool-kit were active in North America up to 8000 years before the extinctions, with no signs of overkill. To be sure, another wave of invading Palaeo-Indians with Clovis technology could, as specified in Martin's theory, have advanced out of Asia but there is no archaeological evidence of similar tool types which connects them with Siberia.

There are strong demographic reasons for questioning the Overkill hypothesis. The extinction model presented by Mosiman and Martin (1975) depends on high rates of hunting associated with a rapidly growing human population which itself impelled the advance through the continent. However there must be considerable doubt that the rates of population growth envisaged in their simulation could be achieved by non-agricultural peoples. Annual growth rates of between three and four per cent, causing the population to double in 20 to 40 years, would have been necessary to have enforced the rapid forward migration by the hunters in the time specified. But this rate of increase is close to record levels of population growth found among modern peoples whose access to secure food supplies and medical support has meant low death rates. In fact sustained rates of population growth above 3.5 per cent over a 20 year period are rare even in poor Third World countries where dramatic improvements in life expectancy have occurred only in the past four decades (World Bank 1991). In a hunting and gathering society, however, mortality levels are known to be high and will often have the effect of causing zero rates of population increase even when no restraints are placed on fertility such as infanticide, abstinence or prolonged breast-feeding. Deaths from hunting accidents, particularly if megafauna are pursued, would have been high among Paleo-Indians and fertility would have been depressed by the stresses placed on women by high mobility. For these reasons it is difficult to conceive of a sustained rise in population at levels close to a theoretical maximum over the course of several centuries.

Another, and perhaps more important, line of evidence which is anthropologically informed can also be applied to Martin's thesis. The Overkill hypothesis relies on an uncritical acceptance of the 'man the hunter' image in which big-game was the principal and preferred food source. Ethnographic evidence has now drastically modified this view as we have already noted. In place of the earlier caricature has come the view of hunting society as one with a balanced foraging economy in which most energy is expended in gaining small game and gathering vegetable food. The Overkill hypothesis therefore becomes much less plausible when it is reviewed against the normal subsistence strategies of hunter-gatherers.

Lee's general research and his empirical work with the !Kung (1968, 1979) have shown that hunting usually plays a lesser part in food provision in foraging societies than gathering plant food. This finding, which has established a new consensus in anthropology, is based upon firm ecological and economic principles. The most efficient mode of extracting food energy from a landscape is a comprehensive subsistence strategy in which a broad range of both vegetable and animal products are utilised. This type of mixed foraging is likely to be much more reliable than a specialised strategy, based upon a single species of animal for instance, because different items of diet may be substituted in times of shortage. If big-game were relied upon exclusively, much greater band mobility would be demanded and this would impose severe limitations on fertility and on the size and age/sex composition of bands. But such reliance on big-game would also be inherently wasteful in that males would be burdened as the sole procurers of food, and the labour of females as potential suppliers of two-thirds of a band's food not utilised. The benefits of a sexual division of labour have been seen by Sahlins (1972) to be highly efficient in that all the subsistence activities of foragers can be confined to a few days of a week. Foraging societies typically enjoy leisure time in excess of that found in modern industrial and peasant-based economies.

It has been observed that in hunting societies where big-game is abundant, plant food is still gathered by choice. The idea of an optimum foraging strategy, which was described in Chapter 8, may perhaps explain this observation. Foraging behaviour will be oriented towards an optimisation of energy and time which in turn will necessitate choices about which varieties of food resource to pursue and what degree of specialisation is appropriate. The earliest American optimum foraging strategy would be unlikely to have privileged more than a few animal species, and would not explain the wholesale extinction of megafauna since hunters would leave intact populations of less desirable animals (Webster 1981). But would these peoples have adopted an inefficient strategy of saturation hunting or would they have attempted to achieve an equilibrium with their resources? It would be idealistic to envisage hunting peoples as natural conservationists with an economy that is always environmentally friendly. There are many instances of wasteful subsistence strategies in foraging societies in both modern and prehistoric times. Game-drives into marshland or over cliffs have killed more animals than can be consumed; tree-cover has been lost by bush-burning to provide more grazing for a prey species, and some animals killed only for their hides or horns. Hunters will frequently be unable to use all parts of an animal. Inevitably the entire carcass of a large animal will be subject to some wastage by scavengers or will spoil before the meat can be preserved or stored; hunters may be unable to carry more than selected parts to camp and will abandon the remains to carnivores. Historical examples of animal extinction through human action may be easily found, yet most of these extinctions have occurred among agriculturally based societies where habitat change, rather than hunting, was the cause. Only in non-foraging societies, where food is produced, can extinction occur without serious economic loss. Hunter-gatherers, however, have no such impunity and the motivation for indiscriminate slaughter as a part of a very inefficient strategy is hard to envisage.

The food yields of North American megafauna would have been prodigious. A single bison bull for instance would provide 250 kg of edible meat (Wheat 1967) and would have easily fed an active adult for six months. The total food yields per person which would result from the level of hunting which Mosiman and Martin (1975) envisage are vastly in excess of nutritional requirements. Killing at even the lower rate required in their simulation would have provided a daily personal meat ration of 10 lb or 4.5 kg for each Paleo-Indian regardless of age, and would have meant enormous wastage if the estimated animal population was to be driven to extinction. The rate of meat consumption is four times higher in Martin's estimates for Clovis peoples than for modern hunters with high meat diets and 30 times higher than among the G/wi Bushmen. Webster calculates that as much as 80 per cent of edible meat would have been wasted even in an all-meat diet. But this rate of killing would also have meant a phenomenal investment of energy, time and risk on the part of the hunters and would depart totally from the foraging strategies of all known hunting societies.

Among modern foragers living with abundant resources, leisure for social activities is preferred to extra labour expenditure and higher levels of consumption. The costs of increasing the rate of predation would include considerable additional personal dangers for hunters and would necessitate the frequent movement of whole bands in pursuit of elusive prey species. In correctly perceiving these costs, hunting societies have chosen to exploit a broad range of food resources and enjoy long occupancy of their camp-sites. In the mid-nineteenth century a population of about 100 thousand Indian hunters in the American Great Plains lived among bison herds that were 30 to 40 million strong and yet their rate of hunting, even in these bison-based economies, was well below the animals' rate of reproduction. Like other foragers they frequently substituted other animals and small game for their main food source while they also relied on gathered plant food whenever hunting was uncertain as in the winter, when this was eaten in a processed form. Even in these abundant circumstances hunting was difficult and unreliable as herds migrated unpredictably or outran their predators. The Indian societies here could not afford to be profligate with their animal resources and preserved bison meat in dried form was used as a means of reducing wastage.

Many of these observations also apply to Paleo-Indians at the time of the extinctions and are confirmed by archaeological evidence. At a kill-site in Colorado in which nearly 200 bison died in a stampede caused by Clovis peoples 8000 years ago, Wheat (1967) found that the hunters had carefully butchered almost all of these animals. Only a few carcasses that had become wedged into a gully and were inaccessible to the hunters were left intact. A standardised form of butchery was used by the Clovis hunters that Wheat considers to be more systematically organised and efficient than the practices of Plains Indians in historical times. The herd provided as much as 30 tons of meat yet, in spite of this abundance, there is good evidence that little wastage occurred. Initially the hunters ate from the fresh carcasses at the site. Having cured the bison skins they then preserved the remaining meat and fat as pemmican by drying and pounding.

The extinction of large mammals in North America would have required a large population of specialised hunters whose activities were uncharacteristically dedicated to only a few species. Such carnage would surely have left an archaeological record. In a survey of American and European mammoth sites Olsen (1990) suggests that while these animals may occasionally have been killed they formed no more than a minor part of human diets. Evidence of mammoth deaths in North America shows a high preponderance of animals having become trapped by natural causes. Nearly three-quarters of the 856 known mammoth sites were once stream, marsh, bog or lake-side environments and since all of the skeletons associated with Clovis hunters are found in such deposits it is reasonable to assume that the animals had become entrapped. The possibility that the mammoths were driven into marsh land by hunters, rather than discovered in this state, has to be considered but since only six per cent of all mammoth sites have any association with human hunting there is no archaeological evidence to link Clovis peoples with mammoth extinction. On the other hand evidence for non-human predation in the form of hunting by sabre-tooth cats and wolves is found at these sites and can be seen to have caused far more deaths among mammoths than those attributed to humans.

Beside the spectacular evidence of large-mammal hunting by Clovis peoples there is also other evidence for the consumption of small game which probably provided most of the daily fare. The mundane food remains of vole, muskrat, gopher and fox together with birds, turtles and fish show that a varied fauna was utilised. While these findings are consistent with hunter-gatherer economies in general they also agree with what is known of some specialist prehistoric hunting strategies. At a number of sites in the Ukraine substantial mammoth bone structures have been found made from long bones, mandibles and tusks which formed the walls and dome of dwellings that were built 15000 years ago (Gladkih et al. 1984). As much as 21 tons of bone together with timber and hides were used in one structure alone which the excavators estimate to have taken 56 labour-days to erect. Between 100 and 150 individual animals are present in the structures found at three of these sites. Initially these settlements might suggest the presence of a widespread mammoth-based economy in Southern Russia at the end of the Pleistocene. However even the prolific use of such quantities of bone for building material and fuel should not be taken as reliance on this animal alone. Many of the bones appear to have been collected from animals that were long dead and were not human food remains. Like the Clovis peoples these steppe hunters pursued a range of large animals but they also relied on small game, fish and birds. Even in a rich habitat of large herd animals the logistics of hunting economies demanded a broadly based subsistence strategy that was unlikely to seriously deplete or extinguish an animal population.

There is no onus on those who reject the Overkill hypothesis to provide alternative explanations for the extinctions, but environmental causes would seem to be sufficient. In reviewing the debate, Stanley (1987, 1989) notes that during the Ice Ages a large body size was adaptive for animals in a cold climate. Pleistocene bison and a species of ground sloth were both the size of modern elephants while mammoths were some 30 per cent larger. Beavers attained the size of bears and horses, lions and elephants dwarfed their modern counterparts. Periodic episodes of extinction were not uncommon among the North American fauna. In fact six such crises are known from the past 9 million years. With the climatic and vegetational changes at the end of the Pleistocene, many bird species, that were unlikely to have been affected by human hunting expired and large-bodied animals experienced severe stress. In particular, the more pronounced seasonal contrasts meant that the young of these animals, who were born at a fixed time of the year, were faced with less predictable temperatures. Among the species which survived, a tendency towards dwarfing is found that may indicate that changes in vegetation also adversely affected large animals. Complex multi-causal explanations for the extinctions have also been proposed. A combination of environmental factors, such as long-term changes in plate movements and the circulation of both the atmosphere and oceans, which gradually produced alterations in food supply, are thought to have drastically changed the balance of competition for many animals and plants (Williams 1993). Large groups of animals were thus already vulnerable to new sources of instability that included fire and hunting but it was climatic change at the end of the Pleistocene which remained as a major factor.

A reasonable conclusion here would be that the efficient hunting practices of late Stone Age peoples did not permanently diminish large animal populations. Early hominid forms of animal exploitation would therefore have been even less likely to have disturbed the ecological balance of African faunal communities and we cannot conclude that the practice of mixed foraging marked the beginning of a new form of assault on the environment.

What can be concluded? Many of the themes followed here and in Chapter 8 have given rise to issues that remain contentious: the extent to which hominids at even the *H. erectus* level were true hunters and the question of fire-use in Oldowan times are examples. However, while the foraging economy as we know it in recent times consists of a lengthy and complex adaptation that is removed in a technical and cultural sense from the early Stone Age, there may nonetheless be continuities which can be traced across this temporal gulf. There is a clear distinction between the societies of modern hunter-gatherers and early hominids. Yet whenever it was practised, foraging carried with it a social logic that points to sharing and a division of labour as both the basis of a collective mentality and an optimum form of resource use. We cannot conclude that our ancestors' predatory experience produced either an excessively aggressive nature or the beginnings of environmental degradation. Studies of hunting peoples have frequently found that a calm uncompetitive personality usually prevails in such societies and foraging normally implies a non-abusive accommodation with nature. Our legacy from these times is a large brain and an elaborate use of culture for rapid and changing modes of adaptation. In particular a new type of cohesion had been created between band members whose interdependence was as much a product of reliance upon technology as upon new patterns of social bonding that are distant from those of all other primates.

10

Modern Humans and Human Behaviour

In this chapter we shall explore two issues. First our sketch of hominid evolution must now be brought to a rapid conclusion with a brief description of the final stages which led to the emergence of our own species *Homo sapiens sapiens*. This brevity can be justified by the fact that our appearance during the last phases of the Pleistocene, a mere 100 thousand years ago, represents the smallest part of the human evolutionary journey. Of greatest significance here are the acquisition of human abilities that may be linked with changes in social behaviour. Our second task will be to examine evidence for the emergence of these new abilities along this road to human status. In particular we shall look at three universal aspects of modern behaviour: the awareness of death, language use and artistic expression, as these first appear in the records of prehistory.

NEANDERTHALS AND MODERN HUMANS

With the decline of the dominant hominid form, represented by *H. erectus* from about 1.7 to 0.5 million years ago (Klein 1989), came archaic *Homo sapiens*. Both hominid species shared features which included an overall robustness, large brow ridges, a low flattened skull with thick cranial walls and a massive chinless jaw with large teeth. However the trend in hominid brain expansion, that was noted in earlier chapters, was evident among the archaics whose brain volumes, which ranged between 1000 and 1400 ml (Stringer 1988A), marked a new stage in encephalization. Archaic skull architecture differed accordingly with an expanded cranial vault with more parallel walls and less protrusion of both the jaw and

266

back of the head. Fossil specimens of archaic *H. sapiens*, which mostly feature only skull material, have been found in Europe, Africa and Asia and are dated to between 450 and 40 thousand years ago. Archaeological evidence does not point to any change in cultural practice with the rise of archaic *H. sapiens*. The same Acheulean tool traditions begun by *erectus* populations persisted among archaics and none of the behavioural innovations that are associated with our species are seen.

In Europe, but not in Africa, the evolution of the archaics suggests a trend towards the form taken by Neanderthal Man, the hominid who seems to have slightly preceded and coexisted with *Homo sapiens sapiens* or anatomically modern humans (AMH). Neanderthals were restricted to Europe and western Asia between 130 and 35 thousand years ago when the final Pleistocene glaciation was most severe. The Neanderthals had bodies and behavioural capabilities that were a clear advance over earlier hominids and in some respects, such as brain size and tool making, they anticipated *H. s. sapiens*. However the Neanderthals retained primitive features that distinguish them from our species.

Since 1856 over 275 individual Neanderthal fossil specimens have been recovered from more than 70 sites. This hominid had thick limb bones and a powerful body (Trinkaus and Howells 1979) which, while not quite as robust as either *H. erectus* or archaic *H. sapiens*, was seen in a heavily muscled barrel-chested build. Trinkaus (1981) has argued that this stocky, short-limbed physique was well adapted for severe cold. Stringer (1988B) claims that the Neanderthals were cultural primitives. Lacking both the technology and planned foraging methods of modern peoples their strong, rotund bodies and large teeth represent an adaptation to sustained periods of haphazard movement over irregular terrain. The Neanderthal head had a bun-like back but was dominated by a low flat forehead, while the enlarged protruding face gave way to pronounced arched brow ridges. Neanderthals lacked a modern chin and had larger jaws and teeth. Brain size in this hominid averaged 1520 ml which is about 10 per cent greater than that of AMH. This superiority however was perhaps more a factor of cold adaptation for a body with a powerful musculature than an indication of innate intelligence. Eskimos have the largest brains among AMH. While the Neanderthal brain is fully asymmetrical and signs of right-handedness are found, there are few indications of higher forms of consciousness such as bodily adornment and art, except for red ochre in many Neanderthal sites which may have been used as body paint (Blanc 1962) or merely for an unknown utilitarian function. There are also few signs that Neanderthals appreciated the importance of bone as a raw material for tool making despite its abundance.

Technology in the Middle Stone Age was characterised by tools of the Mousterian tradition which superseded the Acheulean hand-axe. Suggestions that Neanderthals were less dextrous than AMH have been shown by Musgrave (1971) to be unfounded and this new technology marks a significant advance on Acheulean tools. Mousterian tools, which were made by striking blank flakes from a prepared tortoise shell shaped stone core – Mode 3 in Clark's typology – were fully utilised by Neanderthal bands. However this hominid is not associated with this tool-form exclusively since Mousterian tools are found in African and western Asian sites, and shortly before their demise Neanderthals are known to have made Upper Palaeolithic tools. Spears and other composite tools are also known but, in general, Neanderthal technology was standardised and conservative. Compared with Lower Palaeolithic tool-kits Mousterian artifacts seem varied and complex but, in terms of the coming Upper Palaeolithic technology produced by AMH from 40 to 35 thousand years ago, a considerable uniformity is evident and there are far fewer tool types. Neanderthals rarely transported raw material beyond the confines of a small locality, unlike Upper Palaeolithic peoples who traded stone, amber and shells over hundreds of kilometres. This behavioural conservatism is also seen in the lack of modification to cave dwellings which do not show signs of the structures made by AMH who constructed pits, post-holes, walls and pavements and even dug into existing deposits to level floors or enlarge living areas.

However there is no doubt that Neanderthals were successful hunters who perhaps caused the extinction of the European cave bear, and that they were also skilled tool-makers with a technology that included clothing and the making and controlled use of fire (Shackley 1980). Neanderthals possessed powers of adaptability unseen in any prior hominid which is exemplified by their pioneering occupation of Eurasia during the Ice Ages. No other hominid had been able to effectively come to terms with the climatic extremes of those times, and their successful settlement in arctic-like conditions was due as much to intelligence and social organization as to physical robustness. Neanderthals also appear to have begun the practice of burial and evidence from Belgium,

France, Russia, Palestine and Iraq points to a concern for the dead and perhaps ideas of an afterlife that are not found among sapient contemporaries in Africa. In 16 instances at European Mousterian sites bodies were tightly flexed in a near foetal position and in some cases brain removal and defleshing of corpses has been found suggesting a mortuary ritual. At one grave in the Crimea a Neanderthal boy was found surrounded by six pairs of ibex horns; at La Ferrassie in France an adult male and female were laid head to head, and in Iraq there is evidence that the dead were buried with flowers.

In this last instance a series of remarkable discoveries, made at the Iraqi site of Shanidar, have been drawn from cave deposits that contain a comprehensive record of climate, fauna and culture of the past 100 thousand years which include the inhumation of seven Neanderthal individuals and 28 AMH (Solecki 1963). Some 2000 generations of Neanderthals are thought to have occupied Shanidar over a 60 000 year period leaving skeletal evidence of concern for the dead, compassion for the living and physical violence. The body of one 40 year old male, which had been crushed by a roof fall 46 000 years ago (Stewart 1977), had been protected from carnivores by a pile of loose stones. This act, Solecki argues, suggests that a degree of esteem was accorded to the body. This individual had a crippled right arm, which may have resulted from palsy or polio, that had probably been amputated above the elbow in life. Other pathological indications point to arthritis and blindness in one eye following a series of facial injuries (Solecki 1972). The tooth wear pattern suggests that teeth were used in lieu of hands for manipulation. The undeveloped right side of the body, seen in the shoulder blade, collar bone and upper arm imply that this Neanderthal would have been unable to forage and that survival to such an advanced age must have required full social acceptance of his condition and extensive support and protection from contemporaries.

Investigation of a subsequent Neanderthal skeleton found evidence of burial and ceremony 60 000 years ago (Solecki 1975). The most striking feature here was the discovery of flower pollen by Arlette Leroi-Gourhan (1975) who identified at least eight different species of flowering plant including yarrow, hyacinth, cornflower, ragwort, groundsel and hollyhock. Many of these plants are known to have significant medicinal uses in the treatment of wounds and rheumatism and as stimulants and purges. Neanderthal bands may have been aware of these properties and this, together with the use of flowers as cultural objects in a death ritual, challenges the brutish stereotype associated with this hominid. The high concentration of flower pollen was found only in the vicinity of the burial and was deliberately interred with the bodies. This, it is proposed, occurred as the band chose mixed bouquets or wreaths of colourful flowers to place in the grave of their companion (Solecki 1977).

The more human image of Neanderthal behaviour that is implied by these findings has been widely accepted albeit with some recent detractors such as Gargett (1989) who envisages burial as a practice that occurred only in Upper Palaeolithic times. However if Solecki is correct the Shanidar material alone is evidence for what J. G. D. Clark (1970) has called the 'dawn of self-awareness'. Purposeful burial at once implies a universal human trait that involves the separation of the dead from the living in a formal manner (J. G. D. Clark 1975). The possibility that these inhumations represent the first stirring of abstract ideas and patterns of thought that are now seen to be exclusively human abilities has to be taken seriously. It would be absurd to imagine that a fully human mentality, including an awareness of death and notions of spirituality, was in no sense anticipated by these large-brained hominids who, if not our direct ancestors, were at least close cousins. Evidence of burial in Middle Palaeolithic times is certainly restricted with only 36 separate instances being recorded in Europe and Asia (Harrold 1980) yet in surveys made by this author and by Sally Binford (1968) some consistency of mortuary practices can be found. Both authorities point to the preferential treatment, in terms of burial, of adult males, who outnumber females by six to one, which may suggest lower status accorded to women whose gathering abilities, like those of Eskimos, were restricted by an Ice-Age environment. Most bodies were buried in a flexed position which could indicate either the need to dispose of a corpse in the smallest possible burial trench in a cave floor, or a desire to accommodate the soul within the confines of the body to prevent its return to the community living above it. Neanderthal grave goods are meagre and comprise little more than stone tools, meat-bearing bones and occasionally ochre, which as a whole indicate what a family or small group might acquire.

This evidence is of course problematic. Abandonment or the absence of burial, at least among modern peoples, does not imply any lack of concern for the dead or absence of belief in supernatural ideas. African cattle-based societies such as the Masai expose corpses on open ground to allow the deceased a final communion with the stars, who are regarded as celestial herders, and remain unperturbed at the consumption of a useless body by jackals and vultures (Turnbull 1978). In many hunter-gatherer societies such as the Hadza and the !Kung the dead are disposed of with only the most simple procedures that do not always involve burial (Woodburn 1982A). Ceremonies, such as sacred dances, are used to commemorate the dead affectionately and these hold more significance than any mortuary ritual or particular mode of disposal (Lee 1967, Marshall 1962, 1969). Similarly it can be shown that where burial does occur, assumptions that this indicates special respect for human remains are not always well founded. For the Nupe of Nigeria and the Sudanese Nuer, burial is merely an act of disposal in graves that remain unmarked and forgotten (Ucko 1969).

Nor can any specific mortuary practice be said to be typical of a culture. Kroeber (1927) has shown that within a single society the dead may be buried in a variety of positions, preserved on platforms, in trees or caves, embalmed or cremated. Forms of disposal could vary with age, social class, political status and circumstances of death in some African tribal societies. Distinctions were also made between those who had committed crimes, been bewitched, struck by lightning or been enslaved. But regardless of these different practices we can agree with Robert Hertz that the recognition of death as a social fact, rather than just a biological phenomenon, is of central importance to human society. Hertz, who was a student of Durkheim, argued in 1907 that in many societies death as a *social reality* was governed by the belief system and had not occurred until a multiplicity of rituals had been performed. Without these the dead would remain with the living and the necessity of providing ceremonies to enable this transition in status was basic to human society. In as much as Neanderthals seem to have disposed of their dead with a degree of consistency that points to ritual treatment they can be considered to have at least begun to approach the human condition.

SPEECH AND THE RISE OF MODERN HUMANS

The degree to which Neanderthals shared the behavioural repertoire of AMH has been a matter of controversy for over a century and some of this debate has centred on their ability for articulate speech. In AMH, speech is possible because of a number of innate genetically transmitted anatomical and neural mechanisms, evolved for this specialisation, that are activated by social interaction (Lieberman 1991). In particular we have evolved a unique supralaryngeal tract from the upper half of our airway which includes the oral and nasal cavities, tongue, pharynx and larynx which together allow complex vocalisation. These speech structures, which arose from organs that facilitate breathing and eating, are also present in other primates but have been 'overlaid' with an additional function in AMH (Lieberman 1972).

Regardless of their brain structures non-human primates are not able to produce a human speech pattern because their vocal tracts deny them control of both tongue and larynx when making sounds. This of course does not mean that chimpanzees at least have no language abilities. Several long-term projects have succeeded in teaching chimps to use American Sign Language. In experiments with an infant female chimp named Washoe, for instance, over 350 hand signals were learned, but more important was the fact that naming, the transfer of signs to new referents and novel combinations of signs were all performed spontaneously. Once a basic language grounding had been gained Washoe's further progress accelerated. As an adult she taught a younger male chimp to use signs without prompting by her trainers (Gardner and Gardner 1969, 1975).

Primate-like vocal tracts that were inadequate for speech are found among the australopithicenes and Lieberman suggests that only in *H. erectus* are there signs of a movement towards the human form. When the supralaryngeal tract finally took its modern form is unknown although the Zambian Broken Hill fossil, an archaic *H. sapiens* about 125 thousand years old, may have had this structure. But by 100 thousand years ago the modern vocal tract is indicated among early AMH fossil skulls at Jebel Qafez and Skul in Palestine. But among Neanderthals living in these times no such vocal apparatus existed. Lieberman and Crelin (1971) made reconstructions of the Neanderthal tract and conclude that the anatomical basis for human speech is not present. The position of the tongue and larynx would not have allowed the formation of i, a, and u vowel sounds and a high rate of phonetic error would have been likely. They add that this does not mean that Neanderthals were without language: their primate heritage, level of culture and brain size point to an ability to communicate albeit without fully articulate speech.

These arguments, which are based on reconstructions of soft tissue, are made by inference from features found on the base of fossil skulls. Some commentators have attacked this procedure and one critic has suggested that with such a vocal tract Neanderthals would have been unable to breath. Carlisle and Siegel (1974) claim that it is not reasonable to draw conclusions from a single Neanderthal fossil specimen in which clear pathological traits are manifest, that the reconstructed airway falls within the range of modern populations and that therefore a diagnosis of speech abilities is invalid. Le May (1975) also finds many of the distinctive features of Neanderthal skull architecture present among modern humans. She also notes that Neanderthal brain casts show a well defined Sylvian fissure which is an accepted neurological development necessary for language. After all no single language utilizes all of the sounds that the human vocal tract is capable of producing and lucid communication may have been possible even under the limitations proposed by Lieberman and Crelin. Language is of course possible without speech and perhaps Neanderthals used hand signs and facial expressions like Washoe. A gestural origin of language in Lower Palaeolithic times has in fact been suggested by Hewes (1973) but his proposal only highlights a lack of consensus on how, when and why this leading human ability evolved. A major conference on the *Origins and Evolution of Language and Speech* (Harnad et al. 1976) produced a range of theories and evidence but little agreement on these issues. Participants were divided as to the stage of cultural development and species of hominid that was coeval with the emergence of language. Authorities such as Montague and Holloway favour the era before *H. erectus* and envisage a linguistic tradition some 2 to 3 million years old like Hewes. Marshack and Jaynes however side with Lieberman and see language, expressed as full articulate speech, as a late development which arose after the Neanderthal stage.

Both the place of the Neanderthals in human evolution and their destiny remain controversial but by about 30 thousand years ago this hominid had been replaced in Europe and Asia by our species *H. s. sapiens* (Pilbeam 1986) whose morphology was described in Chapter 1. Whether Neanderthal genes were absorbed by fully human populations with whom they coexisted at this time or whether they had reached a dead end is also contentious. The changing status of Neanderthals in anthropological thought during the past century is a fascinating subject in itself (Spencer

1984, Trinkaus and Shipman 1993). In the 1960s it was widely thought that Neanderthals represented a universal stage in human evolution (Brace 1979). But since this time a more sapient view of Neanderthals, that had replaced the previous brute cave-man image, has itself been superceded by a consensus that is sceptical of any attempt to draw a parallel between the intellect and culture of this hominid and modern humans. The 1960s view presented by Bordes urges us to dispense with the dim and brutish half-man image of Neanderthals. Mousterian peoples lived in organized social groups of 30 to 50 individuals, they invented many of the tool types that were to be perfected by fully sapient peoples, they had weapons adequate to deal with both the cave lion and cave bear, they used body paint, buried their dead and were clever flint workers. 'It is obvious that they did not lack inventive powers' (Bordes 1961: 810). By the late 1980s however many authorities denied any evolutionary connection between Neanderthals and ourselves. The core of this issue rests on the designation of this hominid as *Homo sapiens neanderthalensis,* that is as a member of our species and a human ancestor, or as *Homo neanderthalensis,* an evolutionary side branch from which we are not directly descended (Lewin 1988A). This problem is now closely bound up with another controversy that has recently divided anthropology which concerns where and when AMH evolved. In this debate two polar positions have been established.

Evidence from molecular biology (Stoneking and Cann 1989, Wilson and Cann 1992) has been used to suggest that all modern humans have a recent origin that can be traced along maternal lines of descent to a woman who lived in Africa between 180 and 150 thousand years ago. Their work involves an analysis of the mutations found in mitochondrial DNA – a tiny proportion of genes that are passed down only in the female line – among racially distinct populations living in all continents of the Old World. They propose that all AMH derive from a single migrating African population which replaced indigenous archaic hominids in Europe and Asia as well as Africa. AMH could not have simultaneously evolved from pre-sapient hominids in such far flung locations without an unlikely massive flow of genes between continents that would have been necessary to maintain different populations as a single biological species. The genetic homogeneity of modern humans is one product of this recent replacement that did not involve interbreeding with extant populations. This degree of genetic isolation between what was once two distinct species of hominid often living in proximity can be explained, they reason, by the use of some cultural adaptation, such as language, which would have ensured segregation between sapient and archaic peoples.

This 'Out of Africa' or monogenesis theory has been used by Stringer (1989, 1990) to argue that AMH and Neanderthals represent two distinct lines of hominid that diverged from a common ancestor 200 thousand years ago and therefore constitute separate species. In Europe, Neanderthals were the cold-adapted indigenous population, the descendants of archaic hominids, who were eventually replaced by AMH whose migrations from Africa coincided with periods of amelioration in the Pleistocene climate. The fossil record indicates that AMH originated in Africa but shows no signs of a hybrid population.

An opposing theory, the multi-regional hypothesis, holds that hominids originated in Africa and then gradually evolved into their contemporary form in each inhabited area of the Old World (Thorne and Wolpoff 1992). Distinctive regional populations therefore developed into a single species of modern human in separate locations as an ongoing process which involved continual interbreeding. Fossil evidence in Africa supporting the rise of AMH is poor and fragmentary and complete replacement of one hominid species by another is unlikely in every part of the world since indigenous populations would be likely to have an adaptive advantage in their own localities. There is no archaeological evidence showing by what means AMH succeeded in overcoming local archaic communities. In fact modern behaviour is not found for some 70 000 years after the arrival of our species and one can only assume that invading AMH adopted local behaviour. This implies that in the Near East AMH had the same culture as contemporary Neanderthals with whom they coexisted for at least 50 to 60 thousand years (Mellars 1991).

Wolpoff claims that there is good fossil evidence to support the multi-regional thesis which shows a continuous evolutionary transition from archaic species to AMH (Wolpoff 1989). A continuity of features can be seen to link fossil specimens from different continents in a lattice-like series of connections which unite separate populations over a 1.5 million year period. Wolpoff also finds a continuity of features that unite archaic hominids and AMH within each continent. Hominids from the Zhoukoudian cave for instance have shovel-shaped incisors as do modern East Asian peoples and intermediate fossil specimens. There are also features of jaw, nose and skull anatomy that are shared by Neanderthals and modern Europeans which are not present in African populations. Regionally distinct populations therefore were just as typical of the Middle Stone Age as they are today. On this basis it is argued that 'some Neanderthals were ancestral to the post-Neanderthal populations of Europe'. (Wolpoff 1989A: 139). A mixture of interbreeding and natural selection had transformed Neanderthals into the European version of AMH. This conclusion is largely endorsed by Trinkaus and Shipman (1993) who propose that fossil evidence indicates that the first AMH arose from Neanderthal stock soon after this hominid had appeared 100 thousand years ago and thereafter spread unevenly throughout Eurasia in a complex mosaic pattern involving coexistence, absorption, migration and extinction.

Regardless of this debate there can be no doubt that the adaptive advantage of AMH was to be overwhelming, showing that an important behavioural shift had occurred in the transition between the Middle and Upper Palaeolithic eras. Evolutionary change was to bring about rapid cultural change in what was to be as fundamental a threshold for archaeology as the appearance of the first stone tools. However full scale modernity, attested by evidence for behavioural traits typical of modern humans, such as art, regular burial, ritual and advanced stone technology, is not apparent until some tens of thousands of years *after* the first fossils of AMH for reasons that remain unknown. In terms of their behaviour the first AMH about 100 thousand years ago were not significantly different from other Middle Palaeolithic hominids (Bar-Yosef and Vandermeersch 1993). But by 40 to 35 thousand years ago AMH were to produce a new range of tools from stone and organic material that were unprecedented. Blades which were at least twice the length of their width, which had only occasionally been made in a crude form by Middle Stone Age peoples, become a standard feature of the Upper Palaeolithic era.

This Mode 4 stone technology (Clark 1977) continued the tradition of prepared cores found in Mousterian tools but used a bone or antler punch to detach finer and sharper flakes from a cylindrical stone core. New tool types such as burins, awls, punches and needles appear for the first time along with evidence of personal adornment and art. The new technology also included tools made from bone, ivory and antler that could be made into a range of innovative devices. Upper Palaeolithic living sites show much greater variation over time and place, containing built structures, graves and hearths which provide unequivocal evidence of occupation by people leading a recognisably human way of life. However no simple positive link can be made between the emergence of *H. s. sapiens* and fully 'modern' behaviour. The unexplained time-lag between anatomical modernity and fully human behaviour indicates that our physical form arose *before* either our current capacity for culture or its use. Populations of AMH in Africa and the Near East, now dated to about 120 thousand years ago (Stringer 1990), continued to make Mousterian tools and to use no raw material other than stone in their tool-kits until 40 000 years ago, long predating the beginnings of Upper Palaeolithic tools. Mousterian technology between these dates shows little sign of variation over time and space until an acceleration of cultural change about 35 000 years ago.

All Palaeolithic economies were based on mixed foraging but in the Upper Palaeolithic era there is evidence for a more systematic and intensive form of provisioning. Large animals of the European plains are frequently encountered in the archaeological record. In southwest France there was a marked preference for herd animals, like reindeer, in which both sexes have antlers during most of the year. Fish and birds were preyed on for the first time and dangerous animals like pigs and buffalo regularly hunted. The fact that the remains of both shellfish and tortoises in food middens at African sites tend to become smaller has been taken to indicate larger populations of foragers with more efficient provisioning methods. Tool-kits became even more complex and the movement of people into much harsher climatic zones such as Eastern Europe and Siberia could only have occurred with the aid of better clothes and shelters. There is clear evidence of fur and leather garments comprising sewn skins and bone needles from burials of 22 000 years ago, and the greater use of open sites, rather than caves, points to a new level of cultural adaptability. Hunting equipment now included sophisticated antler and bone spear points (Knecht 1994), detachable harpoon heads, fishing tackle and the spear-thrower which, together with the bow and arrow, radically increased productivity. Long-distance movement, by as much as 200 km, of what must have been luxury goods such as tool-making stone, amber and sea shells indicate enhanced spatial conceptualisation and unparalleled forms of social contact.

Societies in Upper Palaeolithic times supported larger populations in larger bands living at higher densities, with consistent forms of interaction across social boundaries for perhaps the first time. This occurred against a background of climatic deterioration causing widespread fluctuations in population. The Upper Palaeolithic in fact began with a decline in both life expectancy and population that was to recover only 20 000 years ago. Skeletal evidence suggests a 17 year life expectancy for the Middle Palaeolithic and 16 years for the early Upper Palaeolithic, which slightly favoured Neanderthals. This later rose to between 26 and 29 years at the end of the Stone Age. Greater foraging efficiency and carrying capacity in the Upper Palaeolithic has led to estimates of a three-fold increase in population and a possible density of one person per 25 sq km in this era against a norm of one person per 83 sq km in Middle Palaeolithic times (Hassan 1981). Stylistic differences in stone tools, which mark regional variation, may have served to emphasize ethnicity, exchange or mating networks. The information exchange that was embodied in these artifacts is also found in the first art objects and in new antler and bone implements and may have communicated the individual or corporate identity of a people living in one territory. The sudden appearance and spread of various modes of personal adornment could also have had this function (Randall White 1982). But despite regional differences in tool style a more homogeneous form of culture is thought to have developed over vast areas of Europe (Sieveking 1979). Nomadic movement of peoples spread similarly designed pieces of portable art from the Atlantic to Central Europe, showing the existence of an extensive trading system that may also have included marriage bonds initiated at seasonal amalgamations of bands. A similar level of cultural unity can also be seen in the cave art of northern Spain and southern France (Graziosi 1960).

These societies contained a series of interrelated bands, linked by a shared culture, who practised the form of egalitarian life-style seen among modern hunter-gatherers. It is possible that in areas of greater abundance, as in southern France and Spain for instance, the emergence of some ranked societies with hereditary chiefs or shamans occurred. Such societies have been well documented during the nineteenth century on the Pacific coast of north America where a rich environment sustained bands in semi-permanent settlements presided over by officials who directed production, distribution, ritual, trade and even warfare (Coon 1976). All of these changes point to a restructuring of social relations by early Upper Palaeolithic peoples in which previously dispersed, isolated groups of foragers became aggregated into societies whose identity transcended the individual hunting band.

THE BEGINNINGS OF ART

The painting, engraving and portable sculpture of the Upper Palaeolithic which suddenly appears in Europe from 35 000 years ago and then across the inhabited world (Bahn and Vertut 1988), provide the first real insight into a mentality that can be bracketed with those of historical peoples. The social function and meaning of art in this era remains enigmatic but it is reasonable to consider that it arose in cultures that experienced the need for new modes of expression at both a personal and social level. Adornment, in the form of pendants and beads and the appearance of female figurines with exaggerated breasts and genitals suggest a new level of self consciousness and introspection on the nature of fertility unimaginable among earlier hominids.

What is signified by the appearance of art in these times? Recognisable artistic products are found in the first culture periods at the opening of the Upper Palaeolithic era between 40 and 30 thousand years ago which is some 20 to 18 thousand years *before* the development of the mature style of French and Spanish cave painting seen at Lascaux and Altamira. The large colourful representations of animals at these sites are long predated by a large amount of less obvious art consisting of ornaments such as beads and scratches on cave walls and pieces of bone. These symbolic objects, which are dated to about 35 000 years ago (White 1989), are usually considered to be the earliest known attempts to record nature in two or three dimensions and to decorate the body. Earlier evidence of symbolism consisting of the use of ochre by Acheulean peoples and Neanderthals (Bordes 1961), pendant beads used as early as 110 thousand years ago, body decoration, as well as the engraving of images by Mousterian cultures and bone musical instruments from these times have been claimed by Marshack (1976A, 1990). Marshack claims to have distinguished a pattern of continuity between Upper Palaeolithic and Mousterian cultures. Symbolic objects associated with Neanderthals may have been used ritually to demarcate culturally determined social relations involving age, sex, rank and role distinctions. Some of these objects had a wear pattern consistent with frequent handling and long devotional use. Though rare, the existence of such objects points to the possible use of symbolic communication since without some form of language artistic products would not have had a common cultural context. On this basis the possibility that Neanderthal culture contained some symbolic elements has to remain open.

This evidence has been largely rejected (Mellars 1991, White 1989B and Conkey 1983). Concepts of self and social display had not emerged in Mousterian societies because the social and ideological foundations of these processes had not been developed. The lack of appropriate technology to perforate materials even in early Upper Palaeolithic times suggests that Mousterian societies were not making prototypes of beads and pendants from perishable materials. Had they in fact developed such techniques the earliest stone, ivory and shell ornaments would show that they had been easily drilled but this is not the case. Indications of symbolic behaviour from the Middle Palaeolithic are exceptional, equivocal or are the result of natural agencies and become insignificant when compared with the apparent 'creative explosion' which accompanies the early phases of the Upper Palaeolithic (Pfeiffer 1982).

While the evidence for symbolic behaviour in the Middle Palaeolithic allows us to reject the uncouth, sub-human caricature of Neanderthals it is not reasonable to conclude that the culture of this hominid functioned in the same way as that of humans today (Chase and Dibble 1987). We may add that this is also true of AMH before about 40 000 years ago and, in terms of behaviour, the transition between the Middle and Upper Palaeolithic is far from being fully understood. It is uncertain whether this transition involved a change in the content and manifestation of what was already potentially a modern cultural system or, conversely, that an evolutionary change transformed existing forms of cultural behaviour by the introduction of language and symbols. What is clear is that the Middle Palaeolithic cultural record lacks firm evidence of symbolism. Mousterian tools do not show a continuous development of form and style. Few new tools are found in a record that is several times as long as that of the Upper Palaeolithic. If Neanderthals and AMH in this era lived in a world containing diverse cultures that changed with time and place, none of these differences were expressed as variations in everyday artifacts as they have been for the past 30 000 years. Tool-making shows that purposeful, complex learned behaviour, which involved planning, was a central part of this adaptation. Similarly, while Neanderthal burials can be accepted as intentional and the presence of strong emotional bonds and attachment behaviour observed, there is a lack of conclusive evidence that links these societies with culturally defined beliefs and values that were expressed symbolically. No form of ritual may be inferred from Neanderthal grave goods which are arguably either a misinterpretation of evidence or remain indistinct from other objects that are abundant at sites. Claims of symbolic behaviour seen in ritual cannibalism by Neanderthals are also unfounded. While some decorations on bone and teeth may crudely resemble the motifs of the Upper Palaeolithic there is no reason to assume a symbolic content. A sense of symmetry and an aesthetic sense are found in the Middle Palaeolithic, but the lack of stylistic variation suggests that these were not linked with a set of culturally diverse ideas. Symbolic forms of behaviour were at best only weakly developed in these times and Chase and Dibble conclude that the societies of all hominids before 40 to 35 thousand years ago were based upon a form of culture that differed significantly from that of modern humanity.

This position differs from the view taken by Bar-Yosef and Vandermeersch (1993) whose excavations at Palestinian cave sites have convinced them that Middle Palaeolithic hominids possessed many elements of a culture-based way of life but that the pace of innovation and change in these times was unaccountably slow. They point to the findings of perforated shells, the use of red ochre, systematic burial and the anatomical potential for fully articulate speech by hominids well before 40 000 years ago. Finally they remind us that the roots of the Upper Palaeolithic cultural revolution, which above all implies direction by symbolic behaviour, lie unequivocally in Mousterian societies.

In the Upper Palaeolithic, tools as well as art provide evidence for the emergence of symbolic behaviour because a much greater degree of *imposed form* is found in artifacts. The greater complexity and variety of Upper Palaeolithic tools can be seen as the products of people with more clearly defined mental templates whose more specialised artifacts were made to conform with preconceived functional forms. These tool-makers produced a range of differentiated tool types each of which was made repetitively to a highly standardised model. Each tool type was produced to fulfil different purposes as if the makers were under compunction to

ensure that every variety of artifact had the same appearance if it
was intended for one particular use. This coherence between tool-
form and function did not occur in Middle Palaeolithic industries.
Tool-makers here worked haphazardly making multi-functional
artifacts and seem to have been untouched by the later convention
that a tool's appearance should denote its use.

Mellars (1991) argues that this is evidence of a more structured
and formalized pattern of thinking that was dictated by a set of
cultural norms behind which lay a fully symbolic complex
language which was a hallmark of Upper Palaeolithic society.
The archaeological evidence of long-term planning, specialised
hunting and increased social contact – which would have included
alliances to colonize some of the more extreme glacial environments in Europe – together with art, points to the existence of
modern linguistic systems. It is unclear whether these basic
distinctions between the Middle and Upper Palaeolithic eras were
the product of cultural change alone or were no more than a new
expression of existing mental abilities, or that an advance in the
neurological capacity of the brain had occurred signifying new
cognitive potential. However it is now known that the earliest
stone industries of the Upper Palaeolithic originated among
preceding Mousterian peoples. Neanderthals were therefore
behaving, technologically at least, like AMH and cannot have
been handicapped by a supposedly inferior mentality. It is to be
wondered whether the new capacities could have evolved so
rapidly since within a few thousand years after the demise of
Mousterian cultures sophisticated art appears.

How can the appearance of art be explained? In an excoriating
methodological review one authority points out that this problem
has been given little attention, and despite a range of theories
which focus on psychological propensities, initiation ceremonies,
social boundary maintenance and as a means of promoting a
unified cosmology, 'we are nowhere near to understanding
Palaeolithic art' (Conkey 1983: 223). Nonetheless we will focus
here on explanations which give prominence to art as a product of
social adaptation.

Davis (1986) suggests that image-making began with the
discovery of the representational capacity by hominids when
lines, blobs of colour or marks were perceived as things. This
highly specialised ability, which implied that symbolic meaning
was imposed on marks, arose as a product of the ongoing
development of the human visual system at a moment in time

when existing forms of perception were found to be insufficient. At what stage symbolism was first imposed is unclear. The Acheulean hand-axe from Swanscombe that contained a fossil echinoid, referred to in Chapter 7, and the use of ochre by Neanderthals suggest that representational capacities may have been present but we have no assurance that in these instances symbolic meaning was implied. Hominids had been making marks for hundreds of millennia but it was only when symbolic meaning connected these marks with things in the environment that true representational image-making began. In the earliest images of the Upper Palaeolithic there is no doubt that the concrete world is represented by silhouettes of animals and their body parts like horns and legs that were engraved on the same blocks of stone as marks that appear to be images of female genitalia.

However is evidence of new mental abilities sufficient to explain why art appeared? Davis does not provide any reason why the capacity for representation arose around 35 000 years ago and, as Marshack notes in the commentary to this paper, Davis sees the issue entirely as a psychological problem of individual cognition that does not require reference to the changing cultural context revealed in the archaeological record. Explanations of a capacity or even evidence of its expression are beside the point Marshack argues. A baby's babbling will not lead to language unless it occurs within a cultural context, and marks made by humans will not create symbols outside a socially supported symbolic system. The capacity and use of symbols 'requires feedback between the individuals in a group and the ongoing cultural structures' (Marshack 1979: 274). Images of vulvas for instance must be seen as part of a set of forms that also include the phallus represented directly or symbolically as an animal. In one French cave an engraved lioness had been splattered with darts as if repeatedly killed but had had the vulva and other body parts replaced on several occasions. Frequent renewal or overmarking of vulvoid and other images was a common practice but this is significant only in the context of surrounding images which together form a symbolic whole that served a new social purpose. Marshack (1970) found a number of previously unrecognised images incised upon a shaft of reindeer antler which together comprise a metaphoric representation of spring and summer. This 'baton' included the outlines of salmon, seals, snakes, ibex and several plants. The seals are a breeding couple; the salmon has a hook on the lower jaw acquired by males in the spawning season; the phallus-like snakes are a species that reappear in springtime, and the plants have new shoots or are in leaf and flower.

Marshack's study of ivory statuettes from the earliest period of European art shows that these objects have a sophistication that is not found in the first two-dimensional etchings onto stone. It has frequently been observed that there appears to be no genesis for art. The first recognisable artistic productions are not elemental or tentative but emerge as the work of fully modern people who can be credited with the ability to use abstraction and symbolism (Sandars 1968). Small portable statuettes carved from ivory or stone, like those carried by Eskimos for ritual purposes, possess the poise and realistic proportions of a classical Greek figure. Ethnographic parallels make it reasonable to consider that these carvings were also part of a sacred tradition that may have included a comprehensive cosmology among peoples whose religious practices were mediated by a shaman.

Close examination of these small statuettes by Marshack (1976A) has shown that they have been worn down by handling over years of use and may have been carried in a pouch as amulets. The presence of 'killing' marks on animal carvings and vulvas on 'venus' figurines suggest that these images were used for specific purposes. Whether these were to ensure hunting success, fertility or cures, or merely that they served as devotional objects is unknown but it is clear that these figures were part of a symbolic system which links prehistoric societies with those of our own age.

A 32000 year old antler plaque from Abri Blanchard in the Dordogne studied by Marshack (1972) appeared to be a symbolic device used for record keeping. The plaque had been engraved with parallel rows of circular cut-marks in a serpentine pattern, as if a form of cumulative notation was intended. It was clear that this object had been made with great care since a number of cutting tools had been used to make different styles of mark. As many as 24 changes of cutting point had been used to leave a sequence of sets. The similarity between this plaque and other curiously marked bone and ivory objects widely distributed over Europe has led Marshack (1988) to suggest that these artifacts are calendars and that the marks represent a tally of the moon's phases and the passage of the lunar months and seasons. Marshack associates these devices with tribal specialists whose functions included a record of seasonal changes that marked the long-distance movement of migrating herd animals.

A study of body ornaments from the earliest part of the Upper Palaeolithic by White (1989, 1989A, 1989B) has shown a strong preference for imported raw materials in European societies. Sea shells, soft stone and mammoth-ivory beads and pendants have been found at sites in France, Germany and Belgium that are several hundred kilometres from their source. This desire for foreign materials suggests an extensive trading network or travel to obtain exotic goods but, more importantly, the sudden appearance of abundant quantities of ornaments at living-sites, rather than in graves, shows that a novel form of self-definition had arisen together with a modern mode of social display. This high value accorded to economically non-essential items of culture indicates a profound change in social relations. A shift away from the family unit to the corporate involvement of whole bands and larger aggregations is signalled here. These larger social units were committed to obtaining and producing ornaments whose display function served to regulate everyday life.

The majority of beads at these sites had been carefully manufactured by a standardized process and only occasionally were animal teeth used as body ornaments. The choice of animal teeth was however restricted to carnivores, with fox teeth predominating. This, White argues, was to evoke the qualities of hunting animals in humans and is the earliest evidence for the use of metonymy, a conceptual process fundamental to art, in which a part is made to stand for a whole. Metonymical processes were also operating in the bead-making of these societies since many of the ivory and stone body ornaments recovered are exact copies of animal teeth. The beads were thus an artificial means of attaining the qualities of natural objects. Tribal societies in modern New Guinea are exemplars of metonymical body decoration (Rappaport 1984). Here highly valued bird feathers are worn by men to assume bird-like ideals of beauty. It is sufficient to wear only this part of a bird to evoke these qualities rather than create a bird mask for instance.

What processes activated this sudden need for social display? It is possible that in these times the body was first used 'as a symbolic stage on which the drama of socialization is enacted' (White 1989: 79) because this coincided with new divisions in society based on age/sex ranking and that individuals gained distinctiveness or attained qualities like beauty from their body decoration. If this is the case, art may have arisen to fulfil social and psychological needs associated with everyday forms of group interaction. A set of shared metaphorical meanings drawn from visual images was certainly part of society by 30 000 years ago. The pattern of cross-hatching and decorative notches found on figurines and plaques, and the figurines themselves, were clearly understood by communities in whole regions of Europe. It is also possible that some of these symbols were in fact representations of natural objects. The gouged cut-marks on bone plaques, as in Marshack's lunar calendar, could equally be attempts to make decorative patterns of much valued sea shells which they resemble. Ivory pendants with shell markings have been found hundreds of kilometres inland showing that the desire to transfer natural patterns to new contexts was a primary aim of art.

These ideas should be seen in the context of a model proposed by Gamble (1982, 1983) which suggests that art and body decoration emerged as a response to social change. It is no longer possible to invoke biological evolution to explain artistic behaviour. *H. s. sapiens* are known to have been extant for some 60 000 years before art and, as we noted in Chapter 7, evidence proposed by Wynn (1979, 1981) points to a level of intelligence among late Acheulean hominids that would not have precluded art. The sudden emergence and development of all forms of art in the period from 35 to 20 thousand years ago should be seen in relation to a series of interconnected social and environmental changes in which hunting bands in Europe required new means of visual communication to enable more effective exchange of information.

The earliest art and ornaments known are from Europe, which in this era was a high latitude region with a climate of increasing severity. The portable 'venus' figurines which are widely distributed over northern Europe from Russia to France show strikingly similar features. These corpulent statuettes, which emphasize female sexual characteristics with little interest shown in either face or limbs, were used as a medium of exchange over vast areas during a period of growing environmental crisis. The similar design features of these figurines served to confirm a common conceptual framework among dispersed populations. By making and exchanging non-utilitarian artifacts that conformed to a distinctive style, separated hunting bands had evolved a mechanism that would provide a means to make social interaction easier and more predictable.

Gamble has based his model on the ecological and demographic work of Wobst (1974, 1976) who demands that late Palaeolithic settlement in Europe should not be seen simply as a disjointed collection of sites in which isolated populations lived in closed societies. The dynamics of social interaction and communication among peoples in the Pleistocene are predictable on the basis of existing anthropological knowledge of foraging strategies and optimum modes of adaptation. A computer simulation shows that in order to maintain a minimum equilibrium size of between 175 and 475 people, social contacts were required with contiguous societies. A necessary number of mate exchanges with neighbours would have been necessary and the negotiations which these involved presuppose frequent social interactions that are likely to also have involved gift exchanges, food-sharing and perhaps ritual to cement agreements. The low population densities produced by the last glaciation prevented the development of ethnographically distinct cultural identities that might have led to closed societies and put a premium on activating communications networks. Exchange of mates was probably buttressed by exchanges of labour, in large-scale game-drives for instance, suggesting the existence of open social boundaries across which inter-societal cooperation and cohesion were highly functional.

Whether the venuses were part of a ritual or gift-exchange system is unknown. However Gamble envisages some form of social coherence around the stylistic messages embodied in these figurines that had a necessary integrative function in times of environmental deterioration. In the last phases of the Pleistocene beginning 29 000 years ago the onset of harsh climatic conditions in Europe produced widespread ecological change. With the gradual extension of the vast Alpine and Scandinavian ice sheets, which were to grow until 18 000 years ago, large areas of northern Europe changed from grassland to polar desert. The spread of these tundra-like conditions with more snow cover and less radiation and plant growth led to a sharp decline in animal and vegetable resources. Parts of Europe became uninhabited and bands declined to levels that were half of those seen in late Middle Palaeolithic times. As the ferocity of the final glaciation abated 20 000 years ago, the more efficient Upper Palaeolithic economy allowed an increase in numbers to reach unsurpassed levels of population.

The worsening periglacial conditions that were coeval with the first art imposed severe adaptive pressures on hunting communities. Where it remained possible to continue a foraging economy these conditions placed renewed emphasis on the need for social cooperation between bands as a means of mutual insurance. It is known that contemporary foraging societies in poor environments employ a host of social mechanisms to ensure redistribution. Sharing and exchange institutions involving fictive kinship and reciprocal partnerships among arctic and desert-based foragers were given attention in Chapter 8. It is equally likely that such mechanisms were used in Ice Age Europe to ensure continued occupation in harsh conditions. One answer to survival here would be to establish and maintain an alliance system among dispersed bands. Gamble envisages chains of connection that united far-flung bands across large areas. Besides trading stone, shells and portable art objects these exchange networks are likely to have also been the locus of mate selection and periodic meetings of aggregated bands. Alliances formalised by marriage would have produced more positive linkages and, if these had also been coupled with expectations of mutual visiting and bride service (residence of young men with the wife's parents), as they have been among extant foraging societies, a firm basis would have been provided for solidarity over long distances and diverse terrain. However the most significant function of such alliances would have involved both the exchange of information and foraging territories. In a high-risk environment of meagre resources, information on the availability of animals and plants, possible mates or specialist services like curers or shamans is of inestimable value, and social mechanisms designed to facilitate this flow of knowledge enhance survival. Art, in the form of body decoration or as venus figures, served to convey a message of shared values and aspirations that signalled potential cooperation across wide geographical and social boundaries. Coherence to a shared style embodied in artistic images made visually explicit the possibility of support and exchange among distant peoples whose infrequent but significant meetings were a source of mutual sustenance. Art, in other words, was a means of achieving sufficient cultural similarity to gain cooperation. Those perceived through their visual display to conform to an expected pattern were more likely to be able to enjoy participation in an open network whose members shared resources.

Having been invented to conserve social contact the functions of art diversified and came to include a means of transmitting information to other generations. The embodiment of rules and traditions that were combined in ceremonies is a theme proposed by Pfeiffer (1982). The later cave paintings, produced deep within inaccessible passages often as a series of superimposed images of animals that are enigmatically crowded on areas of wall besides which are vacant spaces, suggest that art was in the service of ritual.

The idea of art as a necessary means of providing social integration by maintaining a state of equilibrium between an individual and society is well established (Fischer 1981). Anthropologists have observed such a process in contemporary societies. Among the Nuba in the southern Sudan, art functions as a significant form of information transmission which makes religion and magic possible (Faris 1972). The Nuba mine ochre for daily use as body colour and in face paints that are applied as bold and intricate designs by men and as pure colour and oil by women to naked bodies ostensibly to emphasize beauty and health. To maintain this ideal the Nuba wear necklaces, bracelets, ear rings and plugs on the nose, ears and lips, as well as ochre and oil, plus a mandatory belt, without which a person is considered to be naked. However these motives mask a primary function of this art form which is to visually represent social distinctions. The Nuba follow precise socially agreed rules in their mode of body decoration to indicate status. Colours and decoration, together with hair styles and scar patterns, are used to announce age group status, patriclan section membership, physical and ritual status. Throughout the life-span Nuba art is used to display social position. Girls oil their bodies yellow until they are pregnant, and as the rites of passage associated with puberty, menstruation, birth and the menopause occur scarification patterns and body ornamentation are changed. Similarly men advertise their athletic skills and grades, lineage and society membership through colour and hair style. This form of personal art operates apparently to satisfy Nuba aesthetics, which celebrate physical health and strength: the old or sick must be clothed in this society, but the clear demarcation and integrative functions served here suggest that art is a necessary means of maintaining social order.

Art bursts into prehistory with no preceding developmental stages to become a part of the human repertoire for reasons that in the end remain obscure. Yet it seems reasonable to propose that a pre-existing aesthetic sense had become allied with a new social purpose and that these connected functions provided support for art as a regular practice. Art itself is evidence of a decisive shift in consciousness that marks the appearance of the modern human psyche which gives pride of place to aesthetic and sacred motives, but this universal quality may well have been developed by social need and action. The archaeology of the golden age of European cave art shows that considerable social investment was made in painting. In two separate works Andre and Arlette Leroi-Gourhan (1982, 1982) both point to the quantity of lighting, scaffolding and colouring matter that was provided by whole communities to decorate caves 17 000 years ago. Pigments were transported over considerable distances and prepared for paintings sometimes executed over a kilometre underground. Platforms were built to give artists access to cavern vaulting and large numbers of lamps were made to illuminate working areas and finished paintings (de Beaune and White 1993). These artistic projects had a social foundation that was ostensibly justified by religious practices but ultimately their cost was repaid by their integrative value in societies experiencing growing complexity.

The unexplained suddenness with which fully human behaviour emerges during the Upper Palaeolithic transformation suggests that an evolutionary kick-start was probably not responsible. Perhaps it was social practice rather than genetic mutation that came to initiate new psychological abilities which then became established as universal elements of human performance because they became socially indispensable. Pre-existing minds with latent capacities for metaphor, enhanced space and time configuration, representation and personal identity were jolted into realising these additions to consciousness as these and other new mental qualities came to form the foundations of greater social complexity. The ability to reflect upon the nature and quality of social bonds from a variety of perspectives and to distinguish the intricate categories of relatedness and social difference of the kind that we examined in Chapter 8 would not have been possible without this shift in consciousness. Social bonds would have become far stronger if their significance could be mentally elaborated and once this had formed the basis of consensus no society could afford to relinquish this strength.

11

Sex and the Division of Labour

What form of social organisation and emotional attachments were typical of the succession of hominid species that we have encountered in this book? The social systems and affectual bonds of hominids were as much a part of their whole adaptive pattern as their cultural behaviour and physique. Here and in the final chapter we face the task of considering the nature of these social ties and interconnected affective drives. There are important points of continuity between the strong emotional forces that are found among both primates and modern humans which serve to preserve mother–child relations. Since we seem also to have a preference for living in mixed-sex groups based on heterosexual pairing it is also reasonable to ask whether similar affectual bonds between mates have functioned as part of an adaptive strategy during our evolutionary journey?

Let us begin with a brief summary of the circumstances that were formative for the hominid way of life. By at least 2 million years ago early species of *Homo* had developed a complex and unique mixed foraging economy which drew upon the products of a varied collection of habitats. Hominid food refuse in these times is of a far more diverse nature than that of any other primate and could only have been gained with the aid of a material culture. A wealth of nutritious resources that are denied even to chimpanzees were available with tools. Digging and collecting tubers, fungi, fruit and nuts required containers, as well as tools, to remove these products to a base area. Tools used for pulverizing fibrous vegetable food could also be advantageously used for pounding meat and cracking bones. Many of the stone artifacts found at Olduvai were almost certainly hammers and anvils

291

which were used for these purposes. Early stone tool-kits employed in gathering could also be adapted to remove flesh from carcasses. For the *homo* lineage the increased importance of animal protein, gained by various means, was to have enormous long-term consequences for both physical evolution and social development. Brain enlargement, food-sharing and a division of labour may have been the most significant of these. Such changes were both the products and cause of a way of life in which culture had acquired increasing importance (Geertz 1975). The earliest artifacts are in themselves evidence of behaviour and techniques which sprang from culturally determined activity (Hoebel 1971). Well before *H. s. sapiens*, hominids were using culture, albeit lacking in symbolic content, as a primary means of adaptation. Since we have solid evidence for cultural transmission of basic tool-forms between generations (Benedict 1971) it is also reasonable to suppose that culture was a factor in social interaction, communication, norms and sanctions. Proto-human society then consisted of hominid bands living by an integrated set of learned behaviour patterns.

However, as we saw in Chapters 8 and 9, the early hominids were not hunters in any modern sense and the origins of *Homo* and human society cannot be claimed to have coincided with an era of hunting. It is no longer valid to postulate a pristine state in which a male dominated hunting economy came to shape all succeeding social relations. Modern human sexuality, kinship and gender relations are not the simple by-products of male hunting strategies. On the contrary, the origin of numerous human characteristics and technical abilities emanate from a long formative period of omnivorous foraging which is likely to have integrated the activities of both sexes. But this should not be taken to suggest that the economic activity and contribution of either sex were identical or that there were no established roles that barred males or females from particular tasks. It is clear though that in the practice of active scavenging and later hunting, by the time of archaic *H. sapiens* at least, that certain behavioural elements which were foreign to other primates, but common among social carnivores, were incorporated by hominids because of a pre-existing material culture. The omnivorous hominid subsistence pattern is likely to have rested upon cooperation and food-sharing within a socially cohesive band whose members regularly engaged in different tasks. There were forms of interdependence and social solidarity here that transcended any experienced by the apes. The adaptive value of culture was again confirmed and impetus to further cultural elaboration provided.

But this social and economic change is incomprehensible unless it is seen in the context of developing social bonds which became established as social institutions within the hominid band. In modern hunting and gathering societies division of labour and reciprocal exchange occur primarily between the sexes organized as a family group that comprises variously mated male and female partners, in multiple and single unions, and their children. Wider patterns of sharing, based upon recognized relations among extended kin, are also present now but were unlikely in distant prehistory because the elaboration of kinship networks was beyond the consciousness of pre-sapient hominids. Among the early hominids the development of a pair-bond between males and females may have represented the origins of sharing and formed a fundamental part of the new adaptive strategy. Important issues are raised here which relate to the question of a sexually based division of labour with implications for power and authority.

SEXUAL POLITICS AND PREHISTORY

In a simple sense prehistoric foraging societies, like those of contemporary hunting peoples, were classless and equal. Social ranking and coercive relations lay in the future when immovable property, especially in land, together with restricted rights of access to the new quantities of material wealth created by settled cultivation came to define class-based hierarchies. The first signs that such inequality was developing out of a process of growing social complexity occurs in the Middle East only about six to seven thousand years ago which is well after the establishment of the first agrarian societies some four to five thousand years before this date. All of the power relations associated with complex societies, such as the office of chief, the rule of tribal elders, or a managerial bureaucracy or domination by leaders of patriarchal clans and lineages, to say nothing of the hereditary transmission of privilege or stratification patterns that emerged out of state-based societies with an urban sector, are entirely foreign to the social organisation of foragers. Such societies are, as Mann (1986) argues, able to escape from the 'social cage' of both a permanent territory and fixed power relations. Mobility precludes all possessions beyond those which are portable, and authority lacks a power base in societies with small flexible bands whose members regularly interchange as they aggregate and disperse with the seasons.

But are these societies entirely equal and free of coercion? There are, to be sure, fixed patterns of authority between parents and dependent children. First marriages are frequently arranged and older generations deferred to even though this indulgence does not rest upon any simple material advantage. But it is the relationship between women and men that has been given most scholarly attention. Since the appearance of Engels' study, *The Origin of the Family, Private Property and the State*, in 1884, which envisaged a division of labour based on sex as the basis of social relations in prehistory, controversy has ranged over both the accuracy and implications of this observation. Engels had postulated separate economic roles for each sex.

> The division of labour is purely primitive, between the sexes only. The man fights in the wars, goes hunting and fishing, procures the raw materials of food and the tools necessary for doing so. The woman looks after the house and the preparation of food and clothing, cooks, weaves, sews. They are each master of their own sphere: the man in the forest, the woman in the house. Each is the owner of the instruments which he or she uses: the man of the weapons, the hunting and fishing implements; the woman of the household gear.
>
> (Engels 1972: 218)

This description of society at the level of 'high barbarism', on the eve of a social transition to property-based domestication, serves as a contrast with the development of the monogamous family which accompanied this change. Engels conceived of human society beginning with a mode of primitive communism in which collective families produced in common. Either sex had separate economic domains but equal political and legal rights and were united by unpossessive forms of 'group marriage' in which no partner had exclusive sexual access. With the coming of agriculture the growth of a larger social surplus caused an end to the collective economy and group family as property became concentrated in the hands of men who wished to leave wealth to sons they had knowingly fathered. Private property was incompatible with female rights and forced communal marriage to give way to the exclusive bonds of the patriarchal family. The new division of labour went beyond sex alone and in these circumstances women lost control of their own labour power. Henceforth their subordination within the productive process marked the beginnings of a gender hierarchy and female dependence.

Many feminists have endorsed Engels' materialist theory because it relates the social position of women to prevailing economic and political relations and also shows how changes in the productive forces lead to changes in status (Sacks 1975). In particular, feminists have agreed with Engels that the beginnings of sexual inequality lie with concepts of ownership and the rise of the state rather than primitive forms of kinship (Reiter 1977). But feminist critics of Engels have argued that in proposing an original sexual division of labour where men are concerned with productive tasks and women only with domestic ones an unfounded essentialist assumption is made. Similarly no natural male preoccupation with paternity can be assumed (Moore 1988). Some feminists are concerned that any dichotomy of male and female roles automatically leads to 'sexual asymmetry', a state of affairs where the activities of men are always given more importance than those of women. Since mothering is taken as the basis of subordination, a sexual division of labour, that encourages an opposition between male and female activities, serves to isolate the political power of men from the confined domesticity of women (Lamphere 1987). Feminist opinion is divided on the question of whether sexual asymmetry is a universal trait – like the incest taboo – and whether, in the earliest egalitarian societies at least, a sexual division of labour existed at all.

However a major critical onslaught has been raised by feminists against nineteenth century formulations of human development, particularly those that accord with social evolutionary models, which give a privileged position to male hunting as a catalyst for all intellectual, technical and social progress. Darwin's principle of 'sexual selection' is attacked as a blatant form of male bias (Hubbard 1979). Darwin had proposed in his later work, *The Descent of Man, and Selection in Relation to Sex*, which appeared in 1871, that many secondary sexual characteristics, that have no immediate relevance to reproductive fitness, are the result of selection for traits that were attractive to sexual partners. By evoking the principle of sexual selection as a factor in human evolution Darwin assumed that the visible secondary sexual characteristics of women like breasts and buttocks were the result of male selection. A woman's chances of mating with a well endowed man were greatly enhanced if these features were appropriately developed. On the other hand male characteristics like strength and belligerence were not a consequence of female preference; they were the results of inter-male rivalry. Darwin took no account of cultural variation and, worse, he assumed that mate choice was entirely a male prerogative. Feminists argue that this mode of explanation supported Victorian preconceptions of sexual differences which designated males as inventive, coura- geous and active, and females as reclusive, altruistic and passive. Such a model gave full credence to the moral precepts that underlay the values embodied in nineteenth century views of the western family and had the effect of contrasting an intelligent male, who was the exclusive object of selection, with a female who had been pulled along the evolutionary journey holding onto her mate's coat-tails (Fedigan 1986).

Besides being anachronistic these criticisms of Darwin are not entirely accurate or fair. It is easy to use quotations from *The Descent* to illustrate Darwin's apparent misogyny and his fidelity to the prevailing intolerance of Victorian England. Thus:

> The chief distinction in the intellectual powers of the two sexes is shewn by man attaining to a higher eminence, in whatever he takes up, than woman can attain – whether requiring deep thought, reason or imagination, or merely the use of the senses and hands.   (Darwin 1981: ii. 327)

Or:

> Man is more powerful in body and mind than woman, and in the savage state he keeps her in a far more abject state of bondage than does the male of any other animal; therefore it is not surprising that he should have gained the power of selection.   (Darwin 1981: ii. 371)

But *The Descent* does not show Darwin's acceptance of sexual inequality to have been part of an unquestioning attachment to popular prejudice. He reveals himself as a perceptive and original thinker who was pushing against the intellectual constraints of his times. In one passage for instance he states that man is more courageous, pugnacious and energetic than woman and has a larger brain. He then proceeds to note that the question of whether the brain–body ratio of each sex is constant is unknown. A little later Darwin speculates on the possibility that women could become the intellectual equals of men given the appropriate training. In these instances and in his general assertion that humans are not essentially different from any other animal, which was a major theme of *The Descent*, Darwin was both breaking new ground and laying forth an agenda for biology and the social sciences.

Biology today accepts that sexual selection may have a limited influence on the formation of some human traits and that both sexes may exercise selection against a small number of individuals who do not conform to conventional norms of attractiveness and fail to reproduce, but there is also agreement that Darwin greatly exaggerated this principle (Campbell 1975). It has recently been proposed by Diamond (1991), who follows Darwin here, that sexual selection can be used to explain a diversity of human traits that typify and maintain what are taken to characterise different 'racial' groups. Environmental factors, like climate, have proved to be a poor guide to the present geographical distribution of human 'races' (Washburn 1963). Since there is no simple correlation between solar intensity and skin and eye colour or hair type it is more likely that these and some secondary sexual characteristics, such as the different forms taken by breasts and genitals, which are constant in a population, result from mate-choices rather than climatic influences. Human sexual partners overwhelmingly favour an ideal set of visible attributes that conform to a standard, or 'imprinted aesthetic preference' (Diamond 1991: 104), which is drawn from the most common physical features of a regularly interbreeding population. The genes of a founder population are thereby preserved and a 'racial' group's character is maintained in a region as the preferred desire for dark or light skin colour is expressed in the search for sexual partners. Sexual selection is also currently used in biology as a means of explaining some aspects of animal behaviour. In one example provided by Dawkins (1986) females chose to mate with males who displayed certain sexual characteristics such as the long tail feathers of the African widow bird. The female widow bird has a genetic preference for such plumage and this predisposition has the effect of maintaining male tail feathers at an optimum length even though these secondary sexual characteristics have no direct consequence for reproduction or survival.

Darwin's projection of a patriarchal paradigm onto human prehistory is no more valid than his ideas of social progress which were discussed in Chapter 3. Like his other sociobiological asides Darwin's views on sexual differences and male superiority are based on little more than unsubstantiated speculation. However, like any innovator, Darwin's intellectual universe was embedded in the mores of his own times and his lack of ideological correctness, seen from the vantage point of the late twentieth century, in no sense invalidates the use of natural selection as an approach to the origins of human sexuality. Whatever Darwin's views on women and their place in evolution, his explanation of change in nature is through a general and abstract theory which is not vulnerable to the arbitrariness of current ideology. No scientific revolution can exist independently of the society in which it arose and Darwin, like Newton, who saw no inconsistency between his simultaneous beliefs in the laws of physics and the magic of alchemy and astrology (Goodman and Russell 1991), is not to be rubbished because his social views are inconsistent with modern ideals.

Sexual selection among humans and their ancestors is clearly much more complex than Darwin had ever imagined, but it would be absurd to reject this feature of evolutionary explanation entirely since it provides us with a means of understanding aspects of our sexuality, which will be investigated in Chapter 12. At least one comprehensive theory of human origins has been proposed by a feminist who effectively utilizes this principle by arguing that it was female mate-choice which transformed male behaviour (Tanner 1981). It should be added finally that Darwin was undogmatic in his treatment of human evolution: his ideas were offered as the tentative conclusions of a naturalist investigating human origins. He was all too aware of the then inadequate hominid fossil record and the manifest gap between his theory and direct supporting evidence appalled him. In struggling with the complexities posed by the transition from animal to human status Darwin experienced considerable intellectual problems which he confided to correspondents as doubts and growing difficulties with the principle of sexual selection (Bowlby 1991).

Many driving forces behind human evolution have been proposed since Darwin. Earlier conceptions of 'man the thinker' or 'toolmaker' gave way to 'man the hunter' in the 1960s. One major shift away from an over reliance on the male has occurred since the 1970s with the combined understanding that big-game

hunting came relatively late in prehistory and the discovery that
the economic contributions of women to tropical foraging societies
were usually of greater importance than those of men. These
advances in hunter-gatherer studies constitute a U-turn in
anthropology and show that many assumptions made about the
basic facts of foraging life before 1950 were inaccurate: Forde
(1934) should be compared with Lee (1968) for instance. One
feminist has seized on these findings to rail against anthropologists in general for being biased, white western males (Slocum
1975). In particular she attacks Washburn and Lancaster's 'Man
the Hunter' formulation which was outlined in Chapter 8. The
origins of the human adaptation were not coeval with male
hunting: it is rather the gathering strategies of females which
began this move away from primate status. Ironically she presents
an inverted form of male chauvinism by advancing the proposition that female independence in foraging and social life had only
tenuous links to superfluous males. 'food-sharing and the family
developed from the mother-infant bond' (Slocum 1975: 45).
Hominid adaptation here is not seen as a combined achievement
but as the accomplishment of one sex alone.

But if females were of equal importance in provisioning a
foraging band were they in fact also accorded equal status and
were they engaged in tasks similar to those of males? In
contemporary hunter-gatherer societies status varies with both
female productivity and local culture. Hiatt (1970) has shown that
in areas of abundant food resources women are more important
than men as food providers and that the reverse is found in
regions where food is scarce. In a survey of 49 societies she shows
that in tropical regions females contributed 52 per cent of food
while in higher latitudes women provided 33 per cent in
temperate zones and only 13 per cent in the arctic. Her survey
also shows that in all but two of these societies hunting was an
exclusively male task and in the case of these exceptions group
hunting was predominantly a male activity that included women
in a non-hazardous and subordinate role. Some correlation
between levels of productivity and social status can be observed
with greater freedom and independence among women in
Kalahari foraging societies than those in the arctic. But cultural
differences can also influence the status of women. Despite similar
levels of productivity among women in both African and
Australian foraging societies it is the former who enjoy higher
status.

But can modern foragers really provide us with an understanding of this issue in prehistory? The decimation and retreat
forced on these societies by the violent encroachments of
colonisers has certainly changed the territory and settlement
patterns of surviving foragers. Leacock (1978, 1983) also argues
that the social structures and normative patterns of modern
foragers have been distorted in this process and that current
ethnographic observations of sexual relations are no guide to those
of the past. She envisages a lost freedom for women over the
control of their lives in work, social life and domestic role that is
not always apparent in foraging societies today. The historical
accounts of white male explorers, missionaries and anthropologists, who left records of men dominating deferential women, are
not to be trusted. Egalitarian societies were not marked by such
inequalities before the division of labour became corrupted by
trade and conquest. Leacock does not relate her formulation of
deprived female independence to biology. She explicitly rejects
notions that any aspect of reproduction might constitute a reason
for subjugation. In fact no form of female subordination or sexual
hierarchy can be distinguished among egalitarian societies until
the disruption of their economies.

There are considerable problems with these ideas and specialists
with detailed knowledge of regional ethnographies have rejected
Leacock's thesis (Berndt 1978). It is of course naive to pretend that
ethnographic work accumulated during the past two centuries
was produced in conditions of value neutrality but there is every
reason to consider that these accounts of foraging societies
constitute important evidence which we shall review here.
Unfortunately it has become common among feminists who wish
to deny the universality of a sexual division of labour among
hunting peoples or to question the distinct configuration of
economic activities which are associated with each sex to dismiss
such work on the grounds that it encapsulates an androcentric
bias. Hayden (1992) has attacked this belief as unproven and
nihilistic. There are numerous unambiguous descriptions of sex-
associated behaviour relating to production, religion and politics
as well as accounts of rape and violence against women which
show no real difference when the observation has been made by a
female or a male ethnographer. It is also the case that a positive
association between environmental conditions and sex roles –
gathering and latitude for instance – has been clearly demonstrated by anthropologists of both sexes which would not be

apparent if an androcentric bias was distorting the ethnographic record.

Leacock does not provide us with any explanation of how contact actually changed the division of labour beyond the fact that trade relations were involved in some north American societies. But many egalitarian societies deep in the forests of New Guinea or South America had little significant contact with outsiders until the mid-twentieth century, yet frequently it is these societies, that are based on primitive agriculture, that contain extreme forms of female exploitation and violence (Rappaport 1984, Chagnon 1983). On the other hand some of the most persecuted extant hunting peoples, such as the Kalahari !Kung, display an extraordinary degree of sexual equality and female independence (Lee 1984). Leacock's economic determinism leads her to an idealised view of egalitarian societies without conflict or sexual tension which excludes cultural influences that may cause varying levels of female status. It is unclear why men were not exploited more than women in the process of contact and why the status of women was to suffer if biological facts are irrelevant.

Leacock makes frequent reference to women hunting in addition to their normal gathering activities. Cases of mothers and daughters, sisters and even grandmothers hunting among the Canadian Ojibwa are cited as evidence that a sex-based division of labour was once minimal. If such instances were indeed a part of much wider practices the case for any form of sexually based division of labour in prehistory is weakened. In Chapter 9 the absence of women hunters in all foraging societies was noted. Can this difference still be upheld?

A SEXUAL DIVISION OF LABOUR

In dealing with this problem it is important to recognise that social distinctions based upon sex are universal. Any sociological review of recent hunter-gatherer societies cannot fail to uncover strong lines of gender demarcation that are present among such peoples. Male-female identity, sacred-secular divisions in art, myth and ritual, economic role, domestic responsibility, taboos and sanctions and spheres of authority all reach to the heart of social organisation. Frequently the material culture of a society records this division in the form of everyday tools and utensils that are closely associated with either sex. General-purpose knives in Inuit societies or spears and digging sticks in Australia are as fundamental as a symbolic means of maintaining sexual identity as a tobacco pipe and handbag in western culture. Yet, as Berndt (1970) shows, these divisions are not a source of conflict; in fact they serve as a means of social integration for mutual advantage. Sex is a fundamental and crucial social distinction that is basic to sharing. She continues,

> The basic dual contrast . . . between male and female: what I am calling the two-sex model – is one of very long standing in human affairs, and has been widely accepted as a normal, and pivotal model in human relationships.   (Berndt 1970: 46)

A rigorously upheld sexual division of labour, as described by Berndt, that also generally excludes women from hunting in foraging societies is easily demonstrated. In a large-scale survey of nearly two hundred societies based on foraging and primitive farming Murdock and Provost (1973) found the hunting of large sea animals to be an exclusively male activity while this was also the case in 99.4 per cent of societies that hunted large land animals. Work with stone, metal and wood, boat building, lumbering, trapping, mining and quarrying was also found to be almost entirely a male practice in a range of societies. Women on the other hand were predominantly concerned with gathering vegetable food and small animals, making clothes and preserving and cooking food. Because these activities, in contrast to those of men, were more compatible with child care, Murdock and Provost follow Brown (1970) in characterising women's work as generally safer, closer to home, more monotonous and directly concerned with the domestic economy. A re-examination of Murdock and Provost's evidence by Burton et al. (1977) confirmed that men were most likely to engage in more distant and dangerous occupations and that childbirth and infant nursing remained as the constraints on the sexual division of labour.

But in practice the distinction between male hunting and female gathering is too simply drawn. The degree to which foraging peoples enforce these economic roles upon either sex differs between societies, and a number of patterns in the division of labour can be distinguished (Friedl 1975). A common pattern involving separate male hunting and female gathering and full sharing of the proceeds is found in Australia and Africa. Some women foragers provide important amounts of protein gained from small animals, and in other societies, like the Mbuti, women will join male-led hunts as beaters or net holders (Turnbull 1976). In these communal activities seasonal hunts, animal drives, fishing or gathering can involve both sexes of a whole band or unite several bands when game is available. Alternatively men may hunt and gather separately from women who feed themselves and dependent children but not their husbands. This pattern is found among the East African Hadza where men bring meat back to camp only when their own hunger is satisfied (Woodburn 1968). There are also singular instances of women hunting large game but this does not include dangerous animals and rarely are the same weapons as men used. On the other hand men rarely eschew gathering and may take vegetable food either alongside women or during a hunting expedition. It must also be said that there are no examples of unproductive women living solely as dependents of men.

The sexual division of labour among hunter-gatherers may take on a variety of forms but it remains a universal part of this type of society. In no case is hunting by women a regular institutionalised feature and nor are there examples of feminine hunting equipment. Early studies of Inuit peoples, some of whom had remained isolated in the Arctic until the mid-nineteenth century, show all the hallmarks of these distinctions. Men did almost all of the hunting although some women might stalk seal across the ice using clubs rather than spears. Cases of women with rare hunting skills accompanying male deer-hunters and mixed-sex bands mobilised to scare seals from breathing holes are recorded. Women and children made wolf cries to drive animals towards waiting men and, in another case, a barren woman became a noted caribou hunter. But all of these instances are exceptional and the full rigours of harpooning from a kayak or pursuing dangerous animals such as walrus, narwhal or whales in an open boat were experienced by men alone. Male hunting equipment is strongly associated with masculine identity and rarely in the possession of women who had their own gender-specific tool-kit (Boas 1888, Murdoch 1892, Jenness 1922). Women had highly important roles in these societies as the processors of skins, clothes makers and as the preservers of food. These activities were of primary importance to the economy as a whole and should not be seen as merely supportive or incidental to the work of men. Murdoch's expedition to Point Barrow in Alaska in 1881 left him with the impression of an Inuit people who combined both egalitarian sexual relations and a firm division of labour. He writes,

Women appear to stand on a footing of perfect equality with men both in the family and in the community. The wife is the constant trusted companion of the man in everything except the hunt and her opinion is sought in every bargain or other important undertaking.   (Murdoch 1892: 413)

At the other extremity of the Americas the Ona and Yahgan (or Yamana) peoples of Tierra del Fuego present a similar pattern of sex roles. These foraging societies became all but extinct under the combined impact of European diseases and wars of extermination during the late nineteenth century with the loss of over 90 per cent of their populations. Observations made this century by Lothrop (1928) are of people with a simple technology and social life that had survived though on a drastically reduced scale. He describes Ona tools as akin to those of Neanderthals, and those of the Yahgans as little better. Fuegian peoples lived in the most rudimentary huts, they were semi-naked and would give chase totally unclad even though their climate was more extreme than northern Europe. However each society maintained a set of sex-related economic roles. Among the Ona, men alone made bows and arrows and used them to shoot guanaco that women and children had driven on to heights. Killing, skinning and butchering the prey was also a man's job. In both societies women fished and, among the Yahgans, gathered crabs and shell fish in their canoes or from beaches which, together with vegetable food, was a substantial part of the diet.

The German missionary and anthropologist Martin Gusinde pointed to the interdependence of spouses and to, 'the almost complete equality of both sexes', in his comprehensive account of Yahgan culture which he investigated between 1918 and 1924 (Gusinde 1961: 450). He provides evidence from historical sources that relate to times before European genocide when a marked sexual division of labour was noted that involved men in the hardest and most perilous forms of labour. Yahgan men hunted seals and whales from flimsy bark canoes which, as with some Inuit peoples, were paddled by wives who frequently collaborated in hunts. Men also hunted penguins, cormorants, geese and ducks with slings and pursued land animals, including those killed for their skins like otters and foxes, with bows. Besides meat men were committed to provide a family with flint and raw hides, they would build canoes and huts but would never perform tasks designated for women. The Yahgan woman fulfilled a range of foraging and domestic tasks which, in addition to child care, included fire-tending and repairs to her own canoe and the family hut, the processing of hides and the making of clothing, pouches and fishing tackle. This work load, which included the constant gathering of mussels by wading in cold water, seemed to some casual visitors to be more arduous than male occupations. These responsibilities and the demands of Yahgan culture for female submissiveness caused Darwin to describe Fuegian women as the slaves of their husbands during the *Beagle*'s voyage to Cape Horn in 1833. But Gusinde's extensive study of the Yahgans convinced him that the separate but interdependent roles of each spouse comprised a mutual share of the economic burdens faced by this society. In reality men rested between bouts of extreme physical exertion while women were continually occupied, but neither was exploited. He quotes Thomas Bridges, a missionary whose long-term work with the Fuegians began in 1871, 'The wife is not wholly dependent on her husband and has all the freedom that she wants'. Gusinde continues:

> The woman is well aware of her rights, and that is why she leaves a man who treats her badly . . . She gathers mussels or goes fishing, arranges the pelts, and sews clothing whenever the whim strikes her . . . If her husband expresses special wishes, he knows as well as she that she is by no means bound to fulfil them and that she does so only out of love and affection.
>
> (Gusinde 1961: 462–3)

Foraging societies in other continents show the same pattern of sexual division. Women among the Japanese Ainu used their own gender-specific tools for fishing but were tabooed from using spears at salmon spawning grounds or bows and poisoned arrows, which were exclusively male weapons, to shoot deer. Women assisted in these hunts as beaters and fence makers and occasionally directed dogs and used clubs against this prey. But women were not involved in hunts for dangerous animals like bears (Watanabe 1964). In Australia hunting large animals was also a male activity but aboriginal women, who contributed more food than men, sometimes ran down kangaroos with the help of dogs during a gathering expedition. Besides vegetable food their main foraging activities provided meat from small animals, reptiles, snakes and rabbits. Men also gathered and collected fruit and honey and dug for yams (Berndt and Berndt 1964).

Are there any exceptions to this basic social dichotomy? Leacock's own source for the Ojibwa, Ruth Landes, is clear, unlike Leacock herself, that hunting by women was not a usual feature and that a well-established sexual division of labour was basic to this society. 'Hunting is regarded strictly as a man's sphere . . .' (Landes 1938: 131). The male–female worlds of the Ojibwa were separated by economic activities as much as by taboos and values. Women may have fished on equal terms with men but with different equipment: lances being used by men alone. Much hunting by the Ojibwa in fact meant little more than trapping small animals for their fur and Landes does not suggest that large dangerous prey were taken by women. Female hunting among the Ojibwa occurred under the pressure of circumstances such as widowhood, desertion or illness of a husband. In these predicaments a woman was permitted to oppose convention but her hunting went unenvied by other women who saw her as unfeminine and extraordinary.

It has recently been claimed that one society alone, that of the Agta Negritos in north-east Luzon in the Philippines, does appear to break the rule against women hunting in foraging societies. In Agta society women make a substantial contribution to the economy, they are prominent as decision makers and have considerable authority. They are also accomplished hunters and regularly take deer, wild pig and fish from their forest habitat. Both sexes are trained to hunt from an early age and female prowess, seen in archery and tracking skills, are no less effective than those of men (Estioko-Griffin 1987). The Agta establish hunting camps far into the forest from which mixed or single-sex parties venture over several days. One survey of hunted food showed that ultimately men provided 43 per cent of animal body weight against 22 per cent for women and 35 per cent for mixed groups. However while women hunters actually provided less meat by weight they nonetheless had a higher killing rate reflecting the fact that men hunted large heavy animals with more powerful bows but women successfully took more smaller animals.

While there is no doubt that women in this society are highly proficient hunters, a closer look at their economy shows that the Agta are in fact not true foragers at all. They are accordingly not even an exception to the general rule which excludes female hunting and it would be wrong to use them as a model for Palaeolithic societies. For one thing domesticated dogs are used in hunting and for several hundred years the Agta have traded forest products with local farming societies in exchange for food-grains, tubers, metal and cloth (Griffin 1984). Little gathering of wild vegetable food occurs since the Agta mostly consume domesticated crops. But the most significant aspect of this economy is that the Agta are primitive agriculturalists who engage in a shifting slash and burn mode of cultivation. This type of farming creates small swidden plots in the forest on which staple foods such as rice, cassava, corn and sweet potatoes are grown. The Agta are therefore not bound by the logic of a foraging society: their population level, settlement pattern and demographic pressures do not meet with the same constraints found in prehistory and their division of labour is free to be varied accordingly. Hunting is thus an important addition to an agrarian economy but it is significant that even though women are good hunters this occupation is still seen by the Agta as primarily a male activity (Estioko-Griffin and Griffin 1981).

There is then an apparent iron law which universally prohibits women in foraging societies from hunting on a regular basis, from having gender-specific weapons and equipment for hunting and a further total prohibition against the pursuit of large and dangerous game. Deviations from this pattern seem always to be only exceptional instances, mythical unobserved cases or dubious examples of societies in the process of transition. This ban needs to be examined and it is worth asking whether this form of restriction together with a sexual division of labour may also have operated in a pre-hunting economy. Did hominid females engage in active scavenging on the plains of Africa before the emergence of our genus?

For some feminists this primary division of labour was in no sense neutral; it did not imply complimentary roles and reciprocity but served as the basis for sexual asymmetry and marked the drawing of the first political lines in society. Exclusion from hunting was also to lead to a form of subordination, expressed through cultural and symbolic domination, which represented the beginnings of male supremacy. These consequences occurred, as Tabet (1982) argues, because women have always been denied equal access to technology. Women are in a constant state of under-equipment because, while men are able to surpass their physical limits by virtue of their tools and weapons, women are denied access to these artifacts. Reliance on the male and male control of production follow from this. The division of labour is no more than a means of supporting sexual class relations which means that, among foragers, women do more arduous and less exciting work with more rudimentary tools. Technology beyond that made of wood, clay, skin and vegetable matter is controlled by men. Tabet notes that some male hunting implements – knives, bows, spears, axes and spear-throwers – are tabooed for women who may not even touch these items in some societies.

These ideas have an oddly conspiratorial ring to them. How in fact could men have succeeded in barring women from using the most sophisticated tools in so many different societies for so long for political reasons alone? Tabet's theory has the unintended consequence of excluding women from stone tool-making and assumes that their lack of equality was connected with their technological deficiency. This goes against the proposals of some feminists who suggest that female initiative may have actually begun tool-making among hominids (Tanner 1981) and contrasts sharply with the innovative heroine of Jean Auel's anthropologi-cally informed novels which present a saga of Upper Palaeolithic life.

The blanket exclusion of women from hunting cannot be explained by a male plot and there are no simple social, biological or psychological reasons for this prohibition that are adequate either. Alternative forms of child-care are available in hunting camps. In many non-foraging societies women work away from home and children develop surrogate mother figures. The burden of pregnancy and birth, with the average three to four year birth-spacing usual among foragers, is much lower than that of women in agrarian societies and could be accommodated with the pursuit of game. Female strength, though less than that of males, is certainly sufficient for hunting particularly when this is allied with the advanced weaponry of the Upper Palaeolithic. Most women in fact never develop their full muscular strength and their apparent lack of physical power relative to men is more a product of social position than biology. It can also be seen that in non-foraging societies which permit female hunting women are no more squeamish than men and do not lack the fortitude necessary to endure the combat and gore which animal slaughter entails. Such aggressive impulses are not restricted to men alone.

The most fruitful explanation of the hunting/sex dichotomy is a demographic hypothesis presented by Friedl (1975). Her analysis focuses on the different specialisms involved in the division of labour. While meat is the most favoured food among foraging peoples it does not in fact usually constitute more than 20 to 45 per cent of total diet, with about 35 per cent as an average for peoples in all latitudes (Lee 1968). It is also usual for such societies to consist of a number of separate bands of about 25 to 50 people each (Hayden 1981) who may form brief communities of 100 to 300 members as the seasons and game resources allow. In these circumstances specialisation and exchange is the most efficient means of exploiting the diverse resources within a habitat by such small groups of people during the nomadic cycle of movement through a band's territory. Both gathering and hunting are separate skills that may rarely be mixed on a single occasion. The gathering specialism, which requires considerable botanical knowledge and an extensive spatial memory for vegetable food resources dispersed through large tracts of land, also requires that a woman carry heavy loads. Gathering food for a !Kung woman occupies two to three days per week and involves walking up to 12 miles on each day with a burden of tubers, fruit, berries or nuts and fire wood that may weigh more than 30 lb. She will frequently also carry a breast-fed child under four years old weighing another 25 lb in addition to her gathering equipment and supply of food and water (Lee 1987). No forager thus encumbered could also carry hunting weapons, properly track animals or have the mobility to make a rapid response if game was encountered. Gathering may be arduous and demand reserves of stamina for carrying and walking up to 1500 miles a year but it is also usually a reliable means of gaining most of the food eaten by the band. Women alone or in small groups are exposed to a variety of hazards but these are insignificant compared to the dangers faced by hunters. Gathering then produces high returns for low risks and while the supply of staple food does not carry with it the same level of esteem as meat, the economic value of women's work is clear. On this basis though it would still be possible for child care to be shared with men and for both sexes to alternate in providing gathered vegetable food. However the demographic realities of life in small bands of 25 to 50 people imply that only 7 to 10 of these will be women of child-bearing age and that infant mortality levels will result in the deaths of between one-third and one-half of all children. In these circumstances nearly all of a band's women must be either pregnant or breast-feeding throughout their fertile years in order to maintain the population at a stable size. With so few women in a band and no nutritional alternative to breast-milk

(Molleson 1994) there can be no possibility of a wet nurse and a mother is of necessity closely bound to her child making her unavailable for hunting. But the dangers of male hunting provide added reasons for women to remain gatherers. Hunting accidents are common and there are frequent cases of broken limbs or other injuries which lead to death. Hunters may be drowned, poisoned, lost, gored or run down. This high risk activity is also prone to failure if animals escape or cannot be located and the dangers, unreliability and low returns involved here, contrasted with gathering, implies that male lives are more expendable. The loss of a male hunter is therefore not as demographically significant as that of a female gatherer who usually provides more food, cares for and feeds growing infants and who, as a widow, may be made pregnant by a new mate.

There is no essential form of female exploitation nascent in the economy and sexual relations of hunting peoples. The sexual division of labour is not usually an area of conflict and women rarely express a desire to hunt. !Kung women for instance complain forcefully that their men sometimes fail to produce enough meat. One anthropologist has described these women as impressively self-contained with a high sense of self-esteem. She found relations between the sexes to be relaxed and egalitarian and !Kung society in general is described as, 'the least sexist of any we have experienced' (Draper 1975: 77). The bawdy and unruly behaviour of these women, which derives from their greater foraging effectiveness but shorter working hours, is part of an overall state of female independence that includes the right to retaliate and overthrow a marriage. !Kung marriages are normally happy and Lee reports that sexual gratification is a valued aim for both partners in these unions. Women here are not forced to endure battery, chastity, seclusion or double standards. It would be wrong to imply that women in this society enjoy complete equality with men. Participation in band discussions and decision making occurs to a greater extent here than in almost all other types of society but is still less than among men (Lee 1984). There are then no telling reasons to believe that women in prehistory suffered from male domination either politically or economically but there is every reason to consider that the contributions of each sex were different.

A sex based division of labour is a universal feature of all known foraging societies and even though different species of hominid are involved the consistency found in the ethnographic record is a strong argument that such an adaptive mechanism was also present in distant prehistory. This basic relationship, which divides animal forms of subsistence from an emerging hominid economy, was a part of both the evolutionary ecology and cultural package which facilitated the development of tools and other artifacts. Even at the *H. erectus* level it is unlikely that all aspects of material culture were produced by single individuals. Specialisation and exchange, the beginning of economic relations, should be linked to sexual relations. The same demographic pressures observed among foragers would have also applied to early hominids engaged in active scavenging. By competing with carnivores for dead and vanquished animals hominids had entered a dangerous new niche and proto-hunting skills that included the ability to interpret environmental cues to the presence of game and the removal of carcasses to places of safety would have been better performed by males unencumbered by the demands of breast-feeding or carrying. Isaac's home base model, presented in Chapter 7, suggests that retrieving scavenged meat would have been viable only in the context of a mixed foraging strategy that was held together by social cooperation. The form of social bonding which this adaptation implies points to exchange between parties engaged in different aspects of foraging. These cooperative alliances would have had greater significance for longer-term reciprocity if sexual relations had been established. In this sense economic and social bonding had come to coalesce with the needs of reproduction and nurture. A sexual division of labour came to imply distinctive sets of male and female economic roles but there is no necessity to assume that any form of subordination or dominance was to result from this.

12

Sexuality and Social Life

There is no direct evidence for the origins of modern human social and sexual behaviour, the family or kinship, and the reconstruction of hominid evolutionary history in these areas faces exceptional difficulties (Tooby and DeVore 1987). Perhaps the reciprocal relationship between the sexes described in the last chapter constitutes partial evidence if the division of labour was in fact as ancient as scavenging itself. A new adaptation of radiative foraging across far greater distances into more varied ecological zones than any other primate was involved here. Chimpanzees have a home range of some 50 sq km but the annual territory of African hunter-gatherers is ten times greater than this. Efficient use of the resources from such an extensive habitat implies some form of specialisation and exchange.

A variety of social strategies, such as hominids foraging in variously mixed or single-sex groups, would have been compatible with this land-use pattern on economic grounds alone. It is unclear whether the earliest tool-users were organised as a family group or whether this came later with the first tool-makers, fire-users, or with the makers of the more advanced tool-kits in Middle Palaeolithic times. Nor is it apparent when and in what order the most distinctive features of this universal human institution emerged. Typically human sexuality involving continual female sexual receptivity, formal restrictions on mating and incest avoidance, a sex-based division of labour, the social recognition of both parenthood and the heterosexual pair in marriage almost certainly emerged at different times but in a sequence which is unknown. This is equally true of kinship, the most fundamental mode of human social organisation. At what evolutionary stage the pattern of statuses and obligations that connect related individuals was to become realised as a formal social network is also obscure. But it would be unwise to overlook a variety of new

312

evidence of social and sexual behaviour in both primates and humans that has recently been used to construct models and theories of human social origins. In this final chapter these issues will be reviewed in the context of the whole adaptive pattern that our ancestors had adopted. All explanations of the origins of the family are of course vulnerable but a speculative reconstruction, even from imperfect sources, is a valuable undertaking since it allows human social bonds to be placed in an evolutionary context (Gough 1971). This in turn implies that the problem of bridging the behavioural gap between apes and hominids is comprehensible only in terms of an evolutionary transition comprising complementary social and biological factors. The primate nature of hominids together with their material culture and emerging social and economic behaviour would have comprised major forces that powered this transition.

RECONSTRUCTING THE ORIGINS OF HOMINID SOCIAL LIFE

Explaining the emergence of a family unit implies that we attempt to bridge a huge gulf between primate and human behaviour. But in Chapter 4 it was noted that there is no single form of primate mating pattern and the major characteristics of the human family, which were just mentioned, are almost absent in monkey and ape societies. Among the apes, gibbons are monogamous and chimpanzees promiscuous, while gorilla males control 'harems' and orangutans govern ranges, and it is the proceptive female in the latter two species who approaches the dominant male when in estrus. There are then further problems of determining how modern primates can provide a basis for understanding early hominid society and which primate species provides us with the most appropriate model.

The baboon is proposed by Fox (1967) as a prototype for the hominid family. But baboon society consists of hierarchically controlled inbreeding troops which are closed to outsiders. Sexual relations are more a product of a rigidly ordered dominance system than choice and lasting pair-bonds are rare. Troops of baboons move in a coordinated manner as a single corporate group with strictly assigned positions for members according to sex and position in the hierarchy (Washburn and DeVore 1961). Baboon infant–mother relations are intense but the matrifocal subgroup, which among chimpanzees unites brothers and sisters with their mothers in an enduring bond, is less developed. There is nothing here to suggest even a rudimentary division of labour based upon mixed foraging and these monkey species seem inappropriate as a model for the genesis of the family.

There are no grounds for supposing, on the basis of inferences drawn from the apes, that hominids once lived in family hordes that were presided over by a single dominant male who monopolized females at the expense of subordinate males who might be expelled from the horde itself. This form of jealous and aggressive behaviour is characteristic of baboons and other monkeys rather than apes. In neither gorilla nor chimpanzee troops is there any permanent social tension caused by competition for females. In fact these ape societies contain both hierarchies and open mating systems. These recent findings run counter to a long tradition that reaches back to Darwin, which included Morgan, Engels and Freud, that held to the idea of hominids living in primal hordes.

The chimpanzee provides a far better means of understanding the prehistory of hominid society and is by far the most favoured ape. Both McGrew (1981) and Tanner (1987) focus on the plant-gathering and tool-using activities of female chimpanzees which they use as an analogy for the transition to hominid status. Chimpanzee mothers are much more frequent tool-users than males, they have more technical skill and take an active part in the tool-using apprenticeship of their female offspring. Gathering with tools by hominid females, which is a stone's throw away from chimpanzee behaviour, is for Tanner the basis of the hominid adaptation and a spur to innovation in behaviour and technology like food-sharing between mother and non-dependent offspring and carrying devices. She argues that this form of female-led evolution involved sexual selection. Free mate choice led to a reduction in hominid sexual dimorphism, as it has among chimpanzees, because females preferred to consort with less aggressive males and were able to engage in more pleasurable sexual contact. Wild chimpanzees have been seen kissing, and hominid females 'may have come to prefer to mate more often with males who kissed effectively than those who growled at them and displayed large canines' (Tanner 1987: 14). She also considers sexual selection to have been an operative factor in the evolution of bipedalism itself as females opted for males able to display an erect penis more conspicuously while standing.

Whatever the validity of these arguments it is the full social adaptive pattern of the chimpanzee which is most helpful in the reconstruction of hominid society. Chimpanzees have evolved with a forest adaptation that was unlike the mosaic of ecological zones favoured by the early hominids. Nevertheless there are behavioural features that we share with this species which, together with a high degree of genetic correspondence, show both how close these animals are in evolutionary terms and how remote their societies are from those of monkeys.

Chimpanzee social organization is unique among primates in that it comprises neither the regimentation seen in monkey troops nor the dispersed solitary behaviour typical of orangutans. It is primarily cooperative and sociable and is based upon a flexible assembly and dispersal system in which the animals of a troop gather together to feed from fruiting trees or fragment into a number of sub-groups to forage separately when a food source is exhausted. In this sense there is a rudimentary correspondence between nomadic, wide-ranging chimpanzee troops and hunter-gatherer bands. Members of a troop are non-competitive in both foraging and mating. A roving group will announce the presence of a new food source that it has discovered to the rest of the troop. An adult female is normally sexually receptive for only several weeks every few years. Yet during her short periods of estrus she will mate with a variety of males who themselves display a tolerant indifference to other males regardless of their status.

From this baseline of features it is possible to conceive of an evolutionary modification of behaviour as cultural practices grew in importance. The multi-faceted social relations that are engendered here provide the means by which the origins of both a division of labour and the male–female bond, which is basic to the family, can be understood. The permutation of individuals within the changing membership of sub-groups implies a variegated pattern of relationships that is quite distinct from the homogeneity of social life within a monkey horde. The pattern of social roles which underpin human society could have emerged only from a social system that allowed the free access of individuals to different groups.

But there are considerable problems involved here. There is no single living primate species that provides us with an appropriate analogy for the ecology and social life of the first hominids and early *homo*. The Oldowan hominids of 2.5 to 1.5 million years ago may have possessed unique adaptive strategies that were unlike those of modern humans or other primates (Potts 1987). These earliest tool-makers lived 2 to 6 million years after the divergence of hominids from ape stock and some 1.5 to 2 million years before the emergence of our species. In this case there can be no question of a simple evolutionary continuum which joins humans with ape societies. Yet primate models remain valuable as the source of hypotheses for human evolution and as a means of understanding major behavioural differences. In facing these problems Wrangham (1987) proposes that by examining behavioural traits that are common to both modern African ape species and humans it may be possible to reconstruct the behaviour of a common ancestor. A marked similarity among all three species is found with regard to stable and semi-closed social networks, female out-mating and male residence in the troop of origin. But the sexual life of chimpanzees, gorillas and humans differs widely having in common only a degree of polygyny while relations between groups vary from cooperation to hostility.

EVOLUTION AND HUMAN SEXUALITY

If ape social behaviour is of any value in understanding hominid society there still remains a yawning gulf between human sexual and nurturing practices and those of all other primates. Our sexuality and family organisation is highly varied and because most of these differences are culturally constructed there are few direct lines of continuity that may reasonably be extended between our species and ape societies. Human sexuality is in fact the result of both cultural practice and biological evolution. But it is important to stress that the rise of the hominid line required – for its existence as a culturally directed species – what has been termed 'the progressive development of sexuality' (Beach 1978: 151). Of paramount importance here is the fact that the human sexual response is no longer a reflex form of behaviour. Hormonal control of sexuality has lessened and given way to control by the brain which has allowed an abundant array of factors to act as the sources of both sexual stimulation and inhibition. Mental control of sexual life implies that features which have no reproductive significance may cause excitation. There are an innumerable range of personal qualities, and even objects, which have become the source of erotic attention. Because sexual desire is the product of an interaction between biology, culture and mental life, it is impossible to specify any boundaries to erotic experience. Conversely, cerebral dominance can have the apparent effect of extinguishing sexual feeling. Cultural forces can produce strong inhibitory reactions to the most blatant forms of stimuli. The voluntary withdrawal from reproduction or from any form of sexual experience or an obliviousness to erotic expression are all patterns of behaviour that are frequently encountered.

Nevertheless it is possible to identify a collection of physical and behavioural traits that relate to sexual activity in humans that are not culturally determined and which as a whole constitute a species' specific reproductive pattern. Our sexual orientation and practices vary with culture and mentality but this is not true of our sexual equipment. Human genitalia, reproductive organs and associated patterns of sexual response and emotional bonding are all the products of natural selection that took their current form by virtue of their adaptive value. Reproductive fitness was conferred upon hominids through an integrated package of changing anatomical structures and behavioural processes that led our ancestors along a path which diverged sharply from other primates. A new culturally driven way of life gave rise to new forms of mating and caring behaviour. These in turn were interconnected with modifications of both sexual apparatus and physiological functions and came to effect sexuality in ways that were as significant as the expansion of the brain.

Women are known to give birth and raise infants in primitive conditions with an ease that contrasts with the experience of labour and delivery in advanced societies, albeit with higher rates of maternal and infant mortality (Konner 1972). But evolution has produced a set of anatomical changes that have made reproduction and nurture more difficult for our species and new forms of behavioural adaptation have emerged to deal with these problems. The advantages of bipedal walking and larger brains were in themselves also the causes of an obstetrical dilemma (Lovejoy 1988). The lengthening period of helplessness and stage of total infant dependence which occurred with the rise of *Homo*, and the vulnerability of hominid females in pregnancy and the postnatal phase, demanded that the changing reproductive and foraging strategies of hominids remained in accord. Evolutionary change in the context of bipedalism was to incapacitate and restrict females with young. To be sure the first bipeds did not mature at the same slow rate or have such a prolonged infancy as our own species. Beynon and Dean (1988) have shown that the delayed maturity of

H. s. sapiens was not present among the australopithecines who had early growth rates similar to the apes. Nevertheless for the *Homo* lineage offspring were to be a growing encumbrance which restricted mobility. Infants were unable to walk long distances until several years old and their only secure source of nourishment was their mother's milk. The human baby is in fact born at a much earlier stage of foetal development than any other primate and has a brain that must attain almost three-quarters of its final volume after birth. All the crucial postnatal developmental steps made by human infants, which lead away from a state of total helplessness, occur considerably later than among the apes. The time taken by babies to acquire the ability to cling to the mother, move the head and sit up, crawl and stand without support take at least twice the time taken by gorilla and chimpanzee infants. These apes are able to gather food independently after weaning and take a fraction of the time taken by humans to gain foraging skills. There are clear difficulties here for hominid females carrying infants while engaging in hazardous activities away from a base that required constant wariness, such as the retrieval of bones from a carnivore's kill-site.

The long-term dependence of children on adults for food, and the problems posed by learning to forage, can be seen as a further restriction. The potential rate of reproduction among chimpanzees is extremely low (Richard 1985) but among hominids it is even lower (Catton and Gray 1985). With single births, long sterile inter-birth intervals and high rates of child dependency some form of social support for a hominid female – perhaps from a mate – must be seen as essential. Without the long-term reciprocal arrangements that involved food provision and exchange, a female would be posed with an almost impossible energetic burden of foraging for herself and her single offspring until it was mature before conceiving again. Such prolonged periods between births might have led hominids into a demographic cul-de-sac. The emergence of the two distinct foraging patterns that were noted in Chapter 11 can be more easily envisaged in these circumstances. Specialized food-gathering patterns by males and females, distinguished primarily in terms of distance from a home site, and a tendency for scavenging and rudimentary hunting to become a male activity may have initiated the sexual division of labour.

There is of course no reason to suppose that, because reproductive necessity once dictated a sexual division of labour, this has become an intrinsic part of the human condition. It would be absurd to suggest that modern sexual inequality rests on ancient imperatives or that variations in gender roles within the family can be explained in this context. The contemporary family is not a simple projection of a 'natural state' and nor does the family's past limit its future (Gough 1971). It is equally invalid to claim that our 'true' sexual nature will be revealed through investigation of our primate past. Human–ape comparisons provide insight into the evolutionary forces that our ancestors were once subjected to, but they cannot ultimately designate us with an intrinsic nature that is exclusively monogamous, polygynous, promiscuous or polyandrous. It is our nature to be any or none of these and to find a multiplicity of ways of satisfying the appetites and needs that our species has inherited (Wolfe 1991). Flexible, culturally-based forms of social action are the products of our evolutionary history but for the early hominids biological realities did constitute forces which directly conditioned reproductive behaviour. A novel behaviour pattern was being explored by a creature whose cultural sophistication had hardly advanced beyond that of a chimpanzee and, in the face of severe selection pressures, conduct which conferred survival advantage on proto-human hominids and their young became incorporated within a repertoire.

The sexual and nurturing behaviour of modern humans is highly distinctive when compared with other primates and in stark contrast to that of other mammals. It is worth considering whether these differences mark evolutionary developments which were connected with changes in interpersonal behaviour that arose in response to new social and economic relations. A cluster of physiological and behavioural changes, with a direct bearing on sexuality and reproduction, have arisen in the course of human evolution to form a unique social pattern that demands investigation. Profound long-term personal bonding between sexes and the evolution of sexual advertisements or 'epigamic' features by both sexes (Jolly 1985), used to consolidate these bonds, mark us out from the apes. Human males have an unusual parental role, that is also not found among the great apes, which involves protection, socialization and material provisioning. Humans live in enduring mixed-sex groups that include adults and children in which monogamous male–female pairs are by far the most common form of mating arrangement even in societies which permit polygamous marriage. For Lévi-Strauss this arrangement is seen as a product of expediency:

> That monogamy is not inscribed in the nature of man is
> sufficiently evidenced by the fact that polygamy exists in widely
> different forms and in many types of society . . . [however] . . .
> the prevalence of monogamy results from the fact that, unless
> special conditions are voluntarily or involuntarily brought
> about, there is normally, about just one woman available for
> each man.   (Lévi-Strauss 1971: 340)

The human practice of monogamy is therefore not comparable
with the monogamy of other primates such as the gibbon.
Primates have no choice of mating system while human culture
has allowed different arrangements for family organization and
sexual behaviour to accord with different modes of subsistence as
the basis of diverse social systems. While the nuclear family and
monogamy are prevalent in hunting and industrial societies and
also common among peasants, these are not simple equivalents to
the mating and nurturing systems that are imposed upon other
primates. Whatever parameters a society designates to reproduction and family life it must be remembered that adherence is
culturally determined, that such rules are frequently broken and
that human sexual behaviour is part of a social system rather than
a primate mating system (Kinzey 1987).

Comparative anatomy provides some indication of the sexual
behaviour patterns which were adaptive for human ancestors. As
the bodies of successive hominid species grew larger during the
course of evolution, a general tendency for differences between
male and female size to diminish can be found. This lessening of
sexual dimorphism is a telling indication of social change in
prehistory. Sexual dimorphism in primates provides a reliable
guide to the mating pattern for which a species is adapted. Among
monogamous gibbons, and some monkey species, where males
mate only with a single female for life, both sexes have the same
body weight. But for polygynous primates such as the gorilla,
orangutan and hamadryas baboon, females are respectively 58, 53
and 51 per cent of male body weight (Clutton-Brock and Harvey
1977). The epigamic or secondary sexual characteristics of each of
these male primates, like the crest, throat pouch, mane and large
canine teeth, function to emphasize size and strength in threat
displays and also serve to attract females during courtship. Control
of females by a dominant male is sustained only at the cost of
aggressive inter-male competition which has selected a large body
size. In the least common promiscuous breeding systems, that

allow males equal sexual access to all estrus females, sexual dimorphism tends to be minimised, as we find among chimpanzees where females weigh 83 per cent of male body weight. Among ancestral hominids a decrease in dimorphism can be detected but not calculated with the accuracy applied to modern primates. Both *A. afarensis*, represented by 'Lucy', and *H. erectus* were clearly dimorphic, but to a lessening extent. Lucy probably weighed half as much as her male contemporaries, like a gorilla female, suggesting that in early hominid communities competition among males for receptive females occurred (Foley 1987).

What can be said of sexual dimorphism among modern humans and the mating systems that gave rise to this adaptive difference? Unfortunately authorities do not agree on the precise extent of our dimorphism or, more importantly, on its significance. Variations in body weight that favour men by between eight and twenty per cent are commonly cited. We are also highly unusual in having almost no dimorphism in our canine teeth (Shea 1988). However modern humans are similar to chimpanzees in being only slightly or moderately dimorphic and this convergence is likely to have had multiple causes (Campbell 1975). Larger females are a product of the need to forage independently and have sufficient leverage to excavate deep tubers and carry between camps, as much as the obstetrical demand for a large pelvic girdle. For males, however, this proximity with female stature could be interpreted in terms of the replacement of the primate dominance hierarchy, and rivalry between males, with a pair-bond relationship in which mutual commitment was derived from complementary economic roles. But it is not clear when this interdependent pattern arose or, on the basis of our present dimorphism alone, whether we were once adapted for polygyny, monogamy or promiscuity. For most commentators, however, the existing size differences and epigamic features of humans support the contention that we evolved in a polygynous system. For Reynolds (1991) the additional size and strength of the male jaw, shoulders, limbs, hands and feet point to competition for females, while the deeper voice and facial hair are associated with threat displays.

The comparative anatomy of the reproductive organs in primates may also be considered in this context. While the ovaries of apes and humans are all relatively constant with body weight this is not so of either the testicles or penis. The principle that form and function are interrelated is basic to evolutionary biology and a close association between testicular size, mating behaviour and copulatory frequency has been established among non-human primates. Chimpanzees, with a promiscuous multi-male breeding system, have the largest testicles of any primate in absolute and relative terms while the gorilla has the smallest in terms of body-weight ratio (Short 1980). The size difference here can be explained by the amount of sexual activity, and sperm production necessary, for each species. Chimpanzee troops are likely to contain at least one receptive female at any given time whose estrus will last ten days providing mating opportunities for all males. Because the level of forest resources obliges chimpanzee troops to constantly gather and disperse, males must be ready to mate whenever an estrus female is encountered. Copulation is almost a daily experience for male chimpanzees who are able to produce sufficient sperm for at least four fertile ejaculates a day, compared with only three and a half per week by humans. Chimpanzee males face constant competition from other males and require reproductive organs that allow them to rapidly inseminate available females. But for a dominant gorilla male living with three to six females, and no serious rivals, estrus is much more rare with a fertile period of no more than a few days that occurs each three to four years. The gorilla group, which forages and sleeps as a unit, is far less sexually active and it is usually the female who initiates mating. Testicular capacity fits the demands of these two mating systems. Chimpanzees and gorillas have testicles that weigh 120g and 30g respectively – compared with a weight of 40g for humans – in spite of the fact that the gorilla has a body weight that is almost four times as heavy as the chimpanzee and two and a half times that of humans (Harcourt et al. 1981).

The adaptive significance of penis size in primates and humans is much more difficult to interpret. Humans are unique in this aspect of sexual anatomy and in its functioning. Men have exceptionally large penises and the enlargement of this sexual organ has occurred together with the loss of its skeletal rigidity. The erect penis in man is 13 cm long and that of the largest primate, the gorilla, only 3 cm even when fully aroused, while the chimpanzee and orangutan penises are 8 cm and 4 cm respectively (Short 1980). Of all primates, however, the human male alone is without a strengthening bone or cartilage embedded within the penis and must rely solely upon blood pressure to achieve an erection. The absence of this bone suggests to Campbell (1975) that a new practice of face-to-face sexual contact – in a more emotionally tranquil and familiar manner than the mania which characterizes mating by monkeys and apes – was a causal factor in our evolution.

There would seem to be no obvious reproductive reason that justifies the magnitude, conspicuousness and vulnerability of the human penis. Even though bipedalism has caused some realignment of our genitalia and has favoured frontal intercourse, which is known to be the most popular sexual position in the majority of cultures (Beach 1978), there is no clear necessity for such size. Orangutans mate in a variety of positions that include frontal contact while hanging in trees (Wolfe 1991), and engage in coitus for considerably longer on average than humans. Yet the orangutan penis is only one-third of the human length although this ape is 15 per cent heavier. The most commonly cited explanation concerns the role of the penis as an organ of display and its size as a product of sexual selection (Short 1980). The contention here is that the reproductive organs of both males and females have acquired a powerful epigamic or erotic quality that has led to their enlargement. Women are unique in having the largest breasts among primates that, having become established well before pregnancy and birth, remain as permanent features. As such they have no direct reproductive function during a large part of a woman's life and in addition to their energetic costs, in terms of extra food intake, they remain as sites of potential injury and disease. There are no recorded cases of mammary cancer among apes whose breasts develop and recede with the demands of infant feeding. The erotic role of the breast, and our naked skin as a whole, is well established in a range of cultures and the idea that this means of expressing attractiveness serves to cement a pair-bond is often suggested (Anderson 1983). Among animals the most concrete expression of female attractiveness is through a presentation of epigamic features associated with fertility. Frisch (1988) has shown that sexual maturity in the human female is acquired, and fertility retained, only by the acquisition and maintenance of a critical proportion of body fat. Unless a woman attains such fat reserves on her breasts and buttocks, or if she loses fatty tissue through poor diet or excessive exercise, she will be infertile. This threshold has been observed in non-industrial societies where nursing mothers can become infertile if their reserves of body fat are expended in producing breast milk (Van Ginneken 1974), and in hunting bands breast-feeding is known to have a contraceptive effect (Kolata 1974). Frisch's discovery of the physiological mechanisms that sustain fertility marks confirmation of a long-held perception that correctly associates fitness and fecundity with attractiveness.

Permanent breasts and large buttocks are seen by Reynolds (1991) as a specific hominid adaptation to savanna foraging. These fat deposits evolved as a means of ensuring sufficient bodily reserves for reproduction in challenging environmental circumstances and became associated with parental potential. A female requires an average of 10 to 12 per cent extra daily calories to support a successful pregnancy and a further 15 to 25 per cent for breast-feeding (Frisch 1988). In conditions where seasonal variability would threaten such additional food needs the fatty reserves carried by a female were an important insurance for reproduction. As tropical temperatures militated against the storage of a general layer of body fat, since this would preclude foraging in midday heat, localised storage became a preferred option. In fact an extreme form of this adaptation, known as 'steatopygia', which consists of large pointed buttocks, is seen in Palaeolithic art and among Kalahari hunting peoples to whom it is highly erotic.

If permanent breasts as epigamic display organs are a product of sexual selection, the enlarged penis may also be the result of an ancestral female preference for this form of erotic display. Evidence for this thesis is weak and contradictory. Women commonly deny any such exclusive interest in the male organ and feminists have protested at the phallocentrism implied by this explanation although it is explicitly endorsed by feminist anthropologists (Tanner 1981, Jolly 1985). Yet the role of both the breast and the penis as evocative sexual symbols in human social rituals is too widespread to require comment (Wickler 1969). The courtship behaviour of chimpanzee males includes the conspicuous display of the erect penis (Nadler and Phoenix 1991) and, for what it is worth, objective testing has shown that in the human female the genitalia are highly responsive to erotic depictions of the penis (Bancroft 1980).

Ultimately the unique qualities of both these reproductive organs remain enigmatic and at best we can conclude here that our relatively small testicles and sperm reserves, together with the largest penis of all primates, indicate that humans are adapted for a low level of continuous sexual activity which may have evolved to facilitate bonding. A recent survey of sexual behaviour in Britain found that frequency of sexual activity was somewhat less than three and a half occasions a week (Wellings et al. 1994) which comes close to the biological norm of weekly fertile ejaculates already mentioned. The combined evidence of our dimorphism and reproductive organs implies that the early hominids did not live in long-term monogamous pairs or in a purely promiscuous system. But what other form of mating system is implied? Some primary form of single-male polygynous breeding system is indicated for Martin and May (1981). But, as Reynolds (1991) argues, the presence of female epigamic features suggests that males were competing for particular well-endowed females who were able to choose their male partner. The hominid mixed foraging pattern, which implies a degree of female independence, is not compatible with the highly dominant forms of male control found among gorillas and baboons and it should not be assumed that an ancestral polygynous mating system was as coercive. In this case it may be possible that we evolved with a different mating system involving greater sexual interaction as proposed by Wolfe (1991). She suggests that we '. . . probably lived in social units containing more than one resident male member, and both females and males had multiple sex partners' (Wolfe 1991: 139).

But the capacities of human testicles tend to confirm that we are not adapted for a multi-male promiscuous chimpanzee-like society (Martin and May 1981). Of all the apes, human sexual anatomy most closely resembles the orangutan male who defends a territory containing perhaps six females and their young all of whom live in isolation. However sexual dimorphism in modern humans is at most mild – within the range of chimpanzees in fact – and it is therefore reasonable to propose that some form of serial monogamous bonding remains a possibility at the *H. sapiens* stage since this mating system has been found among some animals to provide an optimum solution for cooperation between parents when a large investment in offspring is necessary.

But the most distinctive and mysterious evolutionary aspect of human sexuality is the loss of estrus by women and its replacement by continual sexual receptivity. This absence of any overt visual, behavioural or olfactory signal of ovulation and fertility is without precedent in primates and mammals as a whole (Burley 1979). Among apes and monkeys sexual play and even copulation is known to occur outside the estrus phase. But most mating is usually confined to only a few days of frenzied activity that is triggered by the estrus state and for females this is frequently followed by abstinence during pregnancy and lactation.

However it would be wrong to suggest that sexual activity in primates is entirely under hormonal control or that in humans the brain has completely replaced hormones. Modification of this older conception has occurred with the discovery that castrated primates, like human males, can continue to engage in sexual activity. Some rhesus monkeys, who lacked the male hormones associated with the sexual urge, continued to ejaculate and penetrate females a year after castration. It is also the case that apes, like humans, use sex to cement social relationships and frequently copulate outside the estrus period – on having been reunited after a period of separation, for instance. One general finding here is that female apes initiated mating primarily when they were in estrus and were especially attractive, while males initiated mating at any time in a female's cycle and engaged in sexual activity that had no clear reproductive purpose (Nadler and Phoenix 1991). In the human female a similar monthly rhythm of ovulation and menstruation occurs under the control of the same hormonal secretions – albeit with considerable natural irregularity compared with other mammals – and with the same dramatic change in hormone level, but without the biochemical reactions which instigate a sudden rise of interest in sexual activity in non-human primates or any awareness that ovulation has occurred (Burley 1979). Prevailing opinion accepts that changes in the basic pattern of sexual arousal or responsiveness to stimulation do not occur in any appreciable sense during the course of a woman's menstrual cycle nor does the release of hormones have any apparent effect upon her attractiveness to males. But some evidence which challenges this has been assembled by Wolfe (1991) who finds that human female sexual behaviour is to a degree concordant with other primates since some preliminary studies have shown that hormonal changes at the time of ovulation do in fact lead to greater sexual attraction and intercourse. As a whole this evidence suggests that rigid distinctions between human and animal sexuality are invalid and that both mental and chemical control of sexual activity occur among all primates.

There are clearly mechanisms at work on our sexuality that are not properly understood. It has been shown for instance that social interaction among cohabiting women has a strong influence upon the menstrual cycle and that this is particularly marked among friends in all female groups. Studies by McClintock (1971) and Graham and McGrew (1980) have shown that women who live together in close proximity come to experience nearly simulta-neous cycles. This phenomenon, known as menstrual synchrony, occurs for reasons that remain unknown although its operation is likely to be by exchange of pheromone. Several explanations have been proposed: mutual female care of offspring, competition for the sexual attention of a male or merely a vestigial physiological reaction that has no current function (Jolly 1985). It has been noted that the 29.5 day menstrual cycle of the human female coincides exactly with the lunar cycle. Catton and Gray (1985) propose that the mechanism is related to the need to maintain a stable monogamous pair-bond as a foundation for successful child rearing. If hominid females once all menstruated in synchrony, and this was widely known because it occurred in accordance with the moon's phases, men would gain little by mating with more than one female. Since all females in a band would be fertile at the same moment, a male who left his mate at this time to copulate elsewhere would lose the opportunity to father and rear thriving offspring by a regular partner and would risk the attack of other loyal males for his attempts at infidelity.

The evolution of concealed ovulation is seen by Tanner (1981) as a byproduct of our new form of locomotion. Bipedalism realigned the genitalia of transitional female hominids and the erotic points of focus to favour upright face-to-face sexual contact. The loss of a periodic swelling, seen on the buttocks of chimpanzees and some monkeys when in estrus, has been replaced by permanent epigamic characteristics on the front of the body. Since the female genital area moved from the rear to between the legs visual signals of estrus were inappropriate and bipedalism made indication by scent less accessible. Publicly recognised anatomical signs of estrus were replaced by non-verbal forms of communication and females increasingly used these to exercise a choice of mating partner. In this thesis concealed ovulation arose to accommodate female decision making but despite the persuasiveness of this explanation the question of continual receptivity remains unanswered.

An alternative explanation proposes that the growing intelligence of hominid females led them quite reasonably to avoid the rigours of conception and gestation by deliberately choosing not to mate when ovulating. Burley (1979) argues that the exercise of this natural form of contraception meant a lowering of the reproductive output of the females who were most aware of their ovulation. As a consequence females who showed the least sign of estrus – by scent, swelling or conduct – conceived more readily, produced more offspring and concealed ovulation evolved accordingly. But was the intelligence of early hominids sufficient to make the necessary association between ovulation, copulation and conception, and would such a strategy in fact be turned to by a creature whose reproductive potential was already low?

A theory which associates concealed ovulation with the maintenance of the monogamous pair-bond is proposed by Lovejoy (1981). The new savanna-based way of life, which included more meat in the diet, demanded the inclusion of males in child rearing. Continual sexual receptivity and frequent copulation would increase pair-bond adhesion making food exchange and longer-term care of offspring more likely. The presence of so many sexual display traits in both sexes, making us the most epigamically endowed of all primates, our high level of sexual activity and our lack of dimorphic features related to possessive dominance, such as longer male canine teeth or a pronounced difference in body size, point to the evolution of characteristics intended to enhance and maintain mutual attraction. Females came to replace overt signs of receptivity with concealed ovulation and indicated their continual availability through emotional contact to remain bonded with a male provider. But did females really need such help from males in what we now know was a pre-hunting economy where meat was still a small part of the diet? A female might potentially gain more by exchanging food with other females or by trading sex for meat with a number of males. Similarly why would a male wish to exchange and procreate with only one female in the presence of others of greater attraction?

Ultimately none of these theories of concealed ovulation and continual sexual receptivity provide us with a conclusive explanation. The evolutionary reorganization of our sexual behaviour has involved us in aspects of all three of these theories. Choice of sexual partner, the desire to control our fertility and the pair-bond are basic parts of human sexuality but in the current state of knowledge the sequence of developments which led to the incorporation of these aspects cannot be established with any certainty.

ATTACHMENT AND THE ORIGINS OF THE FAMILY

Our sexuality and its bodily manifestations evolved in tandem with new forms of ecological adaptation that were made possible by innovatory forms of social behaviour marked by exchange and division of labour. But this emerging social matrix is only comprehensible within a species-specific pattern of emotional bonding with which it is closely connected. The origins of family organisation, which depended upon essential shifts in economic behaviour and interpersonal relations, were founded equally upon a unique form of attachment system. Exactitude here with respect to time and content is not possible but it seems unlikely that larger-brained hominids, such as *H. erectus,* could have success- fully reproduced without an enduring alliance between a feeding strategy and a predicable pattern of social bonds. Any prescriptive reconstruction of these institutions would be unwise. We cannot assume the full existence of the modern family in this era but a review of some elements of an attachment system that had begun to materialise among our ancestors is of value.

The existence of a species-specific social structure and attach- ment patterns has been established among apes and monkeys. Primates have an inherited propensity to form general attachment traits while the particular form taken depends on social experience. A huge gap exists between our nurturing behaviour and that of other primates but there are also significant elements of continuity that make the transition to the human family more comprehensible. Individual differences in maternal behaviour, which originate from infant experience, are seen in all primates and maternal deprivation has predictable effects. Offspring continue to be closely associated with a mother long after the end of caring and juvenile female siblings are particularly likely to be attached to infants and act as carers. Warm intimate relations, expressed by frequent acts of close physical contact, are an essential part of all primate social relations. The difficulties faced by all primate mothers in bearing and raising offspring, which involve huge energetic costs and extra feeding time, lead to the conclusion that such sacrifices are borne only because of the evolution of a series of emotional rewards that have allowed mothering to be inherently gratifying and emotionally fulfilling. In parenthesis here it is worth noting that the grasping and clinging reflexes of human infants, the composition of human milk and the maternal behaviour of foraging peoples all point to the fact that human mothers and infants are likely to be adapted for a greater amount of close contact and nursing than is normally found in western societies (Nicholson 1991).

The emergence of the hominid family involved the utilisation of behaviour and attachment systems that were likely to have been part of the primate repertoire. The hominid pair-bond itself may have evolved from a prolonged consort pair relationship, seen among chimpanzees and baboons, and adult food-sharing from the mother–infant relation. Male nurturing behaviour can be found in some primate species and the idea of males becoming emotionally 'captured' by their young to fulfil a fathering role has an echo if not a precedent among primates and some carnivorous animals. Where pair-bonding is found in mammalian species it usually coexists with paternal infant care suggesting that the two are causally related and that infantile and sexual attachment systems evolved together (Reynolds 1976). Among some canids – species of coyotes, jackals, dingoes and wolves – both sexes regurgitate food to cubs after a hunt and postnatal or nursing females are provisioned by male partners. Male parental involvement, bonding and the prohibition on incest are aspects of evolutionary continuity that need further scrutiny.

Primate males regularly engage in troop protection, surveillance for predators and food resources, and the defence of young. But male infant care is very unusual in mammals, and most primates show little systematic *direct* paternal involvement that would incur costs and increase the fitness of offspring except on an occasional basis (Taub and Mehlman 1991). Rhesus monkey males have acted as surrogate mothers in extreme circumstances. Gorilla and chimpanzee males are known to be solicitous, affectionate and tolerant to infants but these relationships are often brief and rarely involve feeding. Of all the higher primates only the monogamous, pair-bonding siamang male, a 'lesser ape' relative of the gibbon, provides primary infant care from the second year until independence. Strangely the gibbon itself does not become involved in paternal care but in several other monogamous, pair-bonding primate species, such as the owl and titi monkeys, males do the bulk of infant carrying, even in the first month of life, and will also play, groom and share food with their offspring.

A cross-cultural study of 80 non-industrial societies found that in more than 90 per cent, human mothers were the principal or exclusive care-givers and that in only four per cent was there a regular and close relationship between father and child (Katz and Konner 1981). Involvement in infant care by the human male is thus highly limited and, in a simple sense, men are comparable with gorillas. It is interesting to note that paternal care was found to be most prevalent in monogamous societies where the nuclear family prevailed, as in hunting societies where a mother's economic contribution was considerable, but the complexities of culture-based ways of life make analogies with primates less meaningful. The role of fathers in human societies is crucially affected by the social division of labour and the nature of the economy as a whole, as much as by the mode of family organisation and the degree to which parents are integrated in or isolated from the rest of society. The importance of male parenting in serving a purpose beyond economic provision has long been recognised by social science. Fathers have a distinctive and necessary function as decisive agents of socialisation and in providing a critical orientation for a child's psyche (Parsons 1954). Ape and monkey behaviour can provide a guide to understanding the origins of our ancestral paternal adaptation but it can in no sense explain the unique features of human fathering or of the family itself.

Humans are unique in having both sexually exclusive and polygamous relations and our marriage practices are not simple equivalents of a primate mating system. But no dignity is lost by recognising that within these diverse cultural patterns – and sometimes in spite of them – strong and enduring male–female bonding frequently develops which might be considered as part of our own species' attachment pattern. We have a deep propensity to make strong heterosexual bonds that are functional for child-rearing even though the sexual code of a culture may specify polygamous marriage. In a world survey of 554 cultures, 75 per cent were found to be polygynous to some extent, 24 per cent monogamous and one per cent polyandrous (Murdock 1957). But several qualifications must be made here. The cultures sampled contain populations of very different sizes and, in reality, the overwhelming majority of humanity live either in monogamous cultures or have monogamous marriages. Even in the most thoroughly polygynous cultures this is very rarely the form of marriage for the majority. In the minority of polygynous marriages the first wife is usually seen as the only legitimate consort with full rights of marital status and her co-wives can be regarded as concubines (Lévi-Strauss 1971). It is also the case that polygyny itself is not necessarily a barrier to pair-bonding. Among the polygynous Yanomamo of Venezuela a strong persisting attachment was observed between a headman and his principal wife. In all the headman had had six wives and numerous other liaisons yet even in this society, marked by violence and sexual domination, a friendly companionate relationship was found (Chagnon 1983). Our propensity to bond, like other universal traits such as smiling and kissing, is seen by Eibl-Eibesfeldt as an aspect of human ethology. He continues:

> Tradition does determine whether a man may have one or many wives. A permanent long-lasting association of the partners, however, is generally the rule and is necessitated by the slow development of the human child.   (Eibl-Eibesfeldt 1970: 442–3)

The human propensity to form bonds should not be seen in mechanistic terms. Bonding and marriage are not synonymous and can vary independently of one another. In some animals pair-bonds have a reproductive purpose alone which serves to guarantee paternity or female fidelity as we find in many thousands of bird species. But bonding in humans and other primates has an emotional context that is questionable in other animals. An ethological approach to human bonding is proposed by Bowlby (1979) who argues that our motivation for this behaviour stems from a pattern of innate drives that are a part of a life-long attachment system. Bonding with a mother figure occurs at a crucially sensitive phase of infant development and differs from more simple imprinting in lower animals because it is facilitated by a complex pattern of species specific 'social releaser' mechanisms, like smiling, which evoke maternal behaviour. As a child grows, and other significant figures like fathers become important, attachment is spread but a preference for a single figure remains. A primary aim for an infant, expressed through the attachment drive, is the maintenance of contact with known trusted figures to avoid strangeness, and a strong bias toward stimuli that reinforce such bonding is part of our inheritance. Throughout life attachment provides psychic stability and protection from danger rather than being simply a means to food and later sex. In a healthy adult the desire for attachment remains as a normal aspect of maturity. The need to form close bonds with a significant figure as a means of both gaining and providing a secure base from which to venture is a common route to a contented emotional equilibrium in humans and does not indicate dependency. While this form of bonding is functional for child rearing it has much wider connotations which go the heart of our existence and generate the strongest emotions.

INCEST AVOIDANCE

All known societies, or almost all, have placed a taboo upon incestuous sexual relations and elaborate social rules to prevent sexual contact among kinsmen exist everywhere. There appears to be a universal horror of incest, and a desire to outlaw its occurrence, which would seem to transcend all forms of cultural diversity. But it is only in the context of human culture, with its consciousness of paternity and kinship relations, that an incest taboo acquires significance as an element of social organization. Animals are usually unaware of both parents, paternity is at best only partially known among primates and it is human society alone which traces uncles, aunts, cousins and more distant relatives. It is frequently argued by social scientists that the avoidance of incest and the practice of exogamous out-mating are crucial factors that distinguish human and animal statuses. It is proposed that these central aspects of human behaviour also constitute functionally adaptive social mechanisms which make the family and a cultural way of life possible. There are supposedly obvious disadvantages to incestuous mating and considerable gains from broadening social networks through exogamous marriages. The avoidance of incest and the taboo placed upon it are therefore claimed to be derived from culture.

It is in fact difficult to show that culture alone acts to cause either the propensity to avoid incest, the associated taboo or the emergence of kinship. It is also just as difficult to confirm that we possess an innate drive which impels us to obey an incest taboo with no reinforcement from culture. We must conclude that both biology and culture are inextricably involved here and with the rise of the family (Leibowitz 1978). If the taboo is purely cultural in origin why is it apparently universal when almost all other facets of social life are variable? Conversely, if incest is normally avoided and held to be naturally repugnant only because of an innate mechanism it is difficult to envisage why it has become explicitly tabooed and why this taboo has given rise to a range of social sanctions. Much activity that is deemed to be universally repulsive is not prohibited or punished by law. Actual cases of parent–child incest and instances of sexual attraction between kinsmen also prompt us to ask why such a mechanism sometimes fails to work?

Proponents of the cultural explanation point to the diversity of social perception that is involved in deciding what constitutes an incestuous relationship. The abhorrence shown to incest and the social rules used to prohibit it are often biologically confused. Whereas sexual relations between parent and child and between siblings are always proscribed, the social definition of what constitutes consanguinity among wider kinsfolk is subject to variation. There are societies which allow marriage between paternal cousins but forbid the marriage of maternal cousins. There are American states which outlaw the marriage of uncle and niece and half-siblings, yet allow the marriage between double first cousins – children produced by two sets of siblings intermarrying – even though the genetic relationship is the same in each of these cases (Morton 1961). People who do not share any blood relationship such as godchildren and godparents or stepfather and stepdaughter may be forbidden to marry. In the Trobriand Islands no biological connection between father and child was traditionally accepted. Malinowski (1982) found this society to be entirely ignorant of the physiological facts of paternity: conception being explained by an extensive mythology and a series of practical fables. But even though the husband of a Trobriand woman was not considered to be the progenitor of his children, he was nonetheless accorded a father role and was strictly barred from having sexual contact with a daughter. An essential part of the Trobriand Islanders' social organisation was the division of the whole community into four totemic clans that contained both relatives and non-relatives. An Islander would refer to his non-related fellow clansmen as pseudo-kindred and, in theory, would observe a prohibition on sexual relations and marriage within his clan. Even though this prohibition was sometimes broken by non-related clan members, the Islanders feared retribution in the form of a wasting disease and such unions were regarded as a violation of a natural law.

In these examples sexual relations and marriage are prohibited between both related and non-related people. Some of those who are in fact related are thought not to be, while others who are not actually related *are* thought to be, at least for the purposes of extending the taboo. Regardless of the muddled conception of consanguinity exemplified by these different social rules, it is often claimed that they all act as protective barriers to relations within the family since they serve as the social mechanism which preserves exogamy or out mating. A variety of cultural solutions are provided as a means of ensuring that this rule, which is central to human society, is adhered to. Highly deleterious social and biological consequences might result should the rule be consistently broken. Social roles both within the family and in society at large would remain unclear if parent–child incest were prevalent. Sexual rivalry among family members would be disruptive and brother–sister mating would prevent the advantages of exchange between families.

But there are formidable difficulties here. Ethnographic examples show that in a variety of societies, which are marked by stability and persistence, fathers, mothers, brothers and sisters do in fact share sexual partners. In polyandrous societies, brothers may share the same wife but not their sister, or father and son share a wife but not the son's mother, and in polygynous societies this practice is reversed (Aberle et al. 1963). Then there is the question of the incest taboo acting as an adaptive mechanism in support of exogamy. Sahlins (1960) argues, on the basis of what was then known from primate studies, that unregulated competition for sexual partners within a primate horde would have been highly maladaptive for hominids. Indiscriminate mating, that could lead to fatal strife, was then thought to be typical of all primates and by contrast it was the repression of unbridled passion that was fundamental to human evolution. The various taboos which came to surround human sexual activity acted to guard the harmony and solidarity of the family unit which was economically linked by mutual aid to other families because of exogamy. But while this argument can logically explain the origins of exogamy and prohibitions upon incest, it cannot be used to demonstrate the *persistence* of either under different conditions and it is not sufficient to explain the revulsion which is aroused from actual cases of incest. Why should the behaviour of others, whether maladaptive or not, excite general condemnation? One may counter ideas which assert that the incest taboo has survived because it is functionally adaptive for the family or kinship system with examples of other features of these institutions that have *also* survived, and which are disruptive and non-adaptive. Family environments have perennially been the centres of repression, and conflict between families has frequently occurred precisely because of marriage ties.

Theories which claim that the incest taboo is a natural behavioural response which serves to prevent inbreeding rest upon arguments that parallels between human and primate behaviour can be established and that the genetic endowment of children who result from such unions will produce morbid and lethal effects. A number of recent studies have shown that other primates besides humans also avoid incest (Jolly 1985). There would appear to be an important continuity here between animal and human societies which has considerable bearing on the formation of social bonds among the early hominids (Passingham 1982). Long-term observations of monkeys, chimpanzees, gorillas and gibbons have shown that among these species females often transfer to other troops when they become sexually mature. Among chimpanzees, who mate promiscuously, only rare instances of matings have been observed between sons and mothers who protested and tried to escape. Matings between siblings were infrequent and occurred only after aggressive resistance by a young female. Because paternity is unknown in this promiscuous breeding system father–daughter incest is unlikely to be inhibited but young females have usually been observed to transfer to or visit other troops when in estrus. Young females also showed reluctance to respond to the courtship of older males within their natal troop (Goodall 1986). Is it reasonable to suggest that this continuity represents a behavioural basis for the institution of exogamy (Reynolds 1980)? Larger and more intelligent animals who mature slowly and are long-lived usually do restrict inbreeding. Humans conform with this pattern but differ in the ability to limit the choice of mate beyond the nuclear family and in having a taboo on incest which reflects a conscious awareness that incest is a possibility.

But if incest avoidance in humans is innate it is difficult to explain the rare but well documented instances of institutionalised incest which compromise the idea of a universal incest taboo. It has long been known that ritual incest between brother–sister and father–daughter occurred in the ancient royal families of Egypt, Persia, Hawaii and among the Inca aristocracy. Recent research has also shown that incestuous marriages between brother and sister regularly occurred among the Greek communities of ordinary citizens of Egypt during the first three centuries of the Christian era (Hopkins 1980). In this study, Roman census returns show brother–sister marriages to have been between 15 and 21 per cent of 113 marriages that were recorded. These were marriages that were celebrated in public and included all the paraphernalia of contracts, dowries and children.

How reasonable is the idea that human beings are unconsciously motivated to avoid incest because of the adverse effects of inbreeding? In this form of explanation, which is often associated with sociobiology (Lumsden and Wilson 1983), humans, like other animals, possess an innate drive which inhibits sexual desire between relatives because of its deleterious genetic effects. Until recently geneticists confirmed the general proposition that the children of incest are biologically less fit. Morton (1961) found that mortality rates doubled among children whose parents were first cousins and in a study by Adams and Neel (1967) of children produced by brother–sister and father–daughter incest, high levels of mental retardation and physical abnormality occurred together with high mortality rates.

But current genetic research shows that while an inbreeding depression, seen in a higher incidence of deaths and defects, does indeed result from incestuous mating, the very high levels found in the studies just quoted had been exaggerated because the findings here come from *retrospective inquiry* into the parentage of children who were already suffering from hereditarily transmitted conditions. However in surveys of large populations with the use of control groups it is reported that the effects of inbreeding are statistically significant but small, resulting in little loss to net fertility (Bittles 1980, Bittles and Makov 1988).

While the incest avoidance mechanisms of chimpanzees are likely to have been present among our ancestors it is also quite possible that some form of inbreeding among more distant kin occurred in the early hominids without this being a source of instability for whole populations (Livingstone 1969). In conditions of high infant mortality and child morbidity, congenital defects would have had only a slight effect on death rates making the real effects of inbreeding difficult to perceive. In fact the idea that biological harm is a consequence of incest is largely unknown in primitive societies and appeared in western culture only after the sixteenth century (Lévi-Strauss 1969).

But why might exogamy have developed at all? If the early hominids lived in bands of perhaps 15 to 20 paired adults and their offspring, an excess of either sex would have posed an economic problem and only in bands as impossibly large as 500, which included 100 couples, would there have been an even chance of a balanced sex ratio. In a hominid band of 20, there was a one in four chance that all babies born in a three year period would be of one sex. Exogamy became an economic necessity and incest was avoided because of the need for balanced foraging (Washburn and Moore 1974). This theory tends to assume a modern foraging economy and fails to explain either the persistence of exogamy or the taboo on incest in subsequent societies. Nevertheless the connection between economic and reproductive behaviour is important.

THE RISE OF HUMAN SOCIETY

These ideas, which are derived from the work of Lévi-Strauss (1969, 1971), link the division of labour between the sexes with the prohibition of incest. While the interdependence of women and men lies in their different productive roles, the outlawing of incest provides the basis for mutual dependency between families since this rule compels new family units to be formed. Consequently when family formation first occurred a transition between nature and culture was marked. Besides being the smallest viable economic unit the family was also a vehicle through which both human status and society are mutually realised. It is only by this form of continual social extension that society can survive and it is only through the agency of society itself that a human way of life is sustained. For these reasons the family has become a universal phenomenon in all societies even though it is often enveloped in disparate sexual and social customs. At either end of the human scale of development – among foragers and in the advanced societies – similar forms of family organisation are found. As Lévi-Strauss notes, it is possible to distinguish monogamy, independent establishment of the newly formed couple, warm relations between parents and children, and married couples united by strong emotional bonds in the societies of both hunters and industrial peoples. Frequently these elements of the nuclear family can also be detected beneath the complex web of social rules governing the lineage systems and sexual practices of polygamous societies.

For Lévi-Strauss the rise of human society is synonymous with both exogamy, as a means of avoiding incest, and reciprocity between families. It is inconceivable that society could exist without such exchanges which usually involved goods for marriageable women. This position is close to that of Service (1971) who also stresses sharing and reciprocity. Sharing was basic to hominid social life and was a condition of its survival that had become an established norm with associated rewards and sanctions. From sharing, and the rules surrounding it, came the formation of a series of social bonds, alliances and dependencies.

Reciprocity gave impetus to the development of symbolic communication since both social rules and concepts of time, which are the context of these exchanges, flow from this ability. Exchanges of females between patrilineally bonded hordes, which controlled hunting territories, had the effect of transforming simple mating or pair-bonding into marriage because such unions were the result of reciprocity and were now marked by social recognition and a degree of permanence. Males gained women who joined their husband's lineage and remained domiciled with his kinsmen. These new social bonds were ratified by society and went beyond the predilections of partners. Service, like Lévi-Strauss, envisages the exchange of females between bands as the mode in which reciprocity was instituted. These exchanges operated as a means of reducing social conflict by ensuring wider alliances and a necessary part of this institution was to be a permanent sexual division of labour.

The idea of exchange as a solution to intergroup aggression, and as a means of structuring social relations beyond the family, is unobjectionable. But did the rise of exogamy simultaneously imply a trade in women? This notion has met the same set of objections from feminists as the 'trading sex for meat' explanation of concealed ovulation proposed by Lovejoy (1981) that was discussed earlier. While the principles of sharing and reciprocity as major foundations of human society are well founded, the conviction that mate exchanges between families must involve females as gifts for male receivers is not properly substantiated in the ethnographic record. Observation of modern foragers shows that patrilineal control of territory is not typical and that marriage and kinship are more likely to have originated from economic and social organisation that was not dominated by either sex. Lee (1974) found that among !Kung bands no consistency of male or female based genealogy formed the core of these social units. Since !Kung men frequently had difficulty in gaining a wife, a woman's parents could be selective about their potential son-in-law and might demand that he render 'bride service' to their family for as much as a decade or even a lifetime and, in this case, the couple would reside with the natal band of the wife. Once again no clear pattern of sexual domination or essential difference in the distribution of power between male and female is apparent.

What processes instigated the evolutionary change which provided the impetus for ancestral hominids to adopt social institutions that would eventually lead toward human status? By the end of the 1970s the desire to abandon explanations that relied upon stereotypical masculine themes, such as the 'man the hunter' image, were given reinforcement by new departures in anthropology and archaeology. However the comparatively new wealth of findings that have come from these disciplines and from evolutionary biology have allowed for the construction of widely disparate models of our origins. Two competing theories that have recently attempted to make a comprehensive explanation of human prehistory are the 'woman the gatherer' model and the pair-bonding hypothesis.

The idea that a specifically female initiative produced the main stimulus to the formation of the hominid adaptation and associated pattern of social bonds that led to human status is central to the work of Tanner and Zihlman (Tanner and Zihlman 1976, Zihlman 1978, Tanner 1981). Our common primate ancestors, at the point of the initial ape–hominid divergence, are assumed to have had behaviour patterns akin to the chimpanzee. It was maternal investment in a new adaptive strategy that was decisive in making the successful shift to the open-country environment. This move towards bipedal exploitation of the savanna came about as the first female hominids, on whom the sole care of dependent young rested, experienced greater dietary stress in shrinking forest habitats. Mothering imposed an increased foraging burden on females who instigated a series of social and technical measures to cope with these new challenges to feeding. It was in the interests of females to develop food-sharing norms with other females and, since such behaviour already had a precedent in the mother–infant relationship, its extension to non-dependent adults was facilitated. In this thesis the first form of tool-use and then tool-making occurred, together with carrying devices and storage methods, to meet the food needs of female hominids and their offspring. The stone tool was initially an extension of the mother's teeth, that were overworked by preparing baby food, and represented a means of staving off 'dental death'. Both technical and cultural traditions were begun and sustained by female plant gatherers who were to be the architects of the human condition.

Tanner and Zihlman envisage the exercise of female sexual preferences for a less aggressive male as the means by which sexual dimorphism was reduced. It was females, who shared mutual responsibilities for child care, who were most naturally fitted to be the first teachers and who actively passed on technical and foraging elements of the cultural tradition. Since paternal investment did not exist females were obliged to develop foresight and it was feminine initiative that generated human intelligence. No attachment system, such as a proto-family, united the sexes and their offspring in this era. Consequently in the absence of any stable ties between males and females, it was mother–infant and sister–sister bonds that became the foundations of kinship and it was into these social networks that males were incorporated.

This position has been given widespread support by feminist commentators (Fedigan 1986, Ehrenberg 1989, Leibowitz 1986, Wolfe 1991) who in general remain committed to two main arguments. First, female hominids are seen to have been just as mobile, productive and inventive as males, or even more so. The development of traits which characteristically differentiate us from other primates are seen in principle to have resulted from feminine initiative. The sexual division of labour is believed to have developed only in late prehistory when the hominid adaptive pattern was already well formed. This is envisaged to be in late *H. erectus* times by Leibowitz (1986) but much later by others. Second, there is a tendency to consider mating to have been emotionally superficial and reproduction perfunctory and little consideration is given to the special hazards which birth posed for a bipedal female. Wolfe (1991) envisages promiscuous mating among the early hominids – although not to the same extent as chimpanzees – independent foraging by each sex and child care by cooperating groups of females. No form of bonding or sentimental attachment between female and male is assumed and any form of mutual reliance between the sexes is taken to be synonymous with dependency.

The feminist approach to prehistory has been fully justified in its dismissal of models that have relied heavily upon hunting, or more arcane explanations which favoured violence or male bonding. At which stage in our career the sexual division of labour became as fully operative as it is now observed among modern foragers remains unclear, but it is likely to have arisen for both economic and reproductive reasons and hence to have had some association with our unique form of locomotion. Giving privileged status to female-driven evolution has the combined effect of presenting an inversion of male chauvinism and of violating evolutionary theory. The assumption of a superfluous and inept male in feminist theories is inconsistent with the principles of evolutionary biology which insist that appropriate explanations must involve the interaction of both sexes. Natural selection works on whole populations rather than one sex. Tooby and DeVore (1987) also find the 'woman the gatherer' model to be at odds with the basic rules of optimum adaptation and thus inconsistent with biological practice. The laws of behavioural ecology specify that animals will seek the most efficient means of gaining subsistence. In short while there is no doubt that gathering was a critically important activity for hominids it is insufficient, as only one factor in a complex behaviour pattern, to explain the significant trends in our evolution.

If in fact gathering is more productive than hunting why did males not engage in this most profitable form of provisioning? While there is no bar at all to male gathering there are, as we have noted, considerable problems associated with female hunting. Plant food alone is unlikely to provide sufficient nourishment to be worth gathering, carrying and sharing and is unlikely, in itself, to have produced any fundamental change in behaviour. Only high-energy animal protein food would normally repay its procurement and become a part of a system of exchange. The gatherer model also fails to explain why scavenging and then hunting ever evolved. Nor does it address the fact that the sexual division of labour developed to provide an optimum foraging solution for hominids as whole populations, and makes sense only in the context of sharing and exchanging animal and vegetable products between males and females. Taken in isolation female gathering is similar to the adaptations of other primates and does not explain either why the rise of the hominid lineage was to become so distinctive or why the apes did not also adopt new cultural solutions. If males were essentially parasitic and un-resourceful why have men come to be involved with reciprocal exchange and parenting? If males did not bond with females and their offspring they would have been in constant competition for mates and no decline in sexual dimorphism would have occurred. Finally, there is no compelling reason either why adult females should share food among themselves rather than with their mates.

The comprehensive model of human origins presented by Lovejoy (1981) has already been discussed here, in the context of concealed ovulation, and in Chapter 5. Lovejoy rejects simple prime movers, such as brain expansion or tool-use, as the formative cause of hominid beginnings. Instead he examines growing habitat change and the overall environmental challenge in relation to population dynamics. New reproductive and social strategies were adopted to stave off extinction threats encountered by a slow-breeding ape faced with a new mosaic of different habitats within the spreading savanna ecology in the late Miocene 5 to 7 million years ago. To compensate for the depletion of standardised forest resources, increased parental investment had occurred among the last hominoid ancestors of the first hominids. This had the effect of allowing each developmental stage to become proportionately longer which both increased longevity and delayed the age of reproduction. This was to constitute a new demographic pressure and a stable population could be maintained only if a radical adaptive change promoted lower mortality or shorter intervals between births. Intelligence, social bonding and a longer learning period were already utilised to avoid greater hazards in the changing habitat. 'The most obvious, and perhaps only, additional mechanism available with which to meet this "demographic dilemma" is an increase in the direct and continuous participation by males in the reproductive process' (Lovejoy 1981: 346). Monogamous pair-bonding, linked closely with male parental investment, arose to solve this dilemma through mixed foraging that involved females in cooperative groups and males who provisioned their own mates and offspring. This adaptation is consistent with ecological principles provided that an equal number of both sex are involved, that radiative foraging from a base occurs, and that the rate of feeding is constrained only by the time taken to acquire food and not by the need to spend long periods handling or consuming it. Pair-bonding constituted a strong selective pressure for bipedal walking as a means to efficient foraging and carrying to and from more dispersed locations. The sexual division of labour implied that males foraged over greater distances than females who remained closer to a base with young. The decline of sexual dimorphism, our degree of epigamic adornment, continual sexual receptivity and high level of sexual activity all attest to the reality of a pair-bond. The nuclear family then is a product of pair-bonding and is taken to be an ancient and fundamental feature of hominids while the interactive social relations within the family and between other families are seen as major selection pressures for greater intelligence. From this more secure foundation, where reproduction was supported by social behaviour, come further developments such as tool-use, tool-making and brain enlargement.

Lovejoy's theory has been praised for its adherence to evolutionary biology and for the significance of social behaviour as a causal factor in evolution. However the hypothesis has met with vigorous opposition from feminists. Lovejoy is accused by Tanner (1988) of casting males in a role that gives them control over females, and similarly of assuming females to have been passive and economically dependent by Wolfe (1991) who also rejects any notion that early hominid females were limited by their sexual services and reproductive capacities. Both these charges seem unfair since for Lovejoy mutuality and interdependency are an explicit part of his two-sex model of reciprocal provisioning. Similarly the criticism that the theory implies an overreliance on male pioneering and that accordingly bipedalism should have developed in one sex alone is unfounded because females are assumed to have been active foragers as well as the carriers of infants. But can the sexual division of labour and food-sharing really be assumed to have developed as much as some 4 million years before the earliest possible evidence of hominid social behaviour, that is before tool-making and the shift to meat-eating? A new food source with greater nutritive value, the products of scavenging perhaps, would have been more likely as an efficient basis for provisioning. Lovejoy rules out meat as a catalyst in the formation of pair-bonds yet, as we have seen, provisioning by carnivores and meat-sharing by primates is by far the most common form of group feeding. Foraging peoples today consume much vegetable food during gathering and rarely transport and share this beyond the nuclear family. Only high protein vegetable food, like the !Kung mongongo nuts, are routinely shared.

Were the immediate ape ancestors of the first hominids really in a demographic cul-de-sac from which the only viable escape route was monogamous pair-bonding and paternal responsibility? Tooby and DeVore (1987) consider the main selection pressure for hominids to have been competition within the species and find that Lovejoy's model is unable to explain why a greater environmental threat and higher mortality was experienced by early hominids than by other primates. Monogamous mating, as we have already seen, is a correlate of minimal sexual dimorphism and yet the first hominids – the australopithecines – are highly dimorphic indicating a form of polygamous mating system.

The full scenario of family formation in prehistory remains unknown. No single factor that might have instigated the beginnings of this institution can be identified with certainty and no single hypothesis can by defended with complete confidence. What does seem to be clear is the fact that this central part of our existence arose as aspects of our ancestors' reproductive biology interacted with novel forms of social behaviour in the context of changing patterns of adaptation to new ecological conditions. In this case it is better to discard the notion of a prime mover or the relative contribution of either sex, as the means by which a proto-family came into being, and work with an integrative model that combines bio-social factors such as sexuality, infant development, nutrition, technology, cooperation and sharing and the other components of an emerging human life-style that have been introduced in the course of this book.

There is no simple conclusion to be drawn from this study of prehistory. The human evolutionary journey from animal origins to fully sapient status is complex and has occurred in the presence of intricate and progressive cultural development with which it has been closely allied. Darwin concluded *The Origin* with the contention that there was grandeur in a view of life that had been shaped by natural selection. His claim was of course fully justified. In having established both the unity of life and the unity of the human species he had simultaneously placed humans in the context of the natural world. A common human nature, the product of our species inheritance, can now be added to this understanding and these insights should not be overlooked by the social sciences if evolution is to be taken seriously.

Natural selection is clearly of great use in perceiving some aspects of our nature as sophisticated culture-using primates. Our propensity for a gregarious social life and our dependence on a material culture and the biological and social mechanisms on which these fundamental elements of human existence rest are examples. However while an evolutionary explanation is appropriate for the human subject it is of less value in understanding society. Darwin was unaware of the social and historical forces that condition and sustain human existence and he saw no need to provide a holistic view of society.

The familiar patterns that are found repeated in different types of human society are not simply the result of similar biological pressures. Such human universals should be attributed to the fact that all societies face similar complex imperatives as a condition of their existence. The social necessity for common production, reproduction or socialisation, for instance, has produced comparable organisational motifs in a range of profoundly different societies. The cultural solutions to these demands have been a weighty accumulation of diverse institutional practices and meanings that have become represented by powerful historical traditions which have an overwhelming influence on social life. Cultural traditions are flexible and non-deterministic forms of adaptation that are not tied to biological mechanisms, and operate and change with a logic that is amenable only to social explanation.

A natural origin, an evolutionary development and the acquisition of cultural abilities do not mean however that culture and history have gained autonomy from the natural world. During our evolution we have accumulated an inextricable combination of social and biological traits which can only become realised through action in social organisation. A knowledge of prehistory demonstrates that in reality there is no convenient means of disentangling the natural aspects of human abilities and cultural practices from those which are learned in society. The obvious significance of culture in human affairs should not be allowed to obscure the fact that we rely on a symbiotic relationship between social and biological mechanisms which have interacted during human evolution and continue to operate in ordinary social life.

Glossary

Acheulean Mode 2 stone tool industries of the Lower Palaeolithic which were first made in Africa 1.5 million years ago and are most typically represented by bifacial hand-axes and flake tools.

Adaptation A set of characteristics chosen by natural selection which provide an organism with an improved chance of survival by allowing a more effective means to exploit its environment.

Altruism Concern for the welfare of others before that of oneself. Sociobiology takes the explanation of the evolution of altruism as its central problem.

AMH Anatomically modern humans.

Apes Primates closely related to humans who together with hominids comprise the superfamily *Hominoidea* represented by the African apes, which include one species of gorilla and two of chimpanzee, and the Asian apes, who consist of a single species of Orangutan and six species of Gibbon.

Australopithecus A genus of extinct small-brained African hominids from the Pliocene and Pleistocene epochs represented by fossils dated from *c.* 5 to 1 million years ago.

A. afarensis An early African hominid providing the earliest unequivocal evidence of bipedal locomotion from *c.* 4 to 3 million years ago.

A. africanus An African hominid from *c.* 3 to 2.5 million years ago that is likely to have been the immediate ancestor of the earliest species of *Homo*.

A. boisei A hyper-robust hominid species found in East Africa between 2.4 and 1 million years ago. Also known as *Paranthropus* or 'beside man'.

A. ramidus A set of comprehensive fossil remains of the most primitive and apelike hominid ancestor known from 4.4 million years ago were published in September 1994 but the exact status and position of this new species in terms of hominid phylogeny has yet to be fully established.

347

A. robustus A robust hominid species from Southern Africa extant between 1.8 and 1.5 million years ago which, like *A. boisei,* is also known as *Paranthropus* or 'beside man'.

Band The basic economic and social unit of hunter-gatherer society comprising a single camp-based group of interrelated families who forage collectively.

Biologism Explanation of human social behaviour which rests on inherited biological mechanisms.

Bipedalism Locomotion by two feet which is a defining characteristic of the hominid family.

Bonding A widespread behaviour pattern among primates to form and maintain strong attachment relationships between mother and infant and interdependent sexual relationships between human adults.

Catastrophism A pre-Darwinian attempt to explain the newly discovered geological record of mass extinctions by postulating a catastrophic event followed by the divine Creation of new species.

Consciousness An aspect of mind which provides us with an awareness of our perceptions, thoughts and feelings and an ability to reflect upon these.

Cosmology An integrated system of beliefs and ideas embodied in the myths, doctrines and narratives of a culture which relate to its origins, place in the universe and explanation of natural phenomena. Cosmologies frequently contain gods, superior beings and divine ancestors who are thought to play a crucial part in maintaining the relationship between a people and their world order.

Culture Learned behaviour patterns for which there is no genetic basis which are expressed as norms, beliefs and values in ordinary conduct and as collections of artifacts comprising a material culture.

Division of labour This defining human characteristic and basis of economic life implies the separation of productive tasks into a variety of specialised activities as a means to greater productivity.

Ecology The study of the interactive relationship between organisms and their environment.

Environment The entire surroundings or context within which an organism exists and interacts.

Eocene epoch Literally 'dawn of recent': a unit of geological time, within the Tertiary Period, from 55 to 34 million years ago in which significant primate adaptations for tree life, like stereoscopic colour vision and limb-eye coordination were evolving.

Epigamic Secondary sexual characteristics which operate to enhance, attract or stimulate the opposite sex and facilitate reproduction.

Essentialism A doctrine prevalent in biology before the advent of natural selection which held that the form taken by a species remained fixed and that any detectable differences among individuals were the product of superficial variations from an essential modal type.

Ethology The comparative study of animal behaviour in the natural environment.

Ethnography The recording and study of particular cultures and forms of social organisation gained by anthropologists through direct observation and participation.

Evolution A biological theory which explains how in adapting to the environment organisms become modified and diversify into new species.

Exogamy Rules demanding that marriage occurs outside a social group.

Genes Units of inheritance within chromosomes consisting of complex bio-chemical material called DNA (deoxyribonucleic acid) which carry information and instructions to the body of an organism.

Genotype The complete genetic constitution of an organism which is passed on to offspring, including characteristics which may not be actually expressed in an individual phenotype.

Hominid (*Hominidae*) A family of living and extinct bipedal primates consisting of the Australopithicine and *Homo* genera.

Hominoid (*Hominoidea*) A superfamily within the primate order comprising the families of apes and hominids with all common ancestors.

Homo A genus within the hominid family which includes modern humans (*Homo sapiens sapiens*) and extinct species of *Homo* such as *H. habilis* and *H. erectus*.

Homo erectus An extinct species of the *Homo* genus extant from about 2 million to 200 thousand years ago ancestral to *H. sapiens*.

Homo habilis An extinct species of the *Homo* genus from sub-Saharan Africa from 2.1 to 1.6 million years ago who is generally accepted as the ancestor of *H. erectus*.

Homo sapiens (**archaic**) An extinct species of the *Homo* genus extant from *c.* 450 to 40 thousand years ago, also known as *H. heidelbergensis*, who gave rise to our own species.

Homo sapiens sapiens A species of the *Homo* genus represented by all modern humans which evolved from archaic *H. sapiens* populations *c.* 120 thousand years ago but lacking modern behaviour patterns until *c.* 40 thousand years ago.

Hunter-gatherer A type of society in which the economic life of human and pre-human foraging communities is based upon the collecting of wild plants, hunting and fishing.

Kinship A set of interpersonal relations which unite individuals on the basis of descent and marriage and which are maintained by a system of socially recognised obligations, rights and customs.

Lower Palaeolithic The first cultural stage of the old stone age between 3 to 2 million and 130 thousand years ago associated with Mode 1 and 2 Oldowan and Acheulean technology.

Matrifocal subunit A primary social group found within many primate societies consisting of mothers and their offspring.

Mesolithic Final culture stage of the old stone age represented by Mode 5 microlithic technology and an advanced hunter-gatherer economy which developed new and more efficient foraging methods in a period of transition immediately before the advent of domestication 12 to 9 thousand years ago.

Middle Palaeolithic Second culture stage of the old stone age from *c.* 130 to between 40 and 35 thousand years ago typified by Mode 3 Mousterian tools.

Miocene epoch Literally 'fewer recent': a unit of geological time within the Tertiary Period, from 22 to 5 million years ago, in which primitive apes began to evolve adaptations which led to the first hominid species.

Mode of production A particular combination of forces and relations of production which have arisen historically and have come to determine the form of social and economic organisation.

Mousterian Mode 3 stone tool industries of the Middle Palaeolithic made by the Levallois technique from about 130 to 40 thousand years ago.

Mutation An abrupt change in the normal genetic constitution of an individual organism which, if advantageous, may become consolidated by the species as a whole.

Natural selection The central principle of Darwinian evolution which specifies the mechanism of evolutionary change whereby any individual which is better adapted to the environment produces more offspring than other members of the same species and thereby changes the nature of the species as a whole.

Neanderthal An extinct species of the *Homo* genus known from fossil evidence in Europe and western Asia between 130 and 35 thousand years ago. Also known as H. *neanderthalensis* and H. *sapiens neanderthalensis*.

Neolithic The new stone age associated with domestication, farming and ground stone tools which began in the Near East about 10 thousand years ago.

Niche The part of the environment occupied by a particular species which provides all of its ecological requirements including food.

Oldowan Mode 1 stone tool industries of the Lower Palaeolithic associated with H. *habilis* consisting of flaked cobbles and choppers which began between 3 and 2 million years ago.

Oligocene epoch Literally 'few recent': a unit of geological time, within the Tertiary Period, from 34 to 22 million years ago in which monkey-like primates evolved.

Palaeontology The study of fossils.

Phenotype The actual manifestation of an individual organism resulting from interaction of its genetic inheritance, or genotype, with the environment during growth and development.

Phylogeny The evolutionary history or 'family tree' of a group of interrelated species.

Pleistocene epoch Literally 'most recent': a unit of geological time, within the Quaternary Period, from 1.8 million to 10 thousand years ago.

A period of intense glaciation during which major episodes of human evolution occurred including the rise of our own species.

Pliocene epoch Literally 'more recent': a unit of geological time, within the Tertiary Period, from 5 to 1.8 million years ago in which primates adapted to new open country habitats as monkeys, apes and hominids.

Polygamy Plural marriage where one spouse has more than one partner of the opposite sex which takes the form of either **polyandry**, where a woman has more than one husband, or **polygyny** where a man has more than one wife.

Preadaptation An existing physical or behavioural characteristic that becomes modified by natural selection to perform a new function: for example bipedalism may have evolved to facilitate more efficient feeding but allowed the development of material culture.

Primates An order of mammals which have evolved during the past 60 million years marked by their arboreal adaptations, which include hominids, apes, monkeys and prosimians.

Reductionism The analysis of a complex whole by reference to a minimal number of variables which are explained in the terms of a limited range of scientific knowledge on the assumption that the whole consists of no more than the sum of its constituent parts.

Scala Naturae A pre-Darwinian concept which implied that all organisms could be classified according to a progressive system seen as a 'ladder of perfection'.

Sexual dimorphism The characteristic differences in size, structure and other traits found between either sex of the same species.

Social evolution A pre and post-Darwinian social theory which argues that human societies may be likened to organisms which can be classified as a series of progressive adaptive stages that change from simple to complex modes of social organisation that, in turn, embody higher levels of moral, intellectual and aesthetic development.

Social institution A socially recognised pattern of frequently enacted roles, rules, relations and norms which demarcate a standardised area of social behaviour.

Sociobiology A body of theory which attempts to explain the behaviour and social organisation of humans by reference to innate genetic mechanisms chosen by natural selection.

Speciation The emergence of new species from an ancestral lineage which has diversified into distinctly different populations.

Species Populations of naturally interbreeding organisms which are capable of reproducing and are reproductively isolated from other populations.

Sub-species The taxonomic term for geographical races or varieties within a species which has been found to be unviable in classifying human populations and has fallen into disuse.

Symbolism An indirect form of representation.

Teleology Theory or explanation of natural or social phenomena which explains the nature of things in terms of the end-states or purposes which they ultimately achieve – giraffes acquired long necks to allow them to browse from trees.

Taphonomy The study of the processes whereby living organisms are transformed into fossils.

Taxonomy The classification of organisms into a series of classes according to their evolutionary relationships.

Tribe The political, economic and social organisation of primitive agricultural peoples who live in stateless societies which may be ruled by a chief or aristocracy.

Upper Palaeolithic The third culture stage of the old stone age from *c.* 40 to 10 thousand years ago which is marked by a profusion of complex Mode 4 blade-tool industries used by specialist hunter-gatherers.

Further Reading

Readers wishing to approach the issues raised by this book in greater detail will find a rich and intriguing literature. Only a rapid guide to a brief selection of work produced by the usually dispersed sciences which contribute to the study of prehistory can be given here.

A first class introduction to evolution can be found in Dawkins (1986) *The Blind Watchmaker* and in Diamond (1991) *The Rise and Fall of the Third Chimpanzee*. The history of life, fossils and ancient fauna are well discussed by Simpson (1983) *Fossils and the History of Life* and Stanley (1989) *Earth and Life Through Time*. Comprehensive guides to general prehistory are found in Gowlett (1984) *Ascent to Civilisation*, Klein (1989) *The Human Career* and Wenke (1980) *Patterns in Prehistory*.

On Darwin, evolutionary theory and social evolution a first choice must be Bowlby (1991) *Charles Darwin. A New Biography* which provides an excellent account of Darwin's life and work but there can be no substitute for reading *The Origin* itself. Like Dawkins above, the many books by Gould – *Ever Since Darwin* for instance – explore the modern application of natural selection. The archaeological discoveries during Darwin's lifetime which confirmed our palaeolithic roots are well documented by Grayson (1983) *The Establishment of Human Antiquity*. Hirst and Woolley (1982) *Social Relations and Human Attributes* looks at some of the sociological issues raised by a biological and evolutionary perspective.

Works on human biology, sociobiology and behaviour are numerous but Reynolds (1980) *The Biology of Human Action*, Jones (1993) *The Language of the Genes*, Midgley (1980) *Beast and Man* and Sahlins (1977) *The Use and Abuse of Biology* are all impressive. Eibl-Eibesfeldt (1970) *Ethology. The Biology of Behaviour* gives an account of human behaviourial mechanisms while Rose et al. (1984) *Not in Our Genes* provides a good overview and critique of sociobiology.

The best introductions to primate studies are to be found in Passingham (1982) *The Human Primate*, Jolly (1985) *The Evolution of Primate Behaviour*, Richard (1985) *Primates in Nature*, Goodall (1971) *In the Shadow of Man* and (1986) *The Chimpanzees of Gombe*, and Fleagle (1988) *Primate Adaptation and Evolution*. One of the most comprehensive and insightful accounts of our evolutionary journey is contained in the third edition (1985) of Campbell's *Human Evolution*. The studies of human evolution provided by Foley (1987) *Another Unique Species* and Richards (1987) *Human Evolution* are also of great value.

For stone tools, early hominid behaviour and Lower Palaeolithic society there is no better source than the papers of Glynn Isaac which have now been published as a single volume edited by Barbara Isaac (1989) *The Archaeology of Human Origins*. However J.G.D. Clark (1977) *World Prehistory in New Perspective* and Oakley (1972) *Man the Tool Maker* are both classics and Wymer (1984) *The Palaeolithic Age* is a most comprehensive archaeological survey.

Ethnographic works on hunting and gathering and primitive agricultural societies can be found in Balikci (1970) *The Netsilik Eskimo*, Lee (1979) *The !Kung San*, Silberbauer (1981) *Hunter and Habitat in the Central Kalahari Desert*, Rappaport (1984) *Pigs for the Ancestors* and Chagnon (1983) *Yanomamo. The Fierce People*. Coon (1976) *The Hunting Peoples* gives a world view of hunting societies and Sahlins (1972) *Stone Age Economics* is an excellent discussion of the sociological issues presented by these societies.

The evolution of modern humans, human abilities and Neanderthals is covered by Bahn and Vertut (1988) *Images of the Ice Age*, Lieberman (1991) *Uniquely Human: The Evolution of Speech, Thought and Selfless Behaviour*, Pfeiffer (1982) *The Creative Explosion. An Inquiry into the Origins of Art and Religion* and Trinkaus and Shipman (1993) *The Neanderthals*.

On the questions of sex, the division of labour and social bonding Ehrenberg (1989) *Women in Prehistory*, Reynolds (1991) in *Mating and Marriage*, Tanner (1981) *On Becoming Human*, are sources. Connections between primate and human sexual behaviour are dealt with by Loy and Peters (1991) *Understanding Behaviour. What Primate Models Tell us About Human Behaviour* and Kinzey (1987) *The Evolution of Human Behaviour: Primate Models*.

Bibliography

Aberle, D. F. et al. (1963) The Incest Taboo and the Mating Patterns of Animals. *American Anthropologist*, vol. 65, pp. 253–65.

Adams, M. S. and Neel, J. V. (1967) Children of Incest. *Pediatrics*, vol. 40, pp. 55–62.

Aiello, L. C. (1981) Locomotion in the Miocene Hominoidea. In Stringer, C. B. (ed.) *Aspects of Human Evolution. Symposia of the Society for the Study of Human Biology*, vol. 21, pp. 63–97. London: Taylor & Francis.

Alper, J. S. and Lange, R. V. (1981) Lumsden–Wilson Theory of Gene-Culture Coevolution. *Proceedings of the National Academy of Sciences*, vol. 78, no. 6, pp. 3976–79.

Anderson, P. (1983) The Reproductive Role of the Human Breast. *Current Anthropology*, vol. 24, no. 1, pp. 25–45.

Andrews, P. (1988) Hominoidea. In Tattersall, I., Delson, E. and Van Couvering, J. (eds), *Encyclopedia of Human Evolution and Prehistory [EHEP]*. New York & London: Garland.

Ardrey, R. (1961) *African Genesis*. London: Collins.

Ardrey, R. (1976) *The Hunting Hypothesis*. London: Collins.

Arens, W. (1979) *The Man-Eating Myth. Anthropology and Anthropophagy*. New York: Oxford University Press.

Auel, J. M. (1980) *The Clan of the Cave Bear*. London: Hodder & Stoughton.

Auel, J. M. (1982) *The Valley of the Horses*. London: Hodder & Stoughton.

Bahn, P. G. and Vertut, J. (1988) *Images of the Ice Age*. London: Windward.

Balikci, A. (1970) *The Netsilik Eskimo*. New York: The Natural History Press.

Bancroft, J. (1980) Human Sexual Response. In Austin, C. R. and Short, R. V. (eds), *Human Sexuality*. Cambridge: Cambridge University Press.

Banton, M. (1977) *The Idea of Race*. London: Tavistock Publications.

Barbetti, M. (1986) Traces of Fire in the Archaeological Record, Before One Million Years Ago. *Journal of Human Evolution*, vol. 15, pp. 771–81.

Barker, E. (1979) In the Beginning: The Battle of Creationist Science Against Evolutionism. In Wallis, R. (ed.), *On the Margins of Science: The Social Construction of Rejected Knowledge*. Sociological Review Monograph no. 27.

Barnett, A. (1950) *The Human Species. A Biology of Man*. London: MacGibbon & Kee.

Bar-Yosef, O. and Vandermeersch, B. (1993) Modern Humans in the Levant. *Scientific American*, vol. 268, pp. 64–70.

Bates, D.G. and Lees, S.H. (1979) The Myth of Population Regulation. In Chagnon, N.A. and Irons, W. (eds), *Evolutionary Biology and Human Social Behaviour. An Anthropological Perspective.* Mass.: Duxbury Press.

Beach, F.A. (1978) Human Sexuality and Evolution. In Washburn, S.L. and McCown, E.R. (eds), *Human Evolution: Biosocial Perspectives.* California: Benjamin Cummings.

de Beaune, S.A. and White, R. (1993) Ice Age Lamps. *Scientific American*, vol. 266, pp. 74–9.

Beck, B.B. (1975) Primate Tool Behaviour. In Tuttle, R.H. (ed.), *Socioecology and Psychology of Primates.* The Hague: Mouton.

Behrensmeyer, A.K. (1978) The Habitat of Plio-Pleistocene Hominids in East Africa: Taphonomic and Microstratigraphic Evidence. In Jolly, C.J. (ed.), *Early Hominids of Africa.* London: Duckworth.

Benedict, R. (1971) The Growth of Culture. In Shapiro, H.L. (ed.), *Man, Culture and Society.* London: Oxford University Press.

Bergounioux, F.M. (1962) Notes on the Mentality of Primitive Man. In Washburn, S.L. (ed.), *The Social Life of Early Man.* London: Methuen.

Berndt, C.H. (1970) Digging Sticks and Spears, or the Two Sex Model. In Gale, F. (ed.), *Woman's Role in Aboriginal Society. Australian Aboriginal Studies, no. 36.* Canberra: Australian Institute of Aboriginal Studies.

Berndt, C.H. (1978) Comments on Leacock, E.: Women's Status in Egalitarian Society. *Current Anthropology*, vol. 19, no. 2, pp. 256.

Berndt, R.M. and Berndt, C.H. (1964) *The World of the First Australians.* London: Angus & Robertson.

Beynon, A.D. and Dean, M.C. (1988) Distinct Dental Development Patterns in Early Hominids. *Nature*, vol. 335, pp. 509–14.

Bigelow, R. (1975) The Role of Competition and Cooperation in Human Evolution. In Nettleship, M.A. et al. (eds) *War, Its Causes and Correlates.* The Hague: Mouton.

Binford, L.R. (1981) *Bones, Ancient Men and Modern Myths.* New York: Academic Press.

Binford, L.R. (1985) Human Ancestors: Changing Views of Their Behaviour. *Journal of Anthropological Archaeology*, vol. 4, pp. 292–327.

Binford, L.R. (1988) Were there Elephant Hunters at Torralba? In Nitecki, M.H. and Nitecki, D.V. (eds), *The Evolution of Human Hunting.* New York: Plenum.

Binford, L.R. and Ho, C.K. (1985) Taphonomy at a Distance: Zhoukoudian, 'The Cave Home of Beijing Man'? *Current Anthropology*, vol. 26, no. 4, pp. 413–42.

Binford, S.R. (1968) A Structural Comparison of Disposal of the Dead in the Mousterian and Upper Palaeolithic. *Southwestern Journal of Anthropology*, vol. 24, pp. 139–54.

Binford, S. R. and Binford, L. R. (1969) Stone Tools and Human Behaviour. *Scientific American*, vol. 220, no. 4, pp. 70–84.

Bittles, A. H. (1980) Inbreeding in Human Populations. *Journal of Scientific and Industrial Research*, vol. 39, pp. 768–77.

Bittles, A. H. and Makov, E. (1988) Inbreeding in Human Populations: An Assessment of the Costs. In Mascie-Taylor, C. G. N. and Boyce, A. J. (eds), *Human Mating Patterns*. Cambridge: Cambridge University Press.

Blanc, A. C. (1962) Some Evidence for the Ideologies of Early Man. In Washburn, S. L. (ed.), *The Social Life of Early Man*. London: Methuen.

Blumenschine, J. and Cavallo, J. A. (1992) Scavenging and Human Evolution. *Scientific American*, vol. 267, no. 4, pp. 70–6.

Blurton Jones, N. G. (1987) Tolerated Theft, Suggestions About the Ecology and Evolution of Sharing, Hoarding and Scrounging. *Social Science Information*, vol. 26, no. 1, pp. 31–54.

Boas, F. (1888) The Central Eskimo. *Bureau of American Ethnology. Sixth Annual Report for 1884–85*. Washington D.C.: Smithsonian Institution.

Boaz, N. T. (1979) Hominid Evolution in East Africa During the Pliocene and Early Pleistocene. *Annual Review of Anthropology*, vol. 8, pp. 71–85.

Bock, K. (1980) *Human Nature and History. A Response to Sociobiology*. New York: Columbia University Press.

Bonner, J. T. (1980) *The Evolution of Culture in Animals*. Princeton, N.J.: Princeton University Press.

Bordes, F. (1961) Mousterian Cultures in France. *Science*, vol. 134, pp. 803–10.

Bordes, F. (1968) *The Old Stone Age*. London: Weidenfeld & Nicholson.

Bowlby, J. (1979) *The Making and Breaking of Affectual Bonds*. London: Tavistock/Routledge.

Bowlby, J. (1984) *Attachment and Loss*. 2 vols. Harmondsworth: Penguin.

Bowlby, J. (1991) *Charles Darwin. A New Biography*. London: Pimlico.

Boyd, R. and Richerson, P. (1985) *Culture and the Evolutionary Process*. Chicago: University of Chicago Press.

Brace, C. L. (1979) *The Stages of Human Evolution*. Englewood Cliffs: Prentice-Hall.

Braidwood, R. J. (1975) *Prehistoric Men*. Glenview, Ill.: Scott, Foresman & Co.

Brain, C. K. (1970) New Finds at the Swartkrans Australopithecine Site. *Nature*, vol. 225, pp. 1112–19.

Brain, C. K. (1976A) A Re-Interpretation of the Swartkrans Site and its Remains. *South African Journal of Science*, vol. 72, pp. 141–6.

Brain, C. K. (1976B) Some Principles in the Interpretation of Bone Accumulations Associated with Man. In Isaac, G. L. and McCown, E. (eds), *Human Origins. Louis Leakey and the East African Evidence*. Menlo Park, Calif.: W. A. Benjamin.

Brain, C. K. (1978) Some Aspects of the South African Australopithecine Sites and their Bone Accumulation. In Jolly, C. J. (ed.), *Early Hominids in Africa*. London: Duckworth.

Brooks, A. (1988) Ambrona. In Tattersall, I., Delson, E. and Van Couvering, J. (eds), *Encyclopedia of Human Evolution and Prehistory [EHEP]*. New York & London: Garland.

Brown, J. K. (1970) A Note on the Division of Labour by Sex. *American Anthropologist*, vol. 72, pp. 1073–8.

Brues, A. (1959) The Spearman and the Archer – An Essay on Selection in Body Build. *American Anthropologist*, vol. 61, pp. 457–69.

Bunn, H. T. (1981) Archaelogical Evidence for Meat Eating by Plio-Pleistocene Hominids from Koobi Fora and Olduvai Gorge. *Nature*, vol. 291, pp. 574–7.

Bunn, H. T. and Kroll, E. M. (1986) Systematic Butchery by Plio/Pleistocene Hominids at Olduvai Gorge, Tanzania. *Current Anthropology*, vol. 27, no. 5, pp. 431–52.

Burleigh, M. and Wippermann, W. (1991) The Racial State: Germany 1933–1945. Cambridge: Cambridge University Press.

Burley, N. (1979) The Evolution of Concealed Ovulation. *The American Naturalist*, vol. 114, no. 6, pp. 835–58.

Burrow, J. W. (1968) Editor's Introduction to Darwin, C. *The Origin of Species*. Harmondsworth: Penguin.

Burton, M. L. et al. (1977) A Model of the Sexual Division of Labour. *American Ethnologist*, vol. 4, pp. 227–51.

Butzer, K. W. (1977) Environment, Culture and Human Evolution. *American Scientist*, vol. 65, pp. 572–84.

Butzer, K. W. (1982) *Archaeology as Human Ecology*. Cambridge: Cambridge University Press.

Byrne, R. W. and Whiten, A. (eds) (1988) *Machiavellian Intelligence. Social Expertise and the Evolution of Intellect in Monkeys, Apes and Humans*. Oxford: Oxford University Press.

Calvin, W. H. (1994) The Emergence of Intelligence. *Scientific American*, vol. 271, no. 4, pp. 79–85.

Campbell, B. G. (1975) *Human Evolution: An Introduction to Man's Adaptations*. Chicago: Aldine.

Carlisle, R. C. and Siegel, M. I. (1974) Some Problems in the Interpretation of Neanderthal Speech Capabilities. *American Anthropologist*, vol. 76, pp. 319–22.

Carrier, D. R. (1984) The Energetic Paradox of Human Running and Hominid Evolution. *Current Anthropology*, vol. 25, no. 4, pp. 483–95.

Cartmill, M. et al. (1986) One Hundred Years of Paleoanthropology. *American Scientist*, vol. 74, pp. 410–20.

Catton, C. and Gray, J. (1985) *Sex in Nature*. London: Croom Helm.

Chagnon, N. A. (1983) *Yanomamo. The Fierce People.* New York: Holt, Rinehart & Winston.

Chase, P. G. (1987) Scavenging and Hunting in the Middle Paleolithic. The Evidence from Europe. In Dibble, H. L. and Monet-White, A. (eds), *The Upper Pleistocene Prehistory of Western Eurasia.* Philadelphia: Philadelphia University Press.

Chase, P. G. (1989) How Different was Middle Palaeolithic Subsistence? A Zooarchaeological Perspective on the Middle to Upper Palaeolithic Transition. In Mellars, P. and Stringer, C. B. (eds), *The Human Revolution. Behavioural and Biological Perspectives on the Origins of Modern Humans.* Edinburgh: Edinburgh University Press.

Chase, P. G. and Dibble, H. L. (1987) Middle Paleolithic Symbolism: A Review of Current Evidence and Interpretations. *Journal of Anthropological Archaeology,* vol. 6, pp. 263–96.

Chia Lan-po [Jia Lanpo] (1975) *The Cave Home of Peking Man.* Peking: Foreign Language Press.

Clark, J. Desmond, (1960) Human Ecology During Pleistocene and Later Times in Africa South of the Sahara. *Current Anthropology,* vol. 1, no. 4.

Clark, J. Desmond, (1968) Studies of Hunter-Gatherers as an Aid to the Interpretation of Prehistoric Societies. In Lee, R. B. and DeVore, I. (eds), *Man the Hunter.* New York: Aldine.

Clark, J. Desmond, (1976) African Origins of Man the Toolmaker. In Isaac, G. L. and McCown, E. R. (eds), *Human Origins. Louis Leakey and the East African Evidence.* Menlo Park, Calif.: W. A. Benjamin.

Clark, J. Desmond and Harris, J. W. K. (1985) Fire and its Roles in Early Hominid Lifeways. *The African Archaeological Review,* vol. 3, pp. 3–27.

Clark, J. G. D. (1970) *Aspects of Prehistory.* Berkeley: University of California Press.

Clark, J. G. D. (1975) *The Earliest Stone Age Settlement of Scandinavia.* Cambridge: Cambridge University Press.

Clark, J. G. D. (1977) *World Prehistory in New Perspective.* Cambridge: Cambridge University Press.

Clark, J. G. D. (1979) Archaeology and Human Diversity. *Annual Review of Anthropology,* vol. 8.

Clark, W. E. Le Gros (1978) *The Fossil Evidence for Human Evolution.* 3rd edn, revd. B. Campbell. Chicago: Chicago University Press.

Clutton-Brock, T. H. and Harvey, P. H. (1977) Primate Ecology and Social Organisation. *Journal of Zoology,* vol. 183, pp. 1–39.

Conkey, M. (1983) On the Origins of Paleolithic Art: A Review and some Critical Thoughts. In Trinkaus, E. (ed.), *The Mousterian Legacy. Human Biocultural Change in the Upper Pleistocene.* Oxford: BAR International Series.

Coon, C. S. (1976) *The Hunting Peoples.* Harmondsworth: Penguin.

Coppens, Y. (1994) East Side Story: The Origin of Humankind. *Scientific American*, vol. 270, pp. 62–9.

Crook, J. H. (1980) *The Evolution of Human Consciousness*. Oxford: Clarendon Press.

Dalton, G. (1981) Anthropological Models in Archaeological Perspective. In Hodder, I. et al. (eds), *Pattern of the Past*. Cambridge: Cambridge University Press.

Dart, R. A. (1953) The Predatory Transition from Ape to Man. *International Anthropological and Linguistic Review*, vol. 1, no. 4. pp. 201–19.

Dart, R. A. (1957) The Osteodontokeratic Culture of Australopithecus Prometheus. *Transvaal Museum Memoir*, no. 10.

Dart, R. A. (1959) *Adventures with the Missing Link*. London: Hamish Hamilton.

Dart, R. A. (1967) Mousterian Osteodontokeratic Objects from Geula Cave (Haifa, Israel). *Quaternaria*, vol. 9, pp. 105–40.

Darwin, C. (1965) *The Expression of the Emotions in Man and Animals* [1872] Chicago: The University of Chicago Press.

Darwin, C. (1968) *The Origin of Species* [1859]. Edited and introduced by Burrow, J. W. Harmondsworth: Penguin.

Darwin, C. (1981) *The Descent of Man, and Selection in Relation to Sex* [1871] Princeton: Princeton University Press.

Davis, W. (1986) The Origins of Image Making. *Current Anthropology*, vol. 27, no. 3, pp. 193–215.

Dawkins, R. (1986) *The Blind Watchmaker*. Harlow: Longman.

Denham, W. W. (1974) Population Structure, Infant Transport, and Infanticide in Pleistocene and Modern Hunter-Gatherers. *Journal of Anthropological Research*, vol. 30, pp. 191–8.

Diamond, J. D. (1991) *The Rise and Fall of the Third Chimpanzee*. London: Radius.

Dillehay, T. D. (1984) A Late Ice Age Settlement in Southern Chile. *Scientific American*, vol. 251, no. 4, pp. 100–9.

Divale, W. T. (1972) Systematic Population Control in the Middle and Upper Palaeolithic. Inferences based on Contemporary Hunter-Gatherers. *World Archaeology*, vol. 2, pp. 222–43.

Dobzhansky, T. (1963) Cultural Direction of Human Evolution – A Summation. *Human Biology*, vol. 25, no. 3, pp. 311–16.

Dobzhansky, T. (1972) Genetics and the Diversity of Behaviour. *American Psychologist*, vol. 27, pp. 523–30.

Dobzhansky, T., et al. (1977) *Evolution*. San Francisco: W. H. Freeman.

Dobzhansky, T. and Boesiger, E. (1983) *Human Culture. A Moment in Evolution*. New York: Columbia University Press.

Douglas, J. W. B. (1975) Early Hospital Admissions and Later Disturbances of Behaviour and Learning. *Developmental Medicine and Child Neurology*, vol. 17, pp. 456–80.

Douglas, M. (1966) Population Control in Primitive Groups. *British Journal of Sociology*, vol. 17, pp. 263–73.

Dowling, J. H. (1968) Individual Ownership and the Sharing of Game in Hunting Societies. *American Anthropologist*, vol. 70, pp. 502–7.

Draper, P. (1975) !Kung Women: Contrasts in Sexual Egalitarianism in Foraging and Sedentary Contexts. In Reiter, R. R. (ed.), *Toward an Anthropology of Women*. New York: Monthly Review Press.

Driver, J. C. (1990) Meat in Due Season: The Timing of Communal Hunts. In Davies, L. B. and Reeves, B. O. (eds), *Hunters of the Recent Past*. London: Unwin Hyman.

Dunbar, R. I. M. (1976) Australopithecine Diet Based on a Baboon Analogy. *Journal of Human Evolution*, vol. 5, pp. 161–5.

Dunn, L. C. (1956) Race and Biology. In *The Race Question in Modern Science*. Paris: UNESCO.

Durham, W. H. (1976) The Adaptive Significance of Cultural Behaviour. *Human Ecology*, vol. 4, no. 2, pp. 89–121.

Durham, W. H. (1978) Toward a Coevolutionary Theory of Human Biology and Culture. In Caplan, A. L. (ed.), *The Sociobiology Debate*. New York: Harper & Row.

Durham, W. H. (1981) Overview: Optimal Foraging Analysis in Human Ecology. In Winterhalder, B. and Alden Smith, E. (eds), *Hunter-Gatherer Foraging Strategies*. Chicago: University of Chicago Press.

EHEP (1988) *Encyclopedia of Human Evolution and Prehistory [EHEP]*, Tattersall, I., Delson, E. and Van Couvering, J. (eds) New York & London: Garland.

Ehrenberg, M. (1989) *Women in Prehistory*. British Museum Publications.

Eibl-Eibesfeldt, I. (1970) *Ethology. The Biology of Behaviour*. New York: Holt, Rinehart & Winston.

Eibl-Eibesfeldt, I. (1979) Ritual and Ritualization from a Biological Perpective. In Cranach, M. von et al. (eds), *Human Ethology. Claims and Limits of a New Discipline*. Cambridge: Cambridge University Press.

Eimas, P. D. (1985) The Perception of Speech in Early Infancy. *Scientific American*, vol. 252, no. 1, pp. 34–40.

Engels, F. (1940) *Dialectics of Nature* [1872–1882] London: Lawrence & Wishart.

Engels, F. (1972) *The Origin of the Family, Private Property and the State* [1884] London: Lawrence & Wishart.

Estioko-Griffin, A. (1987) Daughters of the Forest. In Whitten, P. and Hunter, D. E. K. (eds), *Anthropology: Contemporary Perspectives*. Boston: Little, Brown & Co.

Estioko-Griffin, A. and Griffin, P. B. (1981) Woman the Hunter: The Agta. In Dahlberg, F. (ed.), *Woman the Gatherer*. New Haven: Yale University Press.

Falk, D. (1980) Hominid Brain Evolution: The Approach From Paleo-neurology. *Yearbook of Physical Anthropology*, vol. 23, pp. 93–107.

Faris, J. C. (1972) *Nuba Personal Art*. London: Duckworth.

Fedigan, L. M. (1986) The Changing Role of Women in Models of Human Evolution. *Annual Review of Anthropology*, vol. 15, pp. 25–66.

Festinger, L. (1983) *The Human Legacy*. New York: Columbia University Press.

Fischer, E. (1981) *The Necessity of Art. A Marxist Approach* [1959] Harmondsworth: Penguin.

Fleagle, J. G. (1988) *Primate Adaptation and Evolution*. San Diego: Academic Press.

Foley, R. (1987) *Another Unique Species. Patterns in Human Evolutionary Ecology*. Harlow: Longman.

Foley, R. (1988) Hominids, Humans and Hunter-Gatherers: An Evolutionary Perspective. In Ingold, T. et al. (eds) *Hunters and Gatherers*, vol. 1, *History, Evolution and Social Change*. Oxford: Berg.

Foley, R. (1991) Investigating the Origins of Human Behaviour. In Foley, R. (ed.), *The Origins of Human Behaviour*. London: Unwin Hyman.

Forde, D. (1934) *Habitat, Economy and Society*. London: Methuen.

Fortes, M. (1983) *Rules and the Emergence of Society*. London: Royal Anthropological Institute Occasional Paper no. 39.

Fossey, D. (1985) *Gorillas in the Mist*. Harmondsworth: Penguin.

Fox, R. (1967) In the Beginning: Aspects of Hominid Behavioural Evolution. *Man*, vol. 2, pp. 415–32.

Frayer, D. W. (1981) Body Size, Weapon Use, and Natural Selection in the European Upper Paleolithic and Mesolithic. *American Anthropologist*, vol. 83, pp. 57–73.

Freeman, L. G. (1968) A Theoretical Framework for Interpreting Archaeological Materials. In Lee, R. B. and DeVore, I. (eds), *Man the Hunter*. New York: Aldine.

Freeman, L. G. (1973) The Significance of Mammalian Faunas from Paleolithic Occupations in Cantabrian Spain. *American Antiquity*, vol. 38, no. 1, pp. 3–44.

Freeman, L. G. (1981) The Fat of the Land: Notes on Paleolithic Diet in Iberia. In Harding, R. S. O. and Teleki, G. (eds), *Omnivorous Primates: Gathering and Hunting in Human Evolution*. New York: Columbia University Press.

Freeman, M. R. (1971) A Social and Ecologic Analysis of Systematic Female Infanticide Among the Netsilik Eskimo. *American Anthropologist*, vol. 73, pp. 1011–18.

Freud, S. (1977) Some Psychical Consequences of the Anatomical Distinction Between the Sexes [1925] In Richards, A. (ed.), The Pelican Freud Library, vol. 7, *On Sexuality*. Harmondsworth: Penguin.

Fried, M. H. (1987) A Four Letter Word that Hurts. In Whitten, P. and Hunter, E. K. (eds), *Anthropology. Contemporary Perspectives*. Boston: Little, Brown & Co.

Friedl, E. (1975) *Women and Men. An Anthropologist's View*. New York: Holt, Rinehart & Winston.

Frisch, K. von (1962) Dialects in the Language of the Bees. *Scientific American*, vol. 207, no. 2, in Wang, W. (ed.), *Human Communication*. San Francisco: W. H. Freeman, 1982.

Frisch, R. E. (1988) Fatness and Fertility. *Scientific American*, vol. 258, no. 3, pp. 70–7.

Gallup, G. G., Jr. (1975) Towards an Operational Definition of Self-Awareness. In Tuttle, R. (ed.), *Socioecology and Psychology of Primates*. The Hague: Mouton.

Gamble, C. (1982) Interaction and Alliance in Palaeolithic Society. *Man*, vol. 17, pp. 92–107.

Gamble, C. (1983) Culture and Society in the Upper Palaeolithic of Europe. In Bailey, G. (ed.), *Hunter-Gatherer Economy in Prehistory. A European Perspective*. Cambridge: Cambridge University Press.

Gardner, R. A. and Gardner, B. T. (1969) Teaching Sign Language to a Chimpanzee. *Science*, vol. 165, pp. 664–72.

Gardner, R. A. and Gardner, B. T. (1975) Early Signs of Language in Child and Chimpanzee. *Science*, vol. 187, pp. 752–3.

Gargett, R. H. (1989) Grave Shortcomings. The Evidence for Neanderthal Burial. *Current Anthropology*, vol. 30, no. 2, pp. 157–90.

Geertz, C. (1975) *The Interpretation of Cultures*. London: Hutchinson.

Geschwind, N. (1979) Specialisations of the Human Brain. *Scientific American*, vol. 241, no. 3, pp. 180–99.

Giddens, A. (1989) *Sociology*. Cambridge: Polity Press.

Ginneken, J. K. Van (1974) Prolonged Breast Feeding as a Birth Spacing Method. *Studies in Family Planning*, vol. 5, no. 6, pp. 201–6.

Gladkih, M. I., Kornietz, N. L. and Soffer, O. (1984) Mammoth-Bone Dwellings on the Russian Plain. *Scientific American*, vol. 251, no. 5, pp. 136–43.

Goodall, J. (1971) *In the Shadow of Man*. London: Collins.

Goodall, J. (1976) Continuities Between Chimpanzee and Human Behaviour. In Isaac, G. L. and McCown, E. R. (eds), *Human Origins: Louis Leakey and the East African Evidence*. Menlo Park, Calif.: W. A. Benjamin.

Goodall, J. (1979) Life and Death at Gombe. *National Geographic*, vol. 155, no. 5, pp. 592–621.

Goodall, J. (1986) *The Chimpanzees of Gombe*. Cambridge, Mass.: The Belknap Press of Harvard University Press.

Goodall, J. (1990) *Through a Window. Thirty Years With the Chimpanzees of Gombe*. London: Weidenfeld and Nicholson.

Goodman, D. and Russell, C. A. (eds) (1991) *The Rise of Scientific Europe 1500–1800*. Sevenoaks: Hodder & Stoughton.

Gough, K. (1971) The Origin of the Family. *Journal of Marriage and the Family*, vol. 33, pp. 760–71.

Gould, J. L. and Marler, P. (1987) Learning by Instinct. *Scientific American*, vol. 256, no. 1, 62–73.

Gould, S. J. (1981) *Ever Since Darwin*. Harmondsworth: Penguin.

Gould, S. J. (1994) The Evolution of Life on Earth. *Scientific American*, vol. 271, no. 4, pp. 62–9.

Gould, S. J. and Eldredge, N. (1977) Punctuated Equilibria: The Tempo and Mode of Evolution Reconsidered. *Paleobiology*, vol. 3, pp. 115–51.

Gowlett, J. A. J. (1984) *Ascent to Civilisation. The Archaeology of Early Man*. New York: Alfred A. Knopf.

Gowlett, J. A. J. (1984A) Mental Abilities of Early Man: A Look at Some Hard Evidence. In Foley, R. (ed.), *Hominid Evolution and Community Ecology*. London: Academic Press.

Gowlett, J. A. J. (1986) Culture and Conceptualisation: The Oldowan-Acheulean Gradient. In Bailey, G. N. and Callow, P. (eds), *Stone Age Prehistory*. Cambridge: Cambridge University Press.

Gowlett, J. A. J. et al. (1981) Early Archaeological Sites, Hominid Remains and Traces of Fire from Chesowanja, Kenya. *Nature*, vol. 294, pp. 125–9.

Graham, C. A. and McGrew, W. C. (1980) Menstrual Synchrony in Female Undergraduates Living on a Coeducational Campus. *Psycho-neuroendocrinology*, vol. 5, pp. 245–52.

Grayson, D. K. (1983) *The Establishment of Human Antiquity*. New York: Academic Press.

Graziosi, P. (1960) *Palaeolithic Art*. London: Faber & Faber.

Griffin, D. R. (1981) *The Question of Animal Awareness*. Los Altos, Calif.: W. Kaufmann.

Griffin, P. B. (1984) Forager Resource and Land Use in the Humid Tropics: The Agta of Northeastern Luzon, the Philippines. In Schrire, C. (ed.), *Past and Present in Hunter-Gatherer Studies*. Orlando: Academic Press.

Grine, F. E. (1988) *Australopithecus*. In Tattersall, I., Delson, E. and Van Couvering, J. (eds), *Encyclopedia of Human Evolution and Prehistory [EHEP]*. New York & London: Garland.

Grine, F. E. (1988A) *Australopithecus boisei*. In Tattersall, I., Delson, E. and Van Couvering, J. (eds), *Encyclopedia of Human Evolution and Prehistory [EHEP]*. New York & London: Garland.

Gusinde, M. (1961) *The Yamana. The Life and Thought of the Water Nomads of Cape Horn*. New Haven: Human Relations Area Files.

Hall, K. R. L. (1963) Tool-Using Performances as Indicators of Behavioural Adaptability. *Current Anthropology*, vol. 4, no. 5, pp. 479–94.

Hallowell, A. I. (1960) Self, Society and Culture in Phylogenetic Perspective. In Tax, S. (ed.), *Evolution after Darwin*, vol. 2. Chicago: University of Chicago Press.

Hallowell, A. I. (1962) The Protocultural Foundations of Human Adaptations. In Washburn, S. L. (ed.), *The Social Life of Early Man*. London: Methuen.

Harcourt, A. H. et al. (1981) Testis Weight, Body Weight and Breeding Systems in Primates. *Nature*, vol. 293, 3/9, pp. 55–7.

Harding, R. S. O. (1975) Meat-Eating and Hunting in Baboons. In Tuttle, R. H. (ed.), *Socioecology and Psychology of Primates*. The Hague: Mouton.

Harding, R. S. O. (1981) Non Human Primate Diets. In Harding, R. S. O. and Teleki, G. (eds), *Omnivorous Primates: Gathering and Hunting in Human Evolution*. New York: Columbia University Press.

Harding, R. S. O. and Teleki, G. (1981) Introduction. In Harding, R. S. O. and Teleki, G. (eds), *Omnivorous Primates: Gathering and Hunting in Human Evolution*. New York: Columbia University Press.

Hardy, A. (1960) Was Man More Aquatic in the Past? *New Scientist*, March 17.

Harlow, H. F. (1959) Love in Infant Monkeys. *Scientific American*, vol. 200, pp. 68–74.

Harlow, H. F. and Harlow, M. K. (1962) Social Deprivation in Monkeys. *Scientific American*, vol. 207, pp. 136–46.

Harnad, S. R., Steklis, H. D. and Lancaster, J. B. (eds) (1976), Origins and Evolution of Language. *Annals of the New York Academy of Sciences*, vol. 280.

Harner, M. (1977) The Ecological Basis for Aztec Sacrifice. *American Ethnologist*, vol. 4, no. 1, pp. 117–35.

Harris, J. W. K. (1983) Cultural Beginnings: Plio-Pleistocene Archaeological Occurrences from the Afar, Ethiopia. *African Archaeological Review*, vol. 1, pp. 3–31.

Harris, M. (1978) *Cannibals and Kings. The Origins of Cultures*. Glasgow: Fontana/Collins.

Harrold, F. B. (1980) A Comparative Analysis of European Palaeolithic Burials. *World Archaeology*, vol. 12, no. 2, pp. 195–211.

Hassan, F. A. (1981) *Demographic Archaeology*. New York: Academic Press.

Hayden, B. (1972) Population Control among Hunter-Gatherers. *World Archaeology*, vol. 4, pp. 205–21.

Hayden, B. (1981) The Subsistence of Modern Hunter/Gatherers. In Harding, R. S. O. and Teleki, G. (eds), *Omnivorous Primates: Gathering and Hunting in Human Evolution*. New York: Columbia University Press.

Hayden, B. (1992) Observing Prehistoric Women. In Claassen, C. (ed.), *Exploring Gender Through Archaeology*. Monographs in World Archaeology no. 11. Madison: Prehistory Press.

Hertz, R. (1960) *Death and the Right Hand* [1907], trans. R. & C. Needham. Aberdeen: Cohen and West.

Hewes, G. W. (1961) Food Transport and the Origins of Hominid Bipedalism. *American Anthropologist*, vol. 63, no. 4, pp. 687–710.

Hewes, G. W. (1973) Primate Communication and the Gestural Origin of Language. *Current Anthropology*, vol. 14, no. 1–2, pp. 5–24.

Hiatt, B. (1970) Woman the Gatherer. In Gale, F. (ed.), *Woman's Role in Aboriginal Society*. *Australian Aboriginal Studies, no. 36.* Canberra: Australian Institute of Aboriginal Studies.

Hill, A. (1988) Taphonomy. In Tattersall, I.,Delson, E. and Van Couvering, J. (eds), *Encyclopedia of Human Evolution and Prehistory [EHEP]* New York & London: Garland.

Hill, K. (1982) Hunting and Human Evolution. *Journal of Human Evolution*, vol. 11, pp. 521–44.

Hindess, B. and Hirst, P. Q. (1975) *Pre-Capitalist Modes of Production*. London: Routledge & Kegan Paul.

Hirst, P. Q. (1976) *Social Evolution and Sociological Categories*. London: G. Allen & Unwin.

Hirst, P. Q. and Woolley, P. (1982) *Social Relations and Human Attributes*. London: Tavistock.

Hirst, P. Q. and Woolley, P. (1985) Nature and Culture in Social Science: The Demarcation of Domains of Being in Eighteenth Century and Modern Discourses. *Geoforum*, vol. 16, no. 2, pp. 151–61.

Hoebel, E. A. (1971) The Nature of Culture. In Shapiro, H. L. (ed.), *Man, Culture and Society*. London: Oxford University Press.

Holloway, R. L. (1969) Culture: A Human Domain. *Current Anthropology*, vol. 10, no. 4, pp. 135–68.

Holloway, R. L. (1976) Paleoneurological Evidence for Language Origins. In Harnad, S. R. et al. (eds), Origins and Evolution of Language. *Annals of the New York Academy of Sciences*, vol. 280.

Holloway, R. L. (1981) Culture, Symbols and Human Brain Evolution: A Synthesis. *Dialectical Anthropology*, vol. 5, pp. 287–303.

Holloway, R. L. (1983) Human Brain Evolution: A Search for Units, Models and Synthesis. *Canadian Journal of Anthropology*, vol. 3, no. 2, pp. 215–30.

Holloway, R. L. (1988) Brain. In Tattersall, I., Delson, E. and Van Couvering, J. (eds), *Encyclopedia of Human Evolution and Prehistory [EHEP]*. New York & London: Garland.

Hopkins, K. (1980) Brother–Sister Marriage in Roman Egypt. *Comparative Studies in Society and History*, vol. 22, pp. 303–54.

Horgan, J. (1993) Trends in Behavioural Genetics. Eugenics Revisited. *Scientific American*, vol. 268, no. 6. pp. 92–100.

Howells, W. W. (1966) *Homo Erectus*. *Scientific American*, vol. 215, no. 11, pp. 46–53.

Howells, W. W. (1980) *Homo Erectus* – Who, When and Where: A Survey. *Yearbook of Physical Anthropology*, vol. 23, pp. 1–23.

Hubbard, R, (1979) Have Only Men Evolved? In Hubbard, R. et al. (eds), *Women Look at Biology Looking at Women*. Boston Mass.: G. K. Hall & Co.

Humphrey, N. (1986) *The Inner Eye*. London: Faber & Faber.

Humphrey, N. (1988) The Social Function of Intellect. In Byrne, R. W. and Whiten, P. (eds), *Machiavellian Intelligence*. Oxford: Oxford University Press.

Huxley, T. H. (1893) *Evolution and Ethics: The Romanes Lecture* 1893. London: Macmillan.

Ingold, T. (1983) The Architect and the Bee: Reflections on the Work of Animals and Man. *Man*, vol. 18, pp. 1–20.

Isaac, B. (1987) Throwing and Human Evolution. *The African Archaeological* Review, vol. 5, pp. 3–17.

Isaac, B. (ed.) (1989) *The Archaeology of Human Origins. Papers by Glynn Isaac*. Cambridge: Cambridge University Press.

Isaac, G. L. (1971) The Diet of Early Man: Aspects of Archaeological Evidence from Lower and Middle Pleistocene Sites in Africa. *World Archaeology*, vol. 2, no. 3, pp. 279–98.

Isaac, G. L. (1975) Stratigraphy and Cultural Patterns in East Africa During the Middle Ranges of Pleistocene Time. In Butzer, K. W. and Isaac, G. L. (eds), *After the Australopithecines*. The Hague: Mouton.

Isaac, G. L. (1976A) Early Stone Tools – an Adaptive Threshold? in Sieveking, G. de G. et al. (eds), *Problems in Economic and Social Archaeology*. London: Duckworth.

Isaac, G. L. (1976B) The Activities of Early African Hominids: A Review of the Archaeological Evidence from the Time Span Two and a Half to One Million Years Ago. In Isaac, G. L. and McCown, E. R. (eds), *Human Origins. Louis Leakey and the East African Evidence*. Menlo Park, Calif.: W. A. Benjamin.

Isaac, G. L. (1976C) Stages of Cultural Elaboration in the Pleistocene: Possible Archaeological Indicators of the Development of Language Capacities. In Harnad, S. R. et al. (eds), Origins and Evolution of Language and Speech. *Annals of the New York Academy of Sciences*, vol. 280.

Isaac, G. L. (1978) The Food-Sharing Behaviour of Protohuman Hominids. *Scientific American*. vol. 238, pp. 90–108.

Isaac, G. L. (1980) Casting the Net Wide: A Review of Archaeological Evidence for Early Hominid Land-Use and Ecological Relations. In Konigsson, L.-K. (ed.), *Current Arguments On Early Man*. Oxford: Pergamon Press.

Isaac, G. L. (1981) Emergence of Human Behaviour Patterns. *Philosophical Transactions of the Royal Society*, series B, vol. 292.

Isaac, G. L. (1981A) Stone Age Visiting Cards. Approaches to the Study of Early Land Use Patterns. In Hodder, I. et al. (eds), *Pattern of the Past*. Cambridge: Cambridge University Press.

Isaac, G. L. (1982) Early Hominids and Fire at Chesowanja, Kenya. *Nature*, vol. 296, p. 870.

Isaac, G. L. (1983) Aspects of Human Evolution. In Bendall, D. S. (ed.), *Evolution from Molecules to Men*. Cambridge: Cambridge University Press.

Isaac, G. L. (1983A) Some Archaeological Contributions Towards Understanding Human Evolution. *Canadian Journal of Anthropology*, vol. 3, no. 2, pp. 233–43.

Isaac, G. L. (1984) The Archaeology of Human Origins: Studies of the Lower Pleistocene in East Africa 1971–1981. *Advances in World Archaeology*, vol. 3, pp. 1–87.

Isaac, G. L. (1986) Foundation Stones: Early Artefacts as Indicators of Activities and Abilities. In Bailey, G. N. and Callow, P. (eds), *Stone Age Prehistory*. Cambridge: Cambridge University Press.

Isaac, G. L. (1989) Cutting and Carrying: Archaeology and the Emergence of the Genus *Homo*. In Durant, J. R. (ed.), *Human Origins*. Oxford: Clarendon Press.

Isaac, G. L. and Crader, D. C. (1981) To What Extent Were Early Hominids Carnivorous? An Archaeological Perspective. In Harding, R. S. O. and Teleki, G. (eds), *Omnivorous Primates: Gathering and Hunting in Human Evolution*. New York: Columbia University Press.

James, S. R. (1989) Hominid Use of Fire in the Lower and Middle Pleistocene. A Review of the Evidence. *Current Anthropology*, vol. 30, pp. 1–26.

Jaynes, J. (1976) *The Origin of Consciousness in the Breakdown of the Bicameral Mind*. Boston: Houghton Mifflin.

Jaynes, J. (1976A) The Evolution of Language in the Late Pleistocene. In Harnad, S. R. et al. (eds), Origins and Evolution of Language. *Annals of the New York Academy of Sciences*, vol. 280.

Jelinek, A. H. (1977) The Lower Paleolithic: Current Evidence and Interpretations. *Annual Review of Anthropology*, vol. 6, pp. 11–32.

Jenness, D. (1922) *The Life of the Copper Eskimos: Report of the Canadian Arctic Expedition 1913–1918*, vol. 13, Ottawa: F. A. Acland.

Jenness, D. (1975) *The People of the Twilight* [1928] Chicago: University of Chicago Press.

Jennings, J. D. (1978) Origins. In Jennings, J. D. (ed.), *Ancient Native Americans*. San Francisco: W. H. Freeman.

Johanson, D. C. (1980) Early African Hominid Phylogenesis: A Reevaluation. In Konigsson, L.-K. (ed.), *Current Arguments on Early Man*. Oxford: Pergamon Press.

Johanson, D. C. and Edey, M. A. (1982) *Lucy. The Beginnings of Humankind.* London: Granada.

Johanson, D. C. and White, T. D. (1979) A Systematic Assessment of Early African Hominids. *Science*, vol. 203, pp. 321–30.

Johnston, F. E. (1982) *Physical Anthropology.* Iowa: W. C. B.

Johnstone, P. (1980) *The Sea-Craft of Prehistory.* London: Routledge & Kegan Paul.

Jolly, A. (1966) Lemur Social Behaviour and Primate Intelligence. *Science*, vol. 153, pp. 501–6.

Jolly, A. (1985) *The Evolution of Primate Behaviour.* New York: Macmillan.

Jolly, C. J. (1970) The Seed-Eaters: A New Model of Hominid Differentiation Based on a Baboon Analogy. *Man*, vol. 5, no. 1, pp. 5–26.

Jolly, C. J. and Plog, F. (1976) *Physical Anthropology and Archaeology.* New York: Alfred A. Knopf.

Jones, P. R. (1981) Experimental Implement Manufacture and Use: A Case Study from Olduvai Gorge, Tanzania. *Philosophical Transactions of the Royal Society*, series B., vol. 292.

Jones, S. (1993) *The Language of the Genes. Biology, History and the Evolutionary Future.* London: HarperCollins.

Katz, M. M. and Konner, M. J. (1981) The Role of the Father: An Anthropological Perspective. In Lamb, M. E. (ed.), *The Role of the Father in Child Development.* New York: John Wiley.

Keeley, L. H. (1977) The Functions of Paleolithic Flint Tools. *Scientific American*, vol. 237, pp. 108–26.

King, J. C. (1980) The Genetics of Sociobiology. In Montague, A. (ed.), *Sociobiology Examined.* Oxford: Oxford University Press.

Kinzey, W. G. (1987) Monogamous Primates: A Primate Model for Human Mating Systems. In Kinzey, W. G. (ed.), *The Evolution of Human Behaviour: Primate Models.* New York: State University of New York Press.

Klein, R. G. (1988) Reconstructing how Early People Exploited Animals: Problems and Prospects. In Nitecki, M. H. and Nitecki, D. (eds), *The Evolution of Human Hunting.* New York: Plenum.

Klein, R. G. (1989) *The Human Career. Human Biological and Cultural Origins.* Chicago: University of Chicago Press.

Knecht, H. (1994) Late Ice Age Hunting Technology. *Scientific American*, vol. 271, no. 1, pp. 66–71.

Kolata. G. B. (1974) !Kung Hunter Gatherers: Feminism, Diet and Birth Control. *Science*, vol. 185, pp. 932–4.

Konner, M. J. (1972) Aspects of the Developmental Ethology of a Foraging People. In Blurton-Jones (ed.), *Ethological Studies of Child Behaviour.* Cambridge: Cambridge University Press.

Kortlandt, A. (1962) Chimpanzees in the Wild. *Scientific American*, vol. 206, no. 5, pp. 128–38.

Kroeber, A. L. (1927) Disposal of the Dead. *American Anthropologist*, vol. 29, no. 3, pp. 308–15.

Kroeber, A. L. (1982) The Super Organic [1957]. In Coser, L. and Rosenberg, B. (eds), *Sociological Theory. A Book of Readings*, 5th edn. New York: Macmillan.

Laitman, J. T. (1988) Origins of Speech. In Tattersall, I., Delson, E. and Van Couvering, J. (eds), *Encyclopedia of Human Evolution and Prehistory [EHEP]* New York & London: Garland.

La Lumiere, L. P. (1981) Evolution of Hominid Bipedalism: A Hypothesis About Where it Happened. *Philosophical Transactions of the Royal Society.* series B, vol. 292.

Lamphere, L. (1987) Feminism and Anthropology. The Struggle to Reshape our Thinking about Gender. In Farnham, C. (ed.), *The Impact of Feminist Research in the Academy*. Bloomington: Indiana University Press.

Lancaster, J. B. (1968) On the Evolution of Tool-Using Behaviour. *American Anthropologist*, vol. 70, no. 1, pp. 56–66.

Lancaster, J. B. (1975) *Primate Behaviour and the Emergence of Human Culture*. New York: Holt, Rinehart & Winston.

Lancaster, J. B. and Whitten, P. (1987) Sharing in Human Evolution. In Whitten, P. and Hunter, D. (eds), *Anthropology. Contemporary Perspectives*. Boston: Little, Brown & Co.

Landau, M. (1984) Human Evolution as Narrative. *American Scientist*, vol. 72, pp. 262–8.

Landes, R. (1938) *The Ojibwa Woman*. New York: Columbia University Press.

Langer, W. L. (1972) Checks on Population Growth: 1750–1850. *Scientific American*, vol. 226.

Laughlin, W. S. (1962) Acquisition of Anatomical Knowledge by Ancient Man. In Washburn, S. L. (ed.), *The Social Life of Early Man*. London: Methuen.

Leach, E. (1982) *Social Anthropology*. Glasgow: Fontana.

Leacock, E. (1978) Women's Status in Egalitarian Society. Implications for Social Evolution. *Current Anthropology*, vol. 19, no. 2, pp. 247–75.

Leacock, E. (1983) Interpreting the Origins of Gender Inequality: Conceptual and Historical Problems. *Dialectical Anthropology*, vol. 7, pp. 263–84.

Leacock, E. and Lee, R. B. (1982) Introduction in Leacock, E. and Lee, R. B. (eds), *Politics and History in Band Societies*. Cambridge: Cambridge University Press.

Leakey, L. S. B. (1968) Bone Smashing by Late Miocene Hominidae. *Nature*, 11 May, vol. 218, pp. 528–30.

Leakey, L. S. B. (1973) Was *Homo Erectus* Responsible for the Hand-Axe Culture? *Journal of Human Evolution*, vol. 2, pp. 493–8.

Leakey, L. S. B., Tobias, P. V. and Napier, J. R. (1964) A New Species of the Genus *Homo* from Olduvai Gorge. *Nature*, vol. 202, pp. 7–9.

Leakey, M. D. (1976) A Summary and Discussion of the Archaeological Evidence from Bed I and Bed II Olduvai Gorge, Tanzania. In Isaac, G. L. and McCown, E. R. (eds), *Human Origins. Louis Leakey and the East African Evidence*. Menlo Park, Calif.: W. A. Benjamin.

Leakey, M. D. (1978) Olduvai Fossil Hominids: their Stratigraphic Position and Associations. In Jolly, C. J. (ed.), *Early Hominids in Africa*. London: Duckworth.

Leakey, M. D. (1980) Early Man, Environment and Tools. In Konigsson, L.-K. (ed.), *Current Arguments on Early Man*. Oxford: Pergamon Press.

Leakey, M. D. (1981) Tracks and Tools. *Philosophical Transactions of the Royal Society*, series B, vol. 292.

Leakey, R. E. (1981) *The Making of Mankind*. London: Michael Joseph.

Leakey, R. E. and Lewin, R. (1979) *Origins*. London: Macdonald and Jane's.

Lee, R. B. (1967) Trance Cure of the !Kung Bushmen. *Natural History*, vol. 76, no. 9, pp. 30–7.

Lee, R. B. (1968) What Hunters do for a Living. In Lee, R. B. and DeVore, I. (eds), *Man the Hunter*. New York: Aldine.

Lee, R. B. (1974) Male–Female Residence Arrangements and Political Power in Human Hunter-Gatherers. *Archives of Sexual Behaviour*, vol. 3, no. 2, pp. 167–73.

Lee, R. B. (1979) *The !Kung San. Men, Women, and Work in a Foraging Society*. Cambridge: Cambridge University Press.

Lee, R. B. (1984) *The Dobe !Kung*. New York: Holt, Rinehart & Winston.

Lee, R. B. (1988) Reflections on Primitive Communism. In Ingold, T. et al. (eds), *Hunters and Gatherers*, vol. 1. *History, Evolution and Social Change*. Oxford: Berg.

Lee, R. B. and DeVore, I. (1968) *Man the Hunter*. New York: Aldine.

Leibowitz, L. (1978) *Females, Males, Families: A Biosocial Approach*. Mass.: Duxbury Press.

Leibowitz, L. (1979) 'Universals' and Male Dominance Among Primates: A Critical Examination. In Hubbard, R. and Lowe, M. (eds), *Genes and Gender II. Pitfalls in Research on Sex and Gender*. New York: Gordian Press.

Leibowitz, L. (1986) In the Beginning . . .' In Coontz, S. and Henderson, P. (eds), *Women's Work, Men's Property. The Origins of Gender and Class*. London: Verso.

Le May, M. (1975) The Language Capability of Neanderthal Man. *American Journal of Physical Anthropology*, vol. 42, pp. 9–14.

Leroi-Gourhan, André (1982) *The Dawn of European Art. An Introduction to Palaeolithic Cave Painting*. Cambridge: Cambridge University Press.

Leroi-Gourhan, Arlette (1975) The Flowers found with Shanidar IV, a Neanderthal Burial in Iraq. *Science*, vol. 190, pp. 562–4.

Leroi-Gourhan, Arlette (1982) The Archaeology of the Lascaux Cave. *Scientific American*, vol. 246, pp. 80–8.

Lévi-Strauss, C. (1969) *The Elementary Structures of Kinship*. London: Eyre & Spottiswoode.

Lévi-Strauss, C. (1971) The Family. In Shapiro, H. L. (ed.), *Man, Culture and Society*. London: Oxford University Press.

Lewin, R. (1981) Ethiopian Stone Tools Are World's Oldest. *Science*, vol. 211, pp. 806–7.

Lewin, R. (1982) How Did Humans Evolve Big Brains? *Science*, vol. 216, pp. 840–1.

Lewin, R. (1984) *Human Evolution. An Illustrated Introduction*. Oxford: Blackwell Scientific Publications.

Lewin, R. (1984A) Man the Scavenger. *Science*, vol. 224, pp. 861–2.

Lewin, R. (1987) *Bones of Contention. Controversies in the Search for Human Origins*. London: Penguin.

Lewin, R. (1988) A New Tool Maker in the Hominid Record? *Science*, vol. 240, pp. 724–5.

Lewin, R. (1988A) *In the Age of Mankind*. Washington D.C.: Smithsonian Books.

Lieberman, P. (1972) Primate Vocalization and Human Linguistic Ability. In Washburn, S. L. and Dolhinow, P. (eds), *Perspectives on Human Evolution*, vol. II. New York: Holt, Rinehart & Winston.

Lieberman, P. (1991) *Uniquely Human: The Evolution of Speech, Thought and Selfless Behaviour*. Cambridge, Mass.: Harvard University Press.

Lieberman, P. and Crelin, E. S. (1971) On the Speech of Neanderthal Man. *Linguistic Inquiry*, vol. 11, no. 2, pp. 203–22.

Livingstone, F. B. (1962) On the Non-Existence of Human Races. *Current Anthropology*, vol. 3, no. 3, pp. 279–81.

Livingstone, F. B. (1969) Genetics, Ecology and the Origins of Incest and Exogamy. *Current Anthropology*, vol. 10, no. 1, pp. 45–61.

Lothrop, S. K. (1928) *The Indians of Tierra Del Fuego*. New York: Museum of the American Indian. Heye Foundation.

Lovejoy, C. O. (1981) The Origin of Man. *Science*, vol. 211, no. 4480, pp. 341–50.

Lovejoy, C. O. (1988) Evolution of Human Walking. *Scientific American*. vol. 259, no. 5, pp. 82–9.

Loy, J. D. and Peters, C. B. (1991) *Understanding Behaviour. What Primate Studies Tell Us About Human Behaviour*. New York: Oxford University Press.

de Lumley, H. (1969) A Paleolithic Camp at Nice. *Scientific American*. vol. 220, no. 5, pp. 42–59.

Lumsden, C. J. and Wilson, E. O. (1983) *Promethean Fire, Reflections on the Origins of Mind*. Cambridge, Mass.: Harvard University Press.

McClintock, M. K. (1971) Menstrual Synchrony and Suppression. *Nature*, vol. 229, pp. 244–5.

McGrew, W. C. (1981) The Female Chimpanzee as a Human Evolutionary Prototype. In Dahlberg, F. (ed.), *Woman the Gatherer*. New Haven: Yale University Press.

MacNeish, R. S. (1971) Early Man in the Andes. *Scientific American*, vol. 224, no. 4, pp. 36–46.

Malinowski, B. (1982) *The Sexual Life of Savages in North Western Melanesia* [1932] 3rd edn. London: Routledge & Kegan Paul.

Mandel, E. (1968) *Marxist Economic History*. London: Merlin.

Mann, A. E. (1981) Diet and Human Evolution. In Harding, R. S. O. and Teleki, G. (eds), *Omnivorous Primates: Gathering and Hunting in Human Evolution*. New York: Columbia University Press.

Mann, M. (1986) *A History of Power from the Beginning to AD 1760*. Cambridge: Cambridge University Press.

Marshack, A. (1970) The Baton of Montgandier. *Natural History*, vol. 79, pp. 57–63.

Marshack, A. (1972) Cognitive Aspects of Upper Palaeolithic Engraving. *Current Anthropology*, vol. 13, pp. 445–77.

Marshack, A. (1976) Some Implications of the Palaeolithic Symbolic Evidence for the Origin of Language. In Harnad, S. R. et al. (eds), Origins and Evolution of Language. *Annals of the New York Academy of Sciences*, vol. 280.

Marshack, A. (1976A) Implications of the Paleolithic Symbolic Evidence for the Origin of Language. *American Scientist*, vol. 64, pp. 136–45.

Marshack, A. (1979) Upper Paleolithic Symbol Systems of the Russian Plain: Cognitive and Comparative Analysis. *Current Anthropology*, vol. 20, pp. 271–311.

Marshack, A. (1988) Paleolithic Calendar. In Tattersall, I., Delson, E. and Van Couvering, J. (eds), *Encyclopedia of Human Evolution and Prehistory [EHEP]*. New York & London: Garland.

Marshack, A. (1990) Early Hominid Symbol and Evolution of the Human Capacity. In Mellars, P. (ed.), *The Emergence of Modern Humans. An Archaeological Perspective*. Edinburgh: Edinburgh University Press.

Marshall, J. K. (1955) *Bitter Melons*. Ethnographic film available through the Royal Anthropological Institute.

Marshall, L. (1960) !Kung Bushman Bands. *Africa*, vol. 4, pt. 4, pp. 325–55.

Marshall, L. (1961) Sharing, Talking, and Giving: Relief of Social Tensions Among !Kung Bushmen. *Africa*, vol. 31, pt. 3, pp. 231–49.

Marshall, L. (1962) !Kung Bushman Religious Beliefs. *Africa*, vol. 32, pt. 3, pp. 221–53.

Marshall, L. (1969) The Medicine Dance of the !Kung Bushmen. *Africa,* vol. 39, pt. 4, pp. 347–81.

Martin, P. S. (1966) Africa and Pleistocene Overkill. *Nature,* vol. 212, pp. 339–42.

Martin, P. S. (1973) The Discovery of America. *Science,* vol. 179, pp. 969–74.

Martin, R. D. and May, R. M. (1981) Outward Signs of Breeding. *Nature,* vol. 293, pp. 7–9.

Marx, K. (1970) *Capital. A Critique of Political Economy,* 3 vols [1867] London: Lawrence & Wishart.

Mayr, E. (1942) *Systematics and the Origin of Species.* New York: Columbia University Press.

Mayr, E. (1970) *Populations, Species and Evolution.* Cambridge, Mass.: Harvard University Press.

Mayr, E. (1974) Behaviour Programs and Evolutionary Strategies. *American Scientist,* vol. 62, pp. 650–9.

Mayr, E. (1978) The Nature of the Darwinian Revolution. In Washburn, S. L. and E. R. McCown (eds), *Human Evolution: Biosocial Perspectives.* Calif.: Benjamin Cummings.

Mayr, E. (1982) *The Growth of Biological Thought.* Cambridge, Mass.: Harvard University Press.

Meillassoux, C. (1973) On the Mode of Production of the Hunting Band. In Alexandre, P. (ed.), *French Perspectives in African Studies.* London: Oxford University Press.

Mellars, P. (1991) Cognitive Changes and the Emergence of Modern Humans in Europe. *Cambridge Archaeological Journal,* vol. 1, no. 1, pp. 63–76.

Merker, B. (1984) A Note on Hunting and Hominid Origins. *American Anthropologist,* vol. 84, pp. 112–3.

Midgley, M. (1980) *Beast and Man.* London: Methuen.

Midgley, M. (1980A) Rival Fatalisms. The Hollowness of the Sociobiology Debate. In Montague, A. (ed.), *Sociobiology Examined.* Oxford: Oxford University Press.

Miller, G. A. and Gildea, P. M. (1987) How Children Learn Words. *Scientific American,* vol. 257, pp. 86–91.

Milton, K. (1993) Diet and Primate Evolution. *Scientific American,* vol. 269, pp. 70–7.

Molleson, T. (1994) The Eloquent Bones of Abu Hureyra. *Scientific American,* vol. 271, no. 2, pp. 60–5.

Montague, A. (1976) Toolmaking, Hunting, and the Origin of Language. In Harnad, S. R. et al. (eds), Origins and Evolution of Language. *Annals of the New York Academy of Sciences,* vol. 280.

Montague, A. (ed.) (1980) *Sociobiology Examined.* Oxford: Oxford University Press.

Moore, H. (1988) *Feminism and Anthropology*. Cambridge: Polity Press.

Morgan, E. (1972) *The Descent of Woman*. London: Corgi.

Morton, N. E. (1961) Morbidity of Children from Consanguineous Marriages. In Steinberg, A. G. (ed.), *Progress in Medical Genetics*, vol. 1, pp. 261–91.

Mosiman, J. E. and Martin, P. S. (1975) Simulating Overkill by Paleoindians. *American Scientist*, vol. 63, pt. 3, pp. 304–13.

Mowat, F. (1989) *People of the Deer* [1951] London: Souvenir Press.

Murdoch, J. (1892) Ethnological Results of the Point Barrow Expedition. *Bureau of American Ethnology. Ninth Annual Report for 1887–88*. Washington D.C.: Smithsonian Institution.

Murdock, G. P. (1957) World Ethnographic Sample. *American Anthropologist*, vol. 59, pp. 664–87.

Murdock, G. P. and Provost, C. (1973) Factors in the Division of Labour by Sex: A Cross-Cultural Analysis. *Ethnology*, vol. 12, no. 2, pp. 203–25.

Musgrave, J. H. (1971) How Dextrous was Neanderthal Man? *Nature*, vol. 233, pp. 538–41.

Musonda, F. B. (1991) The Significance of Modern Hunter-Gatherers in the Study of Early Hominid Behaviour. In Foley, R. A. (ed.), *The Origins of Human Behaviour*. London: Unwin Hyman.

Nadler, R. D. and Phoenix, C. H. (1991) Male Sexual Behaviour: Monkeys, Men and Apes. In Loy, J. D. and Peters, C. B. (eds), *Understanding Behaviour. What Primate Studies Tell us About Human Behaviour*. New York: Oxford University Press.

Napier, J. R. (1967) The Antiquity of Human Walking. *Scientific American*, vol. 216, no. 4, in Isaac, G. L. and Leakey, R. E. (eds) (1979), *Human Ancestors*. San Francisco: W. H. Freeman.

Napier, J. R. (1971) *The Roots of Mankind*. London: G Allen & Unwin.

Nicholson, N. A. (1991) Maternal Behaviour in Human and Non-Human Primates. In Loy, J. D. and Peters, C. B. (eds), *Understanding Behaviour. What Primate Studies Tell us About Human Behaviour*. New York: Oxford University Press.

Oakley, K. P. (1962) On Man's Use of Fire, With Comments on Tool-Making and Hunting. In Washburn, S. L. (ed.), *The Social Life of Early Man*. London: Methuen.

Oakley, K. P. (1972A) *Man the Tool-Maker*. London: British Museum.

Oakley, K. P. (1972B) Skill as a Human Possession. In Washburn, S. L. and Dolhinow, P. (eds), *Perspectives on Human Evolution*, vol. 2. New York: Holt, Rinehart & Winston.

Oakley, K. P. (1981) The Emergence of Higher Thought Patterns. *Philosophical Transactions of the Royal Society*, series B, vol. 292.

Oakley, K. P. et al. (1977) A Reappraisal of the Clacton Spearpoint. *Proceedings of the Prehistoric Society*, vol. 43, pp. 13–30.

Olsen, S. J. (1990) Was Early Man in North America a Big Game Hunter? In Davies, L. B. and Reeves, B. O. (eds), *Hunters of the Recent Past.* London: Unwin Hyman.

Orme, B. (1974) Twentieth-Century Prehistorians and the Idea of Ethnographic Parallels. *Man*, vol. 9, pp. 199–212.

Orme, B. (1981) *Anthropology for Archaeologists.* London: Duckworth.

Ortiz de Montellano, B. R. (1978) Aztec Cannibalism: An Ecological Necessity? *Science*, vol. 200, pp. 611–17.

Parker, S. T. and Gibson, K. R. (1979) A Developmental Model for the Evolution of Language and Intelligence in Early Hominids. *The Behavioral and Brain Sciences*, vol. 2, pp. 367–408.

Parsons, T. (1954) The Incest Taboo in Relation to Social Structure and the Socialization of the Child. *British Journal of Sociology*, vol. 5, pp. 101–17.

Passingham, R. (1982) *The Human Primate.* Oxford: W. H. Freeman.

Patrusky, B. (1987) The First Americans: Who Were They and When Did They Arrive? In Whitten, P. and Hunter, E. K. (eds), *Anthropology. Contemporary Perspectives.* Boston: Little, Brown & Co.

Peukert, D. J. K. (1989) *Inside Nazi Germany. Conformity, Opposition and Racism in Everyday Life.* Harmondsworth: Penguin.

Pfeiffer, J. E. (1982) *The Creative Explosion. An Inquiry into the Origins of Art and Religion.* New York: Cornell University Press.

Pilbeam, D. (1984) The Descent of Hominoids and Hominids. *Scientific American*, vol. 250, no. 3, pp. 60–9.

Pilbeam, D. (1986) The Origin of *Homo sapiens:* The Fossil Evidence. In Wood, B. (ed.), *Major Topics in Human Evolution.* Cambridge: Cambridge University Press.

Potts, R. (1983) Foraging for Faunal Resources by Early Hominids. In Clutton-Brock, J. and Grigson, C. (eds), *Animals in Archaeology*, vol. 1, *Hunters and their Prey.* Oxford BAR International Series.

Potts, R. (1984) Home Bases and Early Hominids. *American Scientist*, vol. 72, pp. 338–47.

Potts, R. (1984A) Hominid Hunters? Problems of Identifying the Earliest Hunter/Gatherers. In Foley, R. (ed.), *Hominid Evolution and Community Ecology.* London: Academic Press.

Potts, R. (1987) Reconstructions of Early Hominid Socioecology: A Critique of Primate Models. In Kinzey, W. G. (ed.), *The Evolution of Human Behaviour: Primate Models.* New York: State University of New York Press.

Potts, R. and Shipman, P. (1981) Cutmarks Made by Stone Tools on Bones from Olduvai Gorge, Tanzania. *Nature*, vol. 291, pp. 577–80.

Price, J. A. (1975) Sharing: The Integration of Intimate Economies. *Anthropologica*, vol. 17, pp. 3–27.

Rappaport, R. A. (1984) *Pigs for the Ancestors. Ritual in the Ecology of a New Guinea People.* New Haven: Yale University Press.

Reader, J. (1988) *Missing Links. The Hunt for Earliest Man.* London: Penguin.

Reiter, R. R. (1977) The Search for Origins: Unravelling the Threads of Gender Hierarchy. *Critique of Anthropology*, vol. 3, nos. 9 & 10, pp. 5–24.

Rennie, J. (1993) How Many Genes and Y. *Scientific American*, vol. 268, no. 1, pp. 7–9.

Rensberger, B. (1987) Racial Odyssey. In Whitten, P. and Hunter, D. E. (eds), *Anthropology. Contemporary Perspectives.* Boston: Little, Brown & Co.

Reynolds, P. C. (1976) The Emergence of Early Hominid Social Organisation: I The Attachment System. *Yearbook of Physical Anthropology*, vol. 20, pp. 73–95.

Reynolds, V. (1966) Open Groups in Hominid Evolution. *Man*, vol. 1, pp. 441–52.

Reynolds, V. (1968) Kinship and the Family in Monkeys, Apes and Man. *Man*, vol. 3, pp. 209–23.

Reynolds, V. (1980) *The Biology of Human Action.* Oxford: W. H. Freeman.

Reynolds, V. (1991) The Biological Basis of Human Patterns of Mating and Marriage. In Reynolds, V. and Kellet, J. (eds), *Mating and Marriage.* Oxford: Oxford University Press.

Richard, A. F. (1985) *Primates in Nature.* New York: W. H. Freeman.

Richards, G. (1987) *Human Evolution. An Introduction for the Behavioural Sciences.* London: Routledge & Kegan Paul.

Rightmire, P. (1988) *Homo habilis.* In Tattersall, I., Delson, E. and Van Couvering, J. (eds), *Encyclopedia of Human Evolution and Prehistory [EHEP].* New York & London: Garland.

Rightmire, P. (1988A) *Homo erectus.* In Tattersall, I., Delson, E. and Van Couvering, J. (eds), *Encyclopedia of Human Evolution and Prehistory [EHEP].* New York & London: Garland.

Rodman, P. S. and McHenry, H. M. (1980) Bioenergetics and the Origin of Hominid Bipedalism. *American Journal of Physical Anthropology*, vol. 52, pp. 103–6.

Romer, A. S. (1958) Phylogeny and Behaviour with Special Reference to Vertebrate Evolution. In Roe, A. and Simpson, G. G. (eds), *Behaviour and Evolution.* New Haven: Yale University Press.

Roper, M. K. (1969) A Survey of the Evidence for Intrahuman Killing in the Pleistocene. *Current Anthropology.* vol. 10, pt. II, pp. 427–59.

Rose, S., Kamin, L. J. and Lewontin, R. C. (1984) *Not in Our Genes.* Harmondsworth: Penguin.

Rudwick, M. (1992) Darwin and Catastrophism. In Bourriau, J. (ed.), *Understanding Catastrophe.* Cambridge: Cambridge University Press.

Sacks, K. (1975) Engels Revisited: Women, the Organization of Production and Private Property. In Reiter, R. R. (ed.), *Towards an Anthropology of Women.* New York: Monthly Review Press.

Sahlins, M. D. (1960) The Origins of Society. *Scientific American.* vol. 203, no. 3, pp. 88–96.

Sahlins, M. D. (1972) *Stone Age Economics.* London: Tavistock.

Sahlins, M. D. (1976) *Culture and Practical Reason.* Chicago: The University of Chicago Press.

Sahlins, M. D. (1977) *The Use and Abuse of Biology. An Anthropological Critique of Sociobiology.* London: Tavistock.

Sahlins, M. D. and Service, E. R. (eds) (1960) *Evolution and Culture.* Ann Arbor: University of Michigan Press.

Sandars, N. K. (1968) *Prehistoric Art in Europe.* Harmondsworth: Penguin.

Sanday, P. R. (1986) *Divine Hunger. Cannibalism as a Cultural System.* Cambridge: Cambridge University Press.

Sarich, V. M. and Wilson, A. C. (1967) Immunological Time Scale for Hominid Evolution. *Science,* vol. 158, pp. 1200–3.

Schaller, G. B. (1972) *The Year of the Gorilla.* London: Collins.

Schaller, G. B. and Lowther, G. R. (1969) The Relevance of Carnivore Behaviour to the Study of Early Hominids. *Southwestern Journal of Anthropology,* vol. 25, no. 4, pp. 307–41.

Schultz, A. H. (1962) Some Factors Influencing the Social Life of Primates in General and of Early Man in Particular. In Washburn, S. L. (ed.), *The Social Life of Early Man.* London: Methuen.

Service, E. R. (1971) *Primitive Social Organisation.* New York: Random House.

Service, E. R. (1979) *The Hunters.* New Jersey: Prentice-Hall.

Shackley, M. (1980) *Neanderthal Man.* London: Duckworth.

Shea, B. T. (1988) Sexual Dimorphism. In Tattersall, I., Delson, E. and Van Couvering, J. (eds), *Encyclopedia of Human Evolution and Prehistory [EHEP].* New York & London: Garland.

Shipman, P. (1983) Early Hominid Lifestyle: Hunting and Gathering or Foraging and Scavenging? In Clutton-Brock, J. and Grigson, C. (eds), *Animals in Archaeology,* vol. 1, *Hunters and their Prey.* Oxford BAR International Series.

Shipman, P. (1984) Scavenger Hunt. *Natural History,* vol. 93, no. 4, pp. 20–7.

Shipman, P. (1986) Scavenging or Hunting in Early Hominids: Theoretical Framework and Tests. *American Anthropologist,* vol. 88, pp. 27–43.

Shipman, P. and Phillips-Conroy, J. (1977) Hominid Tool-Making Versus Carnivore Scavenging. *American Journal of Physical Anthropology,* vol. 46, pp. 77–86.

Short, R. V. (1980) The Origins of Human Sexuality. In Austin, C. R. and Short, R. V. (eds), *Human Sexuality.* Cambridge: Cambridge University Press.

Shostak, M. (1990) *Nisa. The Life and Words of a !Kung Woman.* London: Earthscan.

Sibley, C. G. and Ahlquist, J. E. (1984) The Phylogeny of the Hominoid Primates, as Indicated by DNA–DNA Hybridisation. *Journal of Molecular Evolution*, vol. 20, pp. 2–15.

Sieveking, A. (1979) *The Cave Artists*. London: Thames & Hudson.

Silberbauer, G. B. (1981) *Hunter and Habitat in the Central Kalahari Desert*. Cambridge: Cambridge University Press.

Simons, E. L. (1977) *Ramapithecus*. *Scientific American*, vol. 236, no. 5, pp. 28–38.

Simons, E. L. (1981) Man's Immediate Forerunners. *Philosophical Transactions of the Royal Society*, series B, vol. 292.

Simpson, B. (1984) *Cannibalism and the Common Law*. Harmondsworth: Penguin.

Simpson, G. G. (1983) *Fossils and the History of Life*. New York: Scientific American Books.

Slocum, S. (1975) Woman the Gatherer: Male Bias in Anthropology. In Reiter, R. R. (ed.), *Towards an Anthropology of Women*. New York: Monthly Review Press.

Smith, V. L. (1975) The Primitive Hunter Culture, Pleistocene Extinction, and the Rise of Agriculture. *Journal of Political Economy*, vol. 83, pp. 727–55.

Solecki, R. S. (1963) Prehistory in Shanidar Valley, Northern Iraq. *Science*, vol. 139, pp. 179–93.

Solecki, R. S. (1972) *Shanidar: The Humanity of Neanderthal Man*. London: Allen Lane. The Penguin Press.

Solecki, R. S. (1975) Shanidar IV, a Neanderthal Flower Burial in Northern Iraq. *Science*, vol. 190, pp. 880–1.

Solecki, R. S. (1977) The Implications of the Shanidar Cave Neanderthal Flower Burial. *Annals of the New York Academy of Sciences*, vol. 293, pp. 114–23.

Spencer, F. (1984) The Neanderthals and their Evolutionary Significance: A Brief Historical Survey. In Smith, F. H. and Spencer, F. (eds), *The Origin of Modern Humans. A World Survey of the Fossil Evidence*. New York: Alan R. Liss.

Spencer, H. (1982) *Progress: Its Law and Cause* [1857], in Coser, L. A. and Rosenberg, B. (eds), *Sociological Theory: A Book of Readings*. 5th edn. New York: Macmillan.

Stanley, S. M. (1981) *The New Evolutionary Timetable*. New York: Basic Books.

Stanley, S. M. (1987) *Extinction*. New York: Scientific American Books.

Stanley, S. M. (1989) *Earth and Life Through Time*. New York: W. H. Freeman.

Steele Russell, I. (1979) Brain Size and Intelligence: A Comparative Perspective. In Oakley, D. A. and Plotkin, H. C. (eds), *Brain, Behaviour and Evolution*. London: Methuen.

Stein, G. J. (1988) Biological Science and the Roots of Nazism. *American Scientist*, vol. 76, pp. 50–8.

Steward, J.H. (1968) Causal Factors and Processes in the Evolution of Pre-Farming Societies. In Lee, R.B. and DeVore, I. (eds), *Man the Hunter*. New York: Aldine.

Stewart, T.D. (1977) The Neanderthal Remains from Shanidar Cave, Iraq: A Summary of Findings to Date. *Proceedings of the American Philosophical Society*, vol. 121, no. 2, pp. 121–65.

Stoneking, M. and Cann, R.L. (1989) African Origins Of Human Mitochondrial DNA. In Mellars, P. and Stringer, C. (eds), *The Human Revolution. Behavioural and Biological Perspectives on the Origins of Modern Humans*. Edinburgh: Edinburgh University Press.

Stringer, C.B. (1988A) Archaic *Homo sapiens*, Archaic Moderns. In Tattersall, I., Delson, E. and Van Couvering, J. (eds), *Encyclopedia of Human Evolution and Prehistory [EHEP]*. New York & London: Garland.

Stringer, C.B. (1988B) Neanderthals. In Tattersall, I., Delson, E. and Van Couvering, J. (eds), *Encyclopedia of Human Evolution and Prehistory [EHEP]*. New York & London: Garland.

Stringer, C.B. (1989) Documenting the Origin of Modern Humans. In Trinkaus, E. (ed.), *The Emergence of Modern Humans*. Cambridge: Cambridge University Press.

Stringer, C.B. (1990) The Emergence of Modern Humans. *Scientific American*, vol. 263, no. 6, pp. 68–74.

Strum, S.C. (1987) *Almost Human. A Journey into the World of Baboons*. London: Elm Tree.

Susman, R.L. (1988) Hand of *Paranthropus robustus* from Member I, Swartkrans: Fossil Evidence for Tool Behaviour. *Science*, vol. 240, pp. 781–4.

Suzuki, A. (1975) The Origins of Hominid Hunting: A Primatological Perspective. In Tuttle, R.H. (ed.), *Socioecology and Psychology of Primates*. The Hague: Mouton.

Svoboda, J. (1989) Middle Pleistocene Adaptations in Central Europe. *Journal of World Prehistory*, vol. 3, no. 1, pp. 33–70.

Symons, D. (1979) *The Evolution of Human Sexuality*. New York: Oxford University Press.

Szalay, F.S. (1975) Hunting – Scavenging Protohominids: A Model for Hominid Origins. *Man*, vol. 10, no. 3, pp. 420–9.

Tabet, P. (1982) Hands, Tools, Weapons. *Feminist Issues*, vol. 2, pp. 3–63.

Tanner, N.M. (1981) *On Becoming Human*. Cambridge: Cambridge University Press.

Tanner, N.M. (1987) The Chimpanzee Model Revisited and the Gathering Hypothesis. In Kinzey, W.G. (ed.), *The Evolution of Human Behaviour: Primate Models*. New York: State University of New York Press.

Tanner, N.M. (1988) Becoming Human, Our Links With Our Past. In Ingold, T. (ed.), *What is an Animal?* London: Unwin Hyman.

Tanner, N. and Zihlman, A. (1976) Women in Evolution Part I: Innovation and Selection in Human Origins. *Signs: Journal of Women in Culture and Society*, vol. 1, no. 3, pp. 585–608.

Taub, D. and Mehlman, P. (1991) Primate Paternalistic Investment: A Cross-Species View. In Loy, J.D. and Peters, C.B. (eds), *Understanding Behaviour. What Primate Studies Tell us About Human Behaviour*. New York: Oxford University Press.

Teleki, G. (1973) The Omnivorous Chimpanzee. *Scientific American*, vol. 228, pp. 32–47.

Thorne, A.G. and Wolpoff, M.H. (1992) The Multiregional Evolution of Humans. *Scientific American*, vol. 266, no. 4, pp. 28–33.

Tiger, L. and Fox, R. (1978) The Human Biogram (From The Imperial Animal). In Caplan, A.L. (ed.), *The Sociobiology Debate*. New York: Harper & Row.

Tobias, P.V. (1965) New Discoveries in Tanganika: Their Bearing on Hominid Evolution. *Current Anthropology*, vol. 6, pp. 391–99.

Tobias, P.V. (1976) African Hominids: Dating and Phylogeny. In Isaac, G.L. and McCown, E.R. (eds), *Human Origins. Louis Leakey and the East African Evidence*. Menlo Park, Calif.: W.A. Benjamin.

Tobias, P.V. (1981) The Emergence of Man in Africa and Beyond. *Philosophical Transactions of the Royal Society*, series B, vol. 292.

Tooby, J. and DeVore, I. (1987) The Reconstruction of Hominid Behavioural Evolution Through Strategic Modeling. In Kinzey, W.G. (ed.) *The Evolution of Human Behaviour: Primate Models*. New York: State University of New York Press.

Toth, N. (1985) The Oldowan Reassessed: A Close Look at Early Stone Artefacts. *Journal of Archaeological Science*, vol. 12, pp. 101–20.

Toth, N. (1987) The First Technology. *Scientific American*, vol. 256, pp. 104–13.

Toth, N. and Schick, K.D. (1986) The First Million Years: The Archaeology of Proto Human Culture. *Advances in Archaeological Method and Theory*, vol. 9, pp. 1–96.

Toth, N. and Schick, K.D. (1988) Fire. In Tattersall, I., Delson, E. and Van Couvering, J. (eds) *Encyclopedia of Human Evolution and Prehistory [EHEP]*. New York & London: Garland.

Trevarthen, C. (1979) Instincts for Human Understanding and for Cultural Cooperation: Their Development in Infancy. In Cranach, M. von et al. (eds) *Human Ethology. Claims and Limits of a New Discipline*. Cambridge: Cambridge University Press.

Trevarthen, C. and Grant, F. (1979) Not Work Alone. Infant Play and the Creation of Culture. *New Scientist*, vol. 81, pp. 566–9.

Trigger, B. G. (1966) Engels on the Part Played by Labour in the Transition from Ape to Man: An Anticipation of Contemporary Anthropological Theory. *Canadian Review of Sociology and Anthropology*, vol. 4, pt. 3, pp. 165–76.

Trinkaus, E. (1981) Neanderthal Limb Proportions and Cold Adaptation. In Stringer, C. B. (ed.), *Aspects of Human Evolution. Symposia for the Study of Human Biology*, vol. 21, pp. 187–224. London: Taylor & Francis.

Trinkaus, E. (1988) Bodies, Brawn, Brains and Noses: Human Ancestors and Human Predation. In Nitecki, M. H. and Nitecki, D. V. (eds), *The Evolution of Human Hunting*. New York: Plenum.

Trinkaus, E. and Howells, W. W. (1979) The Neanderthals. *Scientific American*, vol. 241, pp. 118–33.

Trinkaus, E. and Shipman, P. (1993) *The Neanderthals. Changing the Image of Mankind*. London: Jonathan Cape.

Turnbull, C. M. (1976) *The Forest People*. London: Pan Books.

Turnbull, C. M. (1978) *Man in Africa*. Harmondsworth: Penguin.

Tuttle, R. H. (1981) Evolution of Hominid Bipedalism and Prehensile Capabilities. *Philosophical Transactions of the Royal Society*, series B. vol. 292.

Tylor, E. B. (1982) Culture Defined. In *Primitive Culture* [1891], in Coser, L. and Rosenberg, B. (eds), *Sociological Theory. A Book of Readings*, 5th edn. New York: Macmillan.

Ucko, P. J. (1969) Ethnography and Archaeological Interpretations of Funerary Remains. *World Archaeology*, vol. 1, no. 2, pp. 262–80.

van Eysinga, F. W. B. (1978) *Geological Timetable*. Amsterdam: Elservier Science.

Villa, P. et al. (1986) Cannibalism in the Neolithic. *Science*, vol. 233, pp. 431–7.

Villa, P. (1990) Torralba and Aridos: Elephant Exploitation in Middle Pleistocene Spain. *Journal of Human Evolution*, vol. 19, pp. 299–309.

Vrba, E. S. (1985) Ecological and Adaptive Changes Associated with Early Hominid Evolution. In Delson, E. (ed.), *Ancestors: The Hard Evidence*. New York: Alan Liss.

Walker, A. (1981) Dietary Hypotheses and Human Evolution. *Philosophical Transactions of the Royal Society*, series B, vol. 292.

Walker, A. and Teaford, M. (1989) The Hunt for *Proconsul*. *Scientific American*, vol. 260, no. 1, pp. 58–64.

Washburn, S. L. (1957) *Australopithecus*: the Hunters or the Hunted? *American Anthropologist*, vol. 59, pp. 612–14.

Washburn, S. L. (1960) Tools and Human Evolution. *Scientific American*, vol. 203, no. 3, pp. 62–75.

Washburn, S. L. (1963) The Study of Race. *American Anthropologist*, vol. 65, pp. 521–31.

Washburn, S. L. (1978) The Evolution of Man. *Scientific American*, vol. 239, pp. 146–54.

Washburn, S. L. (1980) Human Behaviour and the Behaviour of Other Animals. In Montague, A. (ed.), *Sociobiology Examined*. Oxford: Oxford University Press.

Washburn, S. L. and Avis, V. (1958) Evolution and Human Behaviour. In Roe, A. and Simpson, G. G. (eds), *Behaviour and Evolution*. New Haven: Yale University Press.

Washburn, S. L. and DeVore, I. (1961) The Social Life of Baboons. *Scientific American*, vol. 204, no. 6, pp. 62–71.

Washburn, S. L. and Lancaster, C. S. (1968) The Evolution of Hunting. In Lee, R. B and DeVore, I. (eds), *Man the Hunter*. New York: Aldine.

Washburn, S. L., Jay, P. C. and Lancaster, J. B. (1968) Field Studies of Old World Monkeys and Apes. In Washburn, S. L. and Jay, P. C. (eds), *Perspectives on Human Evolution*, vol. I. New York: Holt, Rinehart & Winston.

Washburn, S. L. and Moore, R. (1974) *Ape Into Man*. Boston: Little Brown & Co.

Watanabe, H. (1964) The Ainu. A Study of Ecology and the System of Social Solidarity Between Man and Nature in Relation to Group Structure. *Journal of the Faculty of Science*, University of Tokyo, section V Anthropology, vol. 2, pt. 6.

Webster, D. (1981) Late Pleistocene Extinction and Human Predation: A Critical Overview. In Harding, R. S. O. and Teleki, G. (eds), *Omnivorous Primates. Gathering and Hunting in Human Evolution*. New York: Columbia University Press.

Wellings, K. et al. (1994) *Sexual Behaviour in Britain. The National Survey of Sexual Attitudes and Lifestyles*. Harmondsworth: Penguin.

Wenke, R. J. (1980) *Patterns in Prehistory. Mankind's First Three Million Years*. New York: Oxford University Press.

Wheat, J. B. (1967) A Paleo-Indian Bison Kill. *Scientific American*, vol. 216, no. 1, pp. 44–52.

Wheeler, M. (1956) *Archaeology from the Earth*. Harmondsworth: Penguin.

Wheeler, P. E. (1984) The Evolution of Bipedality and Loss of Functional Body Hair in Hominids. *Journal of Human Evolution*, vol. 13, pp. 91–8.

White, L. A. (1959A) *The Evolution of Culture. The Development of Civilisation to the Fall of Rome*. New York: McGraw-Hill.

White, L. A. (1959B) The Concept of Culture. *American Anthropologist*, vol. 61, pp. 227–51.

White, L. A. (1982) The Symbol from *The Science of Culture*. [1949], in Coser, L. and Rosenberg, B. (eds) *Sociological Theory. A Book of Readings*. 5th edn. New York: Macmillan.

White, R. (1982) Rethinking the Middle/Upper Paleolithic Transition. *Current Anthropology*, vol. 23, no. 2, pp. 169–92.

White, R. (1989) Visual Thinking in the Ice Age. *Scientific American*, vol. 261, no. 1, pp. 74–81.

White, R. (1989A) Toward a Contextual Understanding of the Earliest Body Ornaments. In Trinkaus, E. (ed.), *The Emergence of Modern Humans*. Cambridge: Cambridge University Press.

White, R. (1989B) Production Complexity and Standardization in Early Aurignacian Bead and Pendant Manufacture: Evolutionary Implications. In Mellars, P. and Stringer, C. B. (eds), *The Human Revolution. Behavioural and Biological Perspectives on the Origins of Modern Humans*. Edinburgh: Edinburgh University Press.

White, T. D. et al. (1994) *Australopithecus ramidus*, A New Species of Early Hominid From Aramis, Ethiopia. *Nature*, vol. 371, pp. 306–12.

Wickler, W. (1969) *The Sexual Code. The Social Behaviour of Animals and Men*. London: Weidenfeld & Nicholson.

Williams, M. A. J. et al. (1993) *Quaternary Environments*. London: Edward Arnold.

Wilson, A. C. and Cann, R. L. (1992) The Recent African Genesis of Humans. *Scientific American*, vol. 266, no. 4, pp. 22–7.

Wilson, E. O. (1975) *Sociobiology: The New Synthesis*. Cambridge, Mass.: Harvard University Press.

Wilson, E. O. (1977) Biology and the Social Sciences. *Daedalus*, vol. 11, pp. 129–40.

Wilson, E. O. (1989) Threats to Biodiversity. *Scientific American*, vol. 261, pp. 60–6.

Wobst, H. M. (1974) Boundary Conditions for Paleolithic Social Systems: A Simulation Approach. *American Antiquity*, vol. 39, no. 2. pp. 147–78.

Wobst, H. M. (1976) Locational Relationships in Paleolithic Society. *Journal of Human Evolution*, vol. 5, pp. 49–58.

Wolfe, L. D. (1991) Human Evolution and the Sexual Behaviour of Female Primates. In Loy, J. D. and Peters, C. B. (eds) *Understanding Behaviour. What Primate Studies Tell us About Human Behaviour*. New York: Oxford University Press.

Wolpoff, M. H. (1989) Multiregional Evolution: The Fossil Alternative to Eden. In Mellars, P. and Stringer, C. B. (eds), *The Human Revolution. Behavioural and Biological Perspectives on the Origins of Modern Humans*. Edinburgh: Edinburgh University Press.

Wolpoff, M. H. (1989A) The Place of Neanderthals in Human Evolution. In Trinkaus, E. (ed.), *The Emergence of Modern Humans*. Cambridge: Cambridge University Press.

Wood, B. (1994) The Oldest Hominid Yet. *Nature*, vol. 371, pp. 280–1.

Woodburn, J. (1968) An Introduction to Hadza Ecology. In Lee, R. B. and DeVore, I. (eds), *Man the Hunter*. New York: Aldine.

Woodburn, J. (1979) Minimal Politics. The Political Organisation of the Hadza of North Tanzania. In Schack, W. A. and Cohen, P. S. (eds), *Politics in Leadership*. Oxford: Clarendon Press.

Woodburn, J. (1982) Egalitarian Societies. *Man*, vol. 17, no. 3, pp. 431–51.

Woodburn, J. (1982A) Social Dimensions of Death in Four African Hunting and Gathering Societies. In Block, M. and Parry, J. (eds), *Death and the Regeneration of Life*. Cambridge: Cambridge University Press.

Woolfson, C. (1982) *The Labour Theory of Culture. A Re-examination of Engel's Theory of Human Origins*. London: Routledge & Kegan Paul.

World Bank (1991) *World Bank Development Report 1991*. Oxford: Oxford University Press.

Wrangham, R. W. (1987) The Significance of African Apes for Reconstructing Human Social Evolution. In Kinzey, W. G. (ed.), *The Evolution of Human Behaviour: Primate Models*. New York: State University of New York Press.

Wright, R. V. S. (1972) Imitative Learning of a Flaked Stone Technology – The Case of an Orangutan. *Mankind*, vol. 8, pp. 296–306.

Wu Rukang and Lin Shenglong (1983) Peking Man. *Scientific American*, vol. 248, no. 6, pp. 78–86.

Wymer, J. (1984) *The Palaeolithic Age*. London: Croom Helm.

Wynn, T. (1979) The Intelligence of Later Acheulean Hominids. *Man*, vol. 14, pp. 371–91.

Wynn, T. (1981) The Intelligence of Oldowan Hominids. *Journal of Human Evolution*, vol. 10, pp. 529–41.

Yellen, J. E. (1977) *Archaeological Approaches to the Present. Models for Reconstructing the Past*. New York: Academic Press.

Yellen, J. E. (1986) Optimization and Risk in Human Foraging Strategies. *Journal of Human Evolution*, vol. 15, pp. 733–50.

Zahn-Waxler, C. (1986) Conclusions: Lessons from the Past and a Look at the Future. In Zahn-Waxler, C. et al. (eds), *Altruism and Aggression. Biological and Social Origins*. Cambridge: Cambridge University Press.

Zihlman, A. L. (1978) Women in Evolution Part II. Subsistence and Social Organization Among Early Hominids. *Signs: Journal of Women in Culture and Society*, vol. 4, no. 1, pp. 4–20.

Zihlman, A. L. (1989) Common Ancestors and Uncommon Apes. In Durant, J. R. (ed.), *Human Origins*. Oxford: Clarendon Press.

Index

Aberle, D. F.   335
aborigines   *see* Australians, native
Acheulean hominids   239, 246,
    247, 257, 267, 268, 279, 283, 286
    see also tools, manufacture and
        use of: Acheulean
Adams, M. S.   337
adaptation, culture and   80–4
adoption   75–6, 347
adornment, personal   279, 283,
    285–6, 289
aggression   73
    see also violence, hunting and
agriculture *see* diet
Agta Negritos people   306–7
Ahlquist, J. E.   93
Aiello, L. C.   147
Ainu people   305
Alper, J. S.   71
altruistic behaviour   66–7, 73–4,
    347
AMH (anatomically modern
    humans)   *see Homo sapiens
    sapiens*
anatomically modern humans
    (AMH)   *see Homo sapiens
    sapiens*
Anderson, P.   323
Andrews, P.   94
animals, extinction of, and
    hunting   256–64
anthropology
    development of   10–11
    evolution and   129–30, 147–9
    racism and   5, 20–1

apes   347
    see also primates
archaeology
    hominids and   23–6, 193–205
    hunting and   241–56
Ardrey, R.   11, 210, 253–4
Arens, W.   86–7
Aridos site   249–50
artistic expression   267–8, 276,
    278, 279–90
attachment, patterns of   328–45
    see also marriage customs
Auel, J. M.   308
Australians, native   165, 212, 302,
    305
australopithecines   5, 132, 136–7,
    139, 143, 144, 166, 171–4,
    176–7, 233, 238, 239, 240,
    253–5, 272, 318, 344, 347
Australopithecus aethiopicus   169
Australopithecus afarensis
    'Lucy'   137, 139, 141, 142,
    168–9, 172, 174, 189–90, 196,
    197, 347
Australopithecus africanus
    168–9, 171, 172–3, 176, 189,
    254, 347
Australopithecus boisei   169,
    171–2, 173–4, 193, 232, 347
Australopithecus ramidus   136,
    347
Australopithecus robustus   169,
    173, 174, 348
Avicenna   27
Avis, V.   96

387

Aztec empire
  cannibalism in 85, 88–90
  human sacrifice in 84–5, 88–90

baboons
  sexual dimorphism of 320–1
  society of 123–5, 313–14
  see also primates
Bahn, P. G. 279, 355
Balikci, A. 214, 222, 355
Bancroft, J. 324
Banton, M. 6
Barbarians 83
Barbetti, M. 228, 229
Barker, E. 2
Barnett, A. 19
Bar-Yosef, O. 276, 281
Bates, D. G. 221
Beach, F. A. 77, 316, 323
Beaune, S. A. de 290
Beck, B. B. 155
behaviour patterns
  culture and 52–8
  evolution and 58–63, 342–4
  genes and 54–5, 58–63, 70–4
  hominids, of 122–3, 152–3, 167,
      178–81, 190–208, 212, 220–1,
      222, 224–5, 264–5, 274
  humans, of 7, 21–3, 266–90
  primates, of *see* primates,
      social behaviour of
Behrensmeyer, A. K. 172
Benedict, R. 292
Bergounioux, F. M. 242
Berndt, C. H. 300, 302, 305
Berndt, R. M. 305
Beynon, A. D. 239, 317–18
Bigelow, R. 254
Binford, L. R. 14, 200–1, 204, 229,
      238–9, 240, 242–4, 247–50, 251
Binford, S. R. 14, 270
biology
  culture and 58–90
  evolutionary 5–7, 35–8, 41

social factors and 15
  see also sociobiology
bipedalism 122, 124, 126, 129,
      132–53, 317–18, 348
  origins of 141–7
  theories of 148–53
bisexuality 71
Bittles, A. H. 337
Blanc, A. C. 268
Blumenschine, J. 205, 232, 235
Blurton Jones, N. G. 224
Boas, F. 303
Bock, K. 79
Boesiger, E. 41, 45
bonding, parent-child 50–1, 56–8,
      348
  see also family units,
      development of
Bonner, J. T. 61–2
Bordes, F. 14, 160–1, 187, 274, 279
Bowlby, J. 5–6, 29, 56–7, 58, 71,
      298, 332, 354
Boyd, R. 64
Brace, C. L. 274
Brain, C. K. 254–6
brain expansion 4, 187–93, 244,
      254, 265, 266–7, 292, 317
Bridges, T. 305
Brooks, A. 244
Brown, J. K. 302
Brues, A. 209
Bunn, H. T. 199, 248–9
burial customs *see* death rituals
Burleigh, M. 6
Burley, N. 325, 326, 327
Burrow, J. W. 26, 39
Burton, M. L. 302
Butzer, K. W. 233–4, 244, 246
Byrne, R. W. 121

Calvin, W. H. 119
Campbell, B. G. 19, 38, 47, 118,
      297, 321, 322–3, 354
Cann, R. L. 20, 274

cannibalism 85–90, 103, 185, 233, 242–3, 254
Carlisle, R.C. 273
Carrier, D.R. 151
Cartmill, M. 5
catarrhines 91, 94, 145
Catton, C. 318, 327
Cavallo, J.A. 205, 232, 235
Chagnon, N.A. 301, 332, 355
Chase, P.G. 250–2, 280–1
Chia, L.-p. 242
chimpanzees
  cannibalism among 103, 233
  diet of 97, 125
  fire and 228
  hierarchy among 102–3, 314
  humans, affinity with 111–12, 122, 155, 207–8, 233, 314–15
  hunting and 125
  incest among 336, 337
  infancy of 105, 314, 318, 330
  language and 23, 272
  learning patterns of 109, 110, 119–20, 121, 125, 272
  sexual behaviour of 108, 313–16, 318, 322, 324, 336, 337
  sexual dimorphism of 106, 171, 314, 320–1
  social life of 99, 125–6, 315–16
  territorial range of 312
  tools and 23, 109, 110, 154–5, 156, 201, 314
  violence among 103
  see also primates
China 75
Clacton site 250
Clark, J.D. 176, 220, 228–9, 237
Clark, J.G.D. 23, 160, 162–3, 215, 270, 276, 355
Clark, W.E. Le Gros 92
class systems 72, 101
clothing, hominids and 268, 277

Clovis peoples 257, 258, 261, 262–3
Clutton-Brock, T.H. 98–9, 320
Combe-Grenal site 251
communism, 'primitive' 217, 218–19
Conkey, M. 280, 282
Constantinople 83
Coon, C.S. 212, 214, 278, 355
Coppens, Y. 141
Crader, D.C. 234
Crelin, E.S. 272, 273
Crook, J.H. 116, 117
culture
  adaptation and 80–4
  animals and 23, 49–50
  behaviour and 52–8
  biology and 58–90
  brain size and 190–3, 265
  civilisation and 48
  definition of 46, 348
  diet and 166, 167
  environment and 47–8, 84–90, 131–2, 176; *see also* environment
  evolution and 10, 12–17, 22–6, 39–63, 80–4, 131–2, 178–81, 345–6; *see also* evolution
  genes and 64–9, 74–8, 113–14
  hominids and 113–14, 123–5, 126–8, 130–2, 264–5; *see also* hominids
  hunting and *see* hunting cultures
  incest and 333–8
  nature of 46–8, 348
  sexual behaviour and 312–346
  social sciences and 48–52
  sociobiological theory of 64–9
  tools and 154–81
  see also human society

Dalton, G. 216
Dart, R.A. 5, 168, 253–6

Darwin, C.   3, 5–6, 8, 23, 26,
    27–35, 37, 39, 41, 44, 79–80, 83,
    91, 95, 149, 295–8, 305, 314,
    345, 354
Davis, W.   282–3
Dawkins, R.   12, 14, 297, 354
Dean, M.C.   239, 317–18
death rituals   56, 268–71, 274, 276,
    281
Denham, W.W.   221
DeVore, I.   106, 124–5, 208, 312,
    313, 342, 344
Diamond, J.D.   77, 297, 354
Dibble, H.L.   280–1
diet   97, 100, 125, 143–5, 148–9,
    151–2, 166, 167, 169, 172, 173,
    176–7, 194, 198–200, 204–5,
    209, 220, 230–1, 232–6, 237–9,
    242, 243, 244–52, 291, 309,
    344
Dillehay, T.D.   258
Divale, W.T.   221
DNA   *see* genes
Dobzhansky, T.   15, 32, 41, 45,
    58–9, 63, 73, 114, 116
Douglas, J.W.B.   57
Douglas, M.   221
Dowling, J.H.   222
Draper, P.   310
Driver, J.C.   224–5
Dryopithecus   26
Dunbar, R.I.M.   233
Dunn, L.C.   19
Durham, W.H.   64–5, 221
Durkheim, E.   271

East African Rift Valley, evolution
    and   141
ecology
    culture and   14, 15, 65, 126–7,
        133–6, 220, 234
    invention of term   5, 348
    see also environment
Edey, M.A.   150

EHEP   *see Encyclopedia of Human
    Evolution and Prehistory*
Ehrenberg, M.   341, 355
Eibl-Eibesfeldt, I.   7, 9, 54, 71, 332,
    354
Eimas, P.D.   61
Eldredge, N.   14
*Encyclopedia of Human Evolution and
    Prehistory (EHEP)*   25, 142,
    169–70
Engels, F.   3–4, 294–5, 314
entrepreneurship   71
environment, culture and   47–8,
    84–90, 126–7, 176, 178, 214–15,
    229, 256–64, 348
    see also ecology
Eocene   133–4, 349
Eskimos   209, 267, 270, 284
    Ihalmiut   212
    Netsilik   222
    see also Inuit peoples
Estioko-Griffin, A.   306, 307
ethology, human behaviour
    and   7–9, 332, 349
evolution
    anthropology and   129–30,
        147–9, 349
    behaviour and   58–63, 187–93
    bipedalism and   *see* bipedalism
    brain expansion in   187–93, 317
    culture and   10, 12–17, 22–6,
        39–63, 80–4, 131–2, 178–81,
        291–346
    diet and   143–5, 148–9, 151–2,
        172, 173, 176, 177, 209, 231,
        232–3, 344
    ecology, environment
        and   126–7, 129–30, 133–6,
        176, 177–8, 234, 342–3
    genes and   73, 274–5
    hominids and   122, 124, 126–53
    human society and   18–63,
        291–346
    hunting and   240

evolution (*cont.*)
labour and 4–5
mental processes and 114–22
primates and 91–5, 313
sexual behaviour and 76–7,
295–301, 316–28, 342–4
social *see* social evolution
social sciences and 2–3, 5, 15,
23, 80–4, 128–9
sociobiology and 78–80
theories of 26–30, 35–8, 129–31,
132, 148–53, 295–301
evolutionary biology 5–6, 342–4

Falk, D. 189, 190
family units, development
of 312–16, 328–45
see also bonding, parent–child;
kinship, systems of
Faris, J. C. 289
farming *see* diet
Fedigan, L. M. 296, 341
Festinger, L. 2
fire, hominids and 185, 226–9,
239, 241–5, 246–7, 264, 268
Fischer, E. 289
Fleagle, J. G. 137, 173, 175, 354
Foley, R. 15, 16, 25, 129–30, 145,
147, 172, 177, 219, 233, 239–40,
321, 355
foraging *see* hominids, diet of
Forde, D. 299
Fortes, M. 1
Fossey, D. 100, 103
fossils, evidence from 14
Fox, R. 65, 313
Frayer, D. W. 211
Freeman, L. G. 216, 244, 245, 248
Freeman, M. R. 221
Freud, S. 77, 314
Fried, M. H. 21
Friedl, E. 302, 308–9
Frisch, K. von 60
Frisch, R. E. 323–4

Gallup, G. G., Jr. 119
Gamble, C. 286–8
Gardner, B. T. 272
Gardner, R. A. 272
Gargett, R. H. 270
Geertz, C. 292
genes
behaviour and 54–5, 58–63,
70–4
culture and 64–9, 74–8, 113–14
human and ape, comparison
of 92–5
monogenesis theory 274–5
Neanderthal 273–4
race and 20–1
sexual selection and 297
society and 78–80
speech and 272
genotype 31–2, 349
Geschwind, N. 156
gibbons *see* primates
Gibson, K. R. 8, 16
Gildea, P. M. 7, 61
Giddens, A. 2
Gigantopithecus 135
Ginneken, J. K. Van 323
Gladkih, M. I. 263
Goodall, J. 8, 96, 99, 103, 105, 108,
120, 122, 125, 154, 155, 233,
234, 336, 354
Goodman, D. 298
gorillas
cannibalism among 103
diet of 100
hierarchy among 102, 314
infancy of 105, 318, 330
infanticide among 103
sexual behaviour of 107, 313,
314, 316, 322
sexual dimorphism of 106, 171,
320–1
social life of 99–100
violence among 103
see also primates

Gough, K.   313, 319
Gould, J. L.   60
Gould, S. J.   14, 21, 35, 354
Gowlett, J. A. J.   157, 159, 178–9,
    185, 186, 202, 227–8, 229, 354
Graham, C. A.   326–7
Grant, F.   9
Gray, J.   318, 327
Grayson, D. K.   30, 354
Graziosi, P.   278
Griffin, D. R.   118–19
Griffin, P. B.   307
Grine, F. E.   168, 171
ground-based living, development
    of   145–7
Gusinde, M.   304–5
G/wi Bushmen   212, 215, 261

Hadza people   271, 303
Haeckel, E.   5–6
Hall, K. R. L.   155
Hallowell, A. I.   123, 126
Harcourt, A. H.   322
Harding, R. S. O.   97, 233
Hardy, A.   148
Harlow, H. F.   50–1
Harlow, M. K.   50–1
Harnad, S. R.   273
Harner, M   84–5, 88–90
Harris, J. W. K.   158, 228–9
Harris, M.   84
Harrold, F. B.   270
Harvey, P. H.   98–9, 320
Hassan, F. A.   278
Hawaiians   75
Hayden, B.   221, 232, 300, 309
Herero people   212
Hertz, R.   271
Hewes, G. W.   151, 273
Hiatt, B.   299
Hill, A.   14
Hill, K.   237–8
Hindess, B.   217
Hirst, P. Q.   42, 50, 77, 83, 217, 354

Ho, C. K.   242–4
Hoebel, E. A.   292
Holloway, R. L.   168, 189–93, 196,
    273
hominids
    archaeology and   23–6, 193–205
    art of   267–8, 286
    behaviour patterns of   167, 340
    bipedalism of   *see* bipedalism
    brain size of   187–90, 244, 254,
        265, 266–7, 292
    camp-sites of   194–205
    cannibalism among   185, 242–3,
        254
    classification of   91, 94, 349
    clothing and   268, 277
    culture of   113–14, 123, 178–81,
        264–5, 278–81, 292, 340
    death rituals and   268–71, 274,
        281
    diet of   143–5, 166, 167, 169–71,
        172, 173, 176–7, 194, 198–200,
        204–5, 220, 233–6, 237–9,
        242, 243, 244–52, 291, 344
    energy, expenditure of   143,
        145–6, 151–2, 226–7
    environment and   126–7,
        129–30, 133–6, 229, 342–3
    evolution of   122, 124, 126–53,
        237–8, 240; *see also* evolution
    family life of   314–19, 325,
        328–30, 337–8, 343–5
    fire and   184, 226–9, 239, 241–5,
        246–7, 264, 268
    hunting and   166, 175, 177,
        199–201, 204, 207–65, 292–3,
        342
    incest among   336, 337–8
    infancy of   138–9, 150, 317–18,
        330, 340
    intelligence of   182–7, 270, 283
    labour, division of, and   291–3,
        310–11, 318–19, 341, 342, 344
    language and   127–8, 271–3, 281

hominids (*cont.*)
  music and 279
  personal adornment and 279,
    283
  scavengers, as 231–6, 239, 240,
    249, 250–1, 342
  sexual behaviour of 292–3,
    310–11, 317–18, 319, 325–8,
    336, 337–8, 341, 343
  sexual dimorphism of 171, 240,
    320, 321, 340–1, 342, 352
  social life of 122–3, 152–3,
    178–81, 190–208, 212, 220–1,
    222, 224–5, 264–5, 274,
    291–3, 310–16, 319, 338,
    339–40
  tools, weapons and 156–206,
    226, 227, 230, 239, 243–4,
    291–2, 309, 340
  see also Acheulean, Mousterian
    and Oldowan hominids;
    australopithecines;
    Neanderthals
hominoids 91, 94, 134–5, 144, 147,
    189, 343, 349
Homo 23, 115, 132, 154, 157, 166,
    169, 171–4, 175, 176–7, 178,
    201, 207, 211, 212, 214, 215,
    222, 225, 233, 243, 291–2,
    315–18, 349
  Homo erectus 'Peking Man' 130,
    172, 173, 175, 177, 184–5,
    193, 205, 208, 214, 222, 226,
    227, 230, 236–8, 239, 240,
    241–4, 264, 266, 267, 272,
    273, 311, 329, 349
  Homo habilis 168, 172, 173, 175,
    185, 190, 193, 205, 232, 239,
    240, 350
  Homo neanderthalensis see
    Neanderthals
  Homo sapiens 1, 113, 128, 141,
    171, 175, 239, 240–1, 244,
    266–7, 272, 292, 325

Homo sapiens neanderthalenis see
    Neanderthals
Homo sapiens sapiens 19, 21,171,
    239, 241, 256, 266, 267, 273,
    277, 286, 350
homosexuality 70, 71
Hopkins, K. 336
Horgan, J. 70
Howells, W.W. 236, 267
Hoxne site 246–7
Hubbard, R. 295
human origin, theories of 10–12,
    14–15
human sacrifice 84–5, 88–90
human society
  animals, comparison with 52–8
  behaviour in 266–90
  concept of 78–9
  culture and 45–63, 205–6,
    264–5, 277–9, 286–90, 333–6
  evolution and 18–44, 129–31,
    132, 205–6, 286–7, 292–3,
    312–46, 338–46, 342–4, 345–6
  genes and 78–80
  hunting in *see* hunting
  incest in 333–8
  labour, division of, in 291–311
  modern humans 266–90
  natural selection and 79–80
  primate societies, comparisons
    with 111–12
  rituals in 56, 268–71, 274, 276,
    279–80, 281, 288–9
  sexual behaviour in 291–346
human species, variability
    and 18–26
humans, modern 266–90
Humphrey, N. 116, 117, 121
hunter-gatherers *see* hunting
    cultures
hunting cultures 207–65, 350
  archaeology and 241–56
  archaic 166, 175, 207–11, 226–7,
    292–3

hunting cultures (*cont.*)
  comparison with modern
      211–17, 219–25, 238–9,
      240–1, 264
  environment and   214–15, 221,
      229, 256–64
  labour, division of, in   291–311
  modern   152–3, 200, 207–8,
      211–25
  society and   208–11, 216–25,
      238, 252–65, 287–8, 292–3,
      298–311
  techniques and equipment
      of 225–31, 277
  violence and   252–6, 264–5
Huxley, T. H.   43–4, 95

Ice Ages   240, 263, 268, 270,
      287–8
incest   333–8
Indians, Dene Forest   212
infanticide   75, 103, 221
Ingold, T.   4
Inuit peoples   212, 221, 301–2,
      303–4
Isaac, B.   197
Isaac, G. L.   124, 144–5, 148, 158,
      159, 167, 173, 176, 193, 197–8,
      199, 200–5, 206, 215, 220–1,
      227, 234, 311, 355

James, S. R.   227, 229
Japan   75
Jay, P. C.   96
Jaynes, J.   116–17, 273
Jelinek, A. H.   16
Jenness, D.   215, 303
Jennings, J. D.   258
Johanson, D. C.   137, 150
Johnston, F. E.   185, 241
Johnstone, P.   214
Jolly, A.   96–7, 110, 121, 319, 324,
      327, 336, 354

Jolly, C. J.   24, 149
Jones, P. R.   164
Jones, S.   20–1, 354

Kamin, L. J.   72
Katz, M. M.   330
Keeley, L. H.   159
King, J. C.   72–3
kinship, systems of   67–8, 74–6,
      123, 125–6, 166, 205–6, 212–13,
      293, 312–3, 328–39, 350
  see also tribalism
Kinzey, W. G.   320, 355
Klein, R. G.   247–8, 266, 354
Knecht, H.   277
Kolata, G. B.   323
Konner, M. J.   317, 330
Kornietz, N. L.   263
Kortlandt, A.   120
Kroeber, A. L.   48–9, 271
Kroll, E. M.   248–9
!Kung Bushmen   211–13, 214, 215,
      218, 221, 222, 259, 271, 301,
      310, 344

labour
  division of   291–311, 318–19,
      330–1, 338, 341–2, 344
  evolution and   4–5
  theory of culture   4
Laitman, J. T.   127
La Lumiere, L. P.   148
Lamarck, J.   27–8
Lamphere, L.   295
Lancaster, C. S.   151, 208–10, 299
Lancaster, J. B.   104, 152, 157
Landau, M.   140
Landes, R.   306
Lange, R. V.   71
Langer, W. L.   221
language, acquisition and
      development of   49–50, 55,
      61, 71, 127–8, 271–3, 281
Laughlin, W. S.   230

Lazaret site 247
Leach, E. 86
Leacock, E. 207, 218–19, 300–1, 306
Leakey, L. S. B. 142, 164, 173
Leakey, M. D. 136–7, 160, 161, 194–5, 196–7, 232
Leakey, R. E. 124, 241
learning, patterns of 7–10, 59–63, 71, 119–21
    see also mental processes
Lee, R. B. 208, 212, 213, 214, 215, 217, 218–19, 259–60, 271, 299, 301, 309, 310, 339, 355
Lees, S. H. 221
Leibowitz, L. 100, 333, 341
Le May, M. 273
Leroi-Gourhan, André 196, 290
Leroi-Gourhan, Arlette 269, 290
Lévi-Strauss, C. 319–20, 331, 337, 338–9
Lewin, R. 14, 124, 140, 143, 157, 173, 174, 177, 189, 274
Lewontin, R. C. 72, 73
Lieberman, P. 272, 273, 355
Lightfoot, Dr 27
Lin, S. 241
Linnaeus, C. 91–2
Livingstone, F. B. 21, 337
locomotion, methods of 145–7
    see also bipedalism
Lothrop, S. K. 304
Lovejoy, C. O. 142, 150–1, 196, 317, 328, 339, 342–4
Lower Palaeolithic 23, 157, 158, 160, 161, 162, 180, 185–7, 244, 254, 268, 273, 350
Lowther, G. R. 235
Loy, J. D. 8
Lumley, H. de 245
'Lucy' *see Australopithecus afarensis*
Lumsden, C. J. 69, 70–1, 337
Lyell, C. 29

MacNeish, R. S. 258
Makov, E. 337
Malinowski, B. 334
Mandel, E. 4
Mann, A. E. 145
Mann, M. 293
marine adaptation theory of evolution 148
Marler, P. 60
marriage customs 56, 75–6, 214
    see also attachment, patterns of; family units, development of
Marshack, A. 273, 279–80, 283–4, 286
Marshall, J. K. 212
Marshall, L. 212, 221, 222, 271
Martin, P. S. 256–9, 261, 263–4
Martin, R. D. 107, 325
Marx, K. 3, 90, 218
Masai people 270–1
May, R. M. 107, 325
Mayr, E. 5, 19, 28, 59, 73
Mbuti people 303
McClintock, M. K. 326–7
McGrew, W. C. 314, 326–7
McHenry, H. M. 146
Mehlman, P. 330
Meillassoux, C. 217
Mellars, P. 275, 280, 282
Mendel, G. J. 32
mental processes, evolution and development of 50–2, 114–22, 270, 283
    see also hominids, intelligence of; language, acquisition and development of; learning, patterns of
Merker, B. 151
Mesoamerica 85, 88–90
Mesolithic 211, 350
Mesozoic 133
Mexico
    cannibalism in 85, 88–90

Mexico (*cont.*)
    human sacrifice in   84–5, 88–90
    see also Aztec empire
Middle Palaeolithic   159, 161, 176,
        238, 244, 250–2, 270, 276, 278,
        280–2, 312, 350
Midgley, M.   53–4, 70, 74, 354
Miller, G. A.   7, 61
Milton, K.   133
Miocene   133, 134–5, 140, 141,
        145, 147, 148, 150, 189, 343,
        350
Molleson, T.   309–10
modern humans   *see* human
    society
monkeys   *see* primates
Mongols   83
monogenesis theory   *see* genes:
    monogenesis theory
Montague, A.   273
Moore, H.   295
Moore, R.   93, 234, 337–8
Morgan, E.   148
Morgan, L. H.   218, 314
Morton, N. E.   334, 337
Mosiman, J. E.   257, 258–9, 261
Mousterian hominids   274, 277,
        279–80, 281
    see also tools, manufacture and
        use of: Mousterian
Mowat, F.   212
multi-regional hypothesis   275–6
Murdoch, J.   303–4
Murdock, G. P.   302, 331
Musgrave, J. H.   268
music   279
Musonda, F. B.   219

Nadler, R.D.   324, 326
Napier, J. R.   36, 173
natural selection   351
    biology of   3, 5–7, 14
    culture and   10, 15–17, 43, 341–2,
        345–6

genes and   73
ground living and   146–7
human society and   79–80,
        341–2
sexual   295–8, 314–15, 317
species and   33–8, 41
nature, man and   3–4
Neanderthals   26, 209, 227,
        266–83, 351
Neel, J. V.   337
Neolithic   176, 212, 351
Newton, I.   90, 298
Nicholson, N. A.   329
Nuba people   289
Nuer people   75, 271
Nupe people   271

Oakley, K. P.   156, 163, 164, 184–5,
        187, 250, 355
Ojibwa people   301, 306
Oldowan hominids   197–9, 200,
        201, 202–4, 227–8, 230, 232,
        234–5, 239, 247, 248–9, 264,
        315–16
    see also tools, manufacture and
        use of: Oldowan
Old World monkeys   *see* primates
Oligocene   133–4, 351
Olsen, S. J.   262–3
Ona people   304
orangutans   *see* primates
Orme, B.   216
Ortiz de Montellano, B. R.   89
Ottoman Empire   83
Ouranopithecus   135
'Out of Africa'   *see* genes:
    monogenesis theory
'Overkill hypothesis'   256–64

palaeoculture   16
Palaeo-Indians   257–9, 261, 262
Palaeolithic era   221, 248
    see also Lower, Middle *and*
                Upper Palaeolithic

Paranthropus 169
Parker, S.T. 8, 16
Parsons, T. 331
Passingham, R. 54, 117, 119, 336, 354
Patrusky, B. 258
'Peking Man' *see Homo erectus*
Peters, C.B. 8
Peukert, D.J.K. 6
Pfeiffer, J.E. 288, 355
phenotype 32, 351
Phillips-Conroy, J. 255
Phoenix, C.H. 324, 326
physical anthropology, racism and 5
Piaget, J. 182–4
Pilbeam, D. 93, 141, 142, 143, 273
Pleistocene 164, 169, 178, 215, 219, 229, 240, 247, 249, 254, 263–4, 266, 267, 275, 287, 351
Pliocene 135, 141, 169, 189, 352
Plog, F. 24
Polynesians 75
population, regulation of 221–2
Potts, R. 199, 201, 202, 204, 205, 235–6, 249, 316
Price, J.A. 224
primates (excluding hominids, q.v.) 352
  brain sizes of 188–9
  cannibalism among 103, 233
  diet of 97, 100, 125, 230–1, 232–3
  ecology of 97–9
  environment and 96–100
  evolution of 134–5
  genetic relationships of 92–3
  hierarchies of 100–1, 102–3
  human evolution and 91–5, 313, 315–16, 319–28
  incest among 336
  infancy of 103–5, 138, 152, 313–4, 329, 330, 331
  infanticide among 103

language and 23, 272
learning patterns of 105, 109–12, 119–20, 121, 125
locomotion, methods of 145–7
sexual behaviour of 106–9, 313–16, 319–26
sexual dimorphism of 106, 171, 314, 320–8
social behaviour of 91–112, 123–6, 315–16, 329
tool-use by 23, 109, 110, 185, 314
violence among 103
primatology 95–7
'primitive communism' 217, 218–19
Proconsul 145, 147
Provost, C. 302

races, human 19–21
racism and natural selection 5–6
Ramapithecus 135, 142
Rappaport, R.A. 285, 301, 355
Reader, J. 122
reductionism 129–31, 352
Reiter, R.R. 295
religious systems 68, 71–2, 88–9
Rennie, J. 20
Rensberger, B. 20
Reynolds, P.C. 330
Reynolds, V. 97, 111–12, 115, 126, 321, 324, 325, 336, 354, 355
Richard, A.F. 97, 318, 354
Richards, G. 148
Richerson, P. 64
Rightmire, P. 173, 175
rituals, social 279–80
  see also death rituals
Rodman, P.S. 146
Romer, A.S. 35–7, 144
Roper, M.K. 254
Rose, S. 72, 354
Rudwick, M. 29
Russell, C.A. 298

Sacks, K.   295
Sahlins, M. D.   74–6, 80–3, 225,
    260, 335, 354, 355
Sandars, N. K.   284
Sanday, P. R.   86
Sarich, V. M.   93
Schaller, G. B.   98, 100, 235, 249
Schick, K. D.   161, 204, 227
Schultz, A. H.   98
Service, E. R.   1, 80–3, 214, 218,
    338
sexual behaviour   56, 71, 75–8,
    291–346
  evolution and   76–7
sexual dimorphism   106, 171, 314,
    320–8, 343, 344, 352
sexual selection   295–8, 314, 323,
    324
Shackley, M.   268
Shanidar site   269–70
Shea, B. T.   321
Shipman, P.   199, 202, 205, 232,
    235–6, 249, 255, 273–4, 276,
    355
Short, R. V.   322, 323
Shostak, M.   218, 221
Sibley, C. G.   93
Siegel, M. I.   273
Sieveking, A.   278
Silberbauer, G. B.   215, 355
Simons, E. L.   134, 135
Simpson, B.   87
Simpson, G.G.   24, 354
Sivapithecine apes   134–5, 136,
    142, 144
Slocum, S.   299
Smith, V. L.   257
sociability   56
  see also primates, social
    behaviour
social evolution, theories of
    39–44, 80–4, 352
social sciences
  culture and   48–52

evolution and   2–3, 5, 15, 23,
    80–4, 128–9
  sociology, concept of   78
society   *see* human society
sociobiology, culture and   64–80,
    353
Soffer, O.   263
Solecki, R. S.   269–70
Solomon Islanders   75
species
  definition of   19
  origin of   33–8
speech   *see* language, acquisition
    and development of
Spencer, F.   273–4
Spencer, H.   40–1, 42–3
Stanley, S. M.   14, 25, 263, 354
Steele Russell, I.   188
Stein, G. J.   6
Steklis, H. D.   273
Steward, J. H.   212
Stewart, T. D.   269
Stoneking, M.   274
Stringer, C. B.   266, 267, 275
Strum, S. C.   125
Sung dynasty   83
Susman, R. L.   174
Suzuki, A.   233
Svoboda, J.   249
symbols, ability to use   *see*
    language, acquisition and
    development of
Symons, D.   76, 77, 78
Szalay, F. S.   149

Tabet, P.   307–8
Tanner, N. M.   298, 308, 314, 324,
    327, 340–1, 344, 355
taphonomy, development of   14,
    254–5, 353
Taub, D.   330
Teaford, M.   145
Teleki, G.   97, 125
Terra Amata site   245–6, 247

territoriality 71

Thorne, A. G. 275

thought *see* mental processes, evolution and development of

Tierra de Fuego 304–5

Tiger, L. 65

Tobias, P. V. 138, 168, 173, 174

Tooby, J. 312, 342, 344

tools, manufacture and use of 4, 16, 49, 62, 109, 110, 142–3, 220, 236–7, 257–8, 276–7, 280–2, 307–8

  Acheulean 157, 160–1, 162–3, 164, 179, 183–4, 185–7, 196, 208, 236, 237, 257, 267, 268, 347

  animals and 154–7

  bone 268

  culture and 154–81

  hominids and 156–206, 226, 227, 230, 239, 243–4

  Lower Palaeolithic 23, 157, 158, 160, 161, 162–3, 180, 185–7

  mental abilities and 182–206

  Middle Palaeolithic 159, 161, 162–3, 176, 196, 268

  Mousterian 162–3, 238–9, 268, 274, 276, 277, 351

  Oldowan 157, 158, 160–1, 162–3, 164, 179, 182–3, 184, 185, 186, 187, 193, 196–8, 199, 200, 201, 202–4, 227–8, 230, 232, 248, 291–2, 351

  Osteodontokeratic 253, 254, 255

  social behaviour and 185, 191–3

  stone 23, 158–65, 173–4, 176, 177, 178, 179, 180, 182–7, 194, 196–8, 199–200, 201–3, 206, 220, 236–7, 248, 253, 268, 308; Clark's typology of, 160–3, 268

Upper Palaeolithic 158, 162–3, 165–6, 176, 196, 268, 277, 308

  wooden 250

Torralba-Ambrona sites 244–5, 246, 247, 248, 249

Toth, N. 161, 164, 165, 179, 184, 204, 227, 248

Trevarthen, C. 9

tribalism 68, 81–2

  see also kinship, systems of

Trigger, B. G. 4

Trinkaus, E. 174, 240, 267, 273–4, 276, 355

Trobriand Islanders 76, 334

Tswana people 212

Turks 83

Turnbull, C. M. 271, 303

Tuttle, R. H. 147

Tylor, E. B. 48

Ucko, P. J. 271

Upper Palaeolithic 158, 165–6, 173, 176, 196, 211, 215, 230, 241, 244, 247, 250–2, 256, 264, 268, 270, 276–90, 308, 353

Usher, J. 27

Vandermeersch, B. 276, 281

van Eysinga, F. W. B. 24

variation, natural 18–26, 31–2

Vertut, J. 279, 355

Villa, P. 86–7, 249–50

violence, hunting and 252–6, 264–5

  see also aggression

Vrba, E. S. 135–6, 176

Walker, A. 143, 145

Wallace, A. 26

Washburn, S. L. 15, 19, 71, 92–3, 96, 106, 123–5, 132, 142, 151, 208–10, 234, 254, 297, 299, 313, 337–8

Watanabe, H. 305

weapons   230, 257
  see also tools, manufacture and
    use of
Webster, D.   257–8, 260
Wellings, K.   324
Wenke, R. J.   241, 354
Wheat, J. B.   261, 262
Wheeler, M.   225
Wheeler, P.E.   146
White, L. A.   49, 96
White, R.   278, 280, 285–6, 290
White, T. D.   136
Whiten, A.   121
Whitten, P.   152
Wickler, W.   324
Williams, M. A. J.   264
Wilson, A. C.   20, 93, 274
Wilson, E. O.   14, 65–73, 74, 78, 80,
  155, 337
Wippermann, W.   6
Wobst, H. M.   286–7
Wolfe, L. D.   319, 323, 325, 326,
  341, 344

Wolpoff, M. H.   275–6
women, role of   *see* labour,
    division of; sexual behaviour
Wood, B.   136
Woodburn, J.   222–3, 271, 303
Woolfson, C.   4
Woolley, P.   50, 77, 354
*World Bank Development Report
  1991*   259
Wrangham, R. W.   316
Wright, R. V. S.   109, 185
Wu, R.   241
Wymer, J.   244, 246, 355
Wynn, T.   182–4, 185, 286

Yahgan (Yamana) people   304–5
Yanomamo people   331–2
Yellen, J. E.   215, 224
Yerkes, R.   122

Zahn-Waxler, C.   74
Zihlman, A. L.   207–8, 340–1